Epigenetics and Human Health

Series editors
Robert Feil
Institute of Molecular Genetics
CNRS
UMR5535
Montpellier Cedex,
France

Mario Noyer-Weidner
Freie Universität Berlin
Berlin,
Germany

Jörn Walter
FR8.4 Biosciences
Universität des Saarlandes
Saarbrücken,
Germany

More information about this series at http://www.springer.com/series/8561

Walter Doerfler • Josep Casadesús
Editors

Epigenetics of Infectious Diseases

Springer

Editors
Walter Doerfler
Institute for Clinical and Molecular Virology
University Erlangen-Nürnberg Medical School
Erlangen, Bayern
Germany

Institute of Genetics
University of Cologne
Köln, Germany

Josep Casadesús
Departamento de Genética
Universidad de Sevilla
Sevilla, Spain

ISSN 2191-2262 ISSN 2191-2270 (electronic)
Epigenetics and Human Health
ISBN 978-3-319-55019-0 ISBN 978-3-319-55021-3 (eBook)
DOI 10.1007/978-3-319-55021-3

Library of Congress Control Number: 2017941638

Printed on acid-free paper

This Springer imprint is published by Springer Nature
The registered company is Springer International Publishing AG
The registered company address is: Gewerbestrasse 11, 6330 Cham, Switzerland

Preface

This book describes epigenetic mechanisms in a variety of human pathogens. The list includes viruses, bacteria, and protists. Two chapters deal with the modifications of the host epigenome induced by bacterial and protist infections. We trust that the eleven chapters in this volume will arouse the interest of researchers in epigenetics, virology, microbiology, and infection biology in general, both in molecular biology and in molecular medicine.

An incentive to arrange this volume has been the consideration that the enormous success of epigenetics in higher eukaryotes and its relevance for human health has somehow overshadowed microbial epigenetics. However, epigenetic mechanisms play crucial roles in the lifestyles of viruses, bacteria, protists, and fungi. Such mechanisms are diverse but have an outcome in common: the generation of nongenetic diversity. In pathogens, phenotypic heterogeneity permits the formation of lineages with distinct properties and contributes to the interaction with the eukaryotic host (e.g., evasion of the immune system, division of work, and preadaptation to host-mediated challenges by bet hedging). An equally relevant, emerging notion is that the interaction of microbial pathogens with their hosts can induce changes in the eukaryotic epigenome. The significance of such changes remains poorly understood in many cases. An appealing speculation is that modulation of the host epigenome might provide a memory mechanism that registers the encounter with a pathogen and transmits this information to daughter cells.

Studies critically directed toward epigenetic alterations of virus-infected or virus-transformed cells have so far received limited attention. In contrast, viral systems have been frequently used as models to document the role of DNA methylation in long-term gene silencing in eukaryotes. However, there is increasing evidence to support the notion that virus infections in general can lead to the destabilization of the host cells' epigenetic profiles, probably early on after virus infections. In this volume, the epigenetic consequences for the host genomes upon infections with human papilloma virus, human herpesvirus type 8, human adenovirus type 12, and the human herpesvirus Epstein–Barr virus have been analyzed by research groups active in these fields.

We are indebted to Anne Clauss of Springer Verlag and to Mario Noyer-Weidner, the series editor, and have appreciated their support in completing this volume.

Erlangen, Germany Walter Doerfler
Sevilla, Spain Josep Casadesús

Contents

Chapter 1
DNA Methylation of Human Papillomavirus Genomes During Infection and Cancer Progression

Hans-Ulrich Bernard

Abstract Human papillomaviruses (HPVs) are important pathogens, as they are the cause of all cervical cancers and of subsets of vulval, anal, oral, and penile cancers. The viral genome is an 8 kb double-stranded circular DNA, which can replicate in various types of epithelial cells. HPV DNA shows changes of its methylation profile during the viral life cycle, namely "sporadic" and "polymorphic" DNA methylation associated with low transcription in basal cells of epithelia, and a lack of methylation in suprabasal cells associated with strong transcription. While these epigenetic changes of HPV DNA during the viral life cycle are still poorly understood, it has emerged that during progression of low-grade precursor lesions to malignant carcinomas, the HPV DNA becomes hypermethylated, probably since the viral genome recombines with the chromosomal DNA of the infected host cell. This methylation signal is intensely studied as a candidate biomarker for the diagnosis of HPV-associated lesions that have the potential to progress to cancer.

Keywords Papillomavirus • HPV • Viral life cycle • Cancer progression • Recombination • Insertion

1.1 Introduction

Papillomaviruses are defined by their (1) non-enveloped capsids, (2) circular double-stranded DNA genomes with sizes close to 8000 bp and highly conserved gene organization, (3) host species specificity, (4) tropism for epithelial cells, and (5) transforming rather than lytic effects on the host cells. They cause neoplastic growth of the infected epithelium or can persist in asymptomatic infections.

Papillomavirus genomes have a noncoding region (long control region, LCR) of about 800 bp, which harbors the replication origin, a transcriptional enhancer, and a

H.-U. Bernard (✉)
Department of Molecular Biology and Biochemistry and Program in Public Health, University of California Irvine, Irvine, CA 92697, USA
e-mail: hbernard@uci.edu

W. Doerfler, J. Casadesús (eds.), *Epigenetics of Infectious Diseases*, Epigenetics and Human Health, DOI 10.1007/978-3-319-55021-3_1

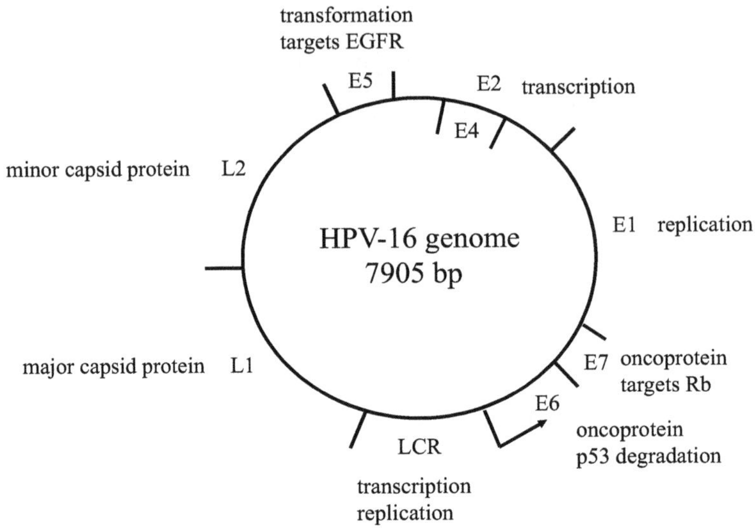

Fig. 1.1 Genome organization of human papillomavirus 16

promoter (Bernard 2013). About half of the genome downstream of the promoter contains the early genes E6, E7, E1, E2, E4, and E5, with the remainder further downstream encoding the late proteins L2 and L1 (Fig. 1.1). For the purpose of this chapter, the following brief summary of the function of these proteins may suffice: E6 and E7 trigger the principal transforming mechanisms, such as interference with p53 and RB cell cycle control (Roman and Munger 2013; vande Pol and Klingelhutz 2013). E1 binds the replication origin and functions as helicase (Bergvall et al. 2013). E2 functions as activating and repressing transcription factor and cooperates with E1 in identification of the replication origin (McBride 2013). And L1 and L2 are the major and the minor capsid proteins (Buck et al. 2013; Wang and Roden 2013).

The papillomavirus life cycle and papillomavirus pathogenesis can be summarized as follows: Papillomaviruses most often infect squamous, i.e., multilayered and differentiating epithelia. In order to establish a stable infection, a papillomavirus particle has to infect the basal layer of such an epithelium, where the circular viral episome persists and replicates. Asymmetric cell divisions of the basal cells lead to suprabasal cells, beginning with the spinous layer. Suprabasal cells normally lack mitotic activity and DNA replication properties. Papillomavirus genomes that are sorted into such cells express E6 and E7 oncoproteins, which target the cellular Rb and p53 proteins, and thereby reestablish an environment of continuing DNA replication and mitoses. The resulting expansion of the suprabasal cell population leads to neoplastic lesions, referred to in the case of skin as "warts." In cell layers close to the epithelial surface, the virus expresses the capsid proteins, which encapsidate the viral DNA into viral particles that are released upon disintegration of terminally differentiated epithelial surface cells. This life cycle also applies during nonmalignant infections to those papillomaviruses that are found in cancer, while it becomes distorted during carcinogenesis. All details of carcinogenesis are not yet understood, but early molecular events frequently involve recombination between papillomaviruses and host cell DNA in a genomic arrangement that leads to stimulation of papillomavirus oncogene transcription. For progression to a malignant phenotype, the affected cell has to undergo numerous additional mutations and epigenetic changes of cellular genes (see Mine et al. 2013, and references therein).

For taxonomic purposes, papillomaviruses are referred to as "types," and their names are abbreviated with the letters PV, preceded by one or two letters that define the host, and followed by a number indicating the historic sequence of isolation (Bernard et al. 2010). Among more than 300 papillomavirus types described so far, only the cottontail rabbit papillomavirus (CRPV1) and those human papillomaviruses (HPVs), which are most prevalent in carcinomas of the cervix uteri (HPV16, 18, 31, 33, 35, 45, 52, 58), were addressed by DNA methylation research. No methylation studies have been done with the well-developed cell culture system for bovine papillomavirus-1 (BPV1) or in situ with those HPVs that cause mostly benign lesions such as HPV2 (common warts) or HPV6 (genital warts). These particular PVs are stably maintained as episomes, while the aforementioned PVs can recombine with cellular DNA, what may influence viral DNA methylation.

1.2 History of Papillomavirus Methylation Research

The first records of PV DNA methylation were garnered with CRPV1 (at that time also called Shope papillomavirus), which was shown to have methylated, chromosomally integrated, multicopy viral DNA in rabbit skin tumors (Wettstein and Stevens 1983; Sugarawa et al. 1983). These observations preceded the modern understanding of the regulatory importance of epigenetic alterations. Unfortunately, the CRPV/rabbit system has never been reinvestigated since then. Subsequently, the potential for transcriptional effects of DNA methylation on HPV16 and HPV18 DNA became established in vitro and in cell culture experiments (Thain et al. 1996; Rösl et al. 1993), but these observations were not extended to a search of methylated HPV DNA in situ. Several years later it turned out that HPV DNA methylation is actually a widespread phenomenon, and became observed in a cell line with episomal HPV16 DNA (Kim et al. 2003), in cell lines with integrated HPV16 and HPV18 DNA as well as in carcinomas and their precursor lesions (Badal et al. 2003, 2004; Kalantari et al. 2004). While these studies opened a rich field of investigation, their analytical power was initially limited by the use of methylation sensitive restriction enzymes, and by a focus on small genomic regions, including E2 binding sites. In the last 10 years DNA methylation studies by bisulfite sequencing have targeted larger parts of the genomes or whole genomes of several HPV types in the context of the viral life cycle and carcinogenesis, and these findings will be reviewed in this chapter.

The following questions emerged as the most challenging research objectives:

1. Does DNA methylation affect HPV biology during the normal viral life cycle and are HPVs unique DNA methylation targets, e.g., as a form of cellular defense against foreign DNA?
2. Are there specific regulatory effects of DNA methylation via CpG dinucleotides in E2 binding sites?
3. Do HPVs affect the cellular epigenome?
4. Does DNA methylation differentially affect HPV genomes during progression of asymptomatic infections through precursor lesions to malignant lesions?
5. Are HPV epigenomes or cellular epigenomic properties in HPV infected cells useful biomarkers in the diagnosis of cancer precursor lesions?

1.3 Methylation of HPV DNA During the Normal Life Cycle

In order to understand a potential role of HPV DNA methylation during the normal life cycle, it would be desirable to investigate the DNA first in the viral capsid, then immediately following infection of the basal layer, and further during epithelial differentiation. Unfortunately, HPV research here and elsewhere has always been hampered by the absence of animal models and by difficulties to establish or reproduce cell culture systems. As a consequence, there are presently only two sources of information about the epigenetics of HPVs during the viral life cycle, namely, the epigenetic properties of HPV DNA from patients likely to harbor only episomal DNA and a stable cell line, W12E, that maintains HPV16 DNA episomally.

HPV16, HPV18 and several related high-risk HPV types infect squamous mucosal cells of the female genital tract subclinically, and these infections can progress through cervical cancer precursor lesions (cervical intraepithelial neoplasia I and III, CIN I and CIN III) to invasive cervical cancer. In subclinical infections and CIN I lesions, HPV genomes exist as episomes, while an increasing portion of them recombines during progression. Consequently, it can be assumed that clinical samples obtained from asymptomatic individuals and CIN I patients contain HPVs during the normal viral life cycle. Studies from numerous labs agree that such clinical samples contain "sporadically" methylated HPV genomes, "sporadically" referring to average methylation frequencies per CpG in the range of 5–10%, and a lack of specificity for the CpG target (Kalantari et al. 2004; Turan et al. 2006; Brandsma et al. 2009; Sun et al. 2011; Mirabello et al. 2013). All of these studies addressed CpGs in the LCR and the L1 gene, but some extended the findings throughout the genome. These data constituted methylation analyses of short PCR amplicons. This is a nontrivial limitation, as the sequencing of multiple cloned amplicons from each patient sample found substantial heterogeneity of methylation between HPV genomes from the same sample (Kalantari et al. 2004).

This investigation of clinical samples became complemented by cell culture studies. Among the few cell culture models in papillomavirus research are W12E cells cloned from a CIN lesions of an HPV16-infected patient. The cells grow on a fibroblast feeder layer and morphologically resemble the basal layer of epithelia. Some differentiation can be observed in confluent cultures, and these differentiated cells can be separated from the undifferentiated cells. Lambert and colleagues (Kim et al. 2003) observed a consistent, but only "sporadic" methylation of the HPV16 LCR in undifferentiated cells, similar to patterns observed in situ in cells harvested from asymptomatic or CIN I patients. Most of this methylation was lost upon differentiation of the W12E cells. It is known from studies addressing different aspects of HPV biology that transcription of HPVs becomes activated upon differentiation, and so it is tempting to hypothesize that the observed epigenetic change is the switch between two different transcription states. It should be noted that methylation was only rarely observed at the E2 binding sites overlapping with the E6 promoter, which would activate rather than repress this promoter (see below).

Another study of W12E cells confirmed the methylation of the HPV16 LCR in undifferentiated cells, as well as the ensuing demethylation upon differentiation (Kalantari et al. 2008a). This study also addressed five clonal derivatives of W12E, where all HPV16 genomes had recombined with the cellular DNA. Three of these clones with few HPV16 copies had nearly no methylation of LCR sequences but some methylation of the L1 gene, which is adjacent to the LCR but not transcriptionally affected by its properties. Two clones with numerous HPV16 copies showed strong methylation of the LCR. In contrast to the W12E cells with episomal DNA, differentiation of these five clones with chromosomally integrated viral DNA did not alter HPV16 DNA methylation, neither in the LCR nor in the L1 gene.

1.3.1 In Summary

Studies of clinical samples as well as the W12E line agree that episomal HPV16 DNA is targeted by DNA methylation. DNA methylation is sporadic, i.e., low, and polymorphic both within an individual sample as well as between comparable samples. Differentiated W12E cells contain completely unmethylated HPV16 LCR segments, and such molecules exist in most clinical samples with episomal DNA. It is therefore likely but not mechanistically understood that HPV16 episomes are methylation targets in undifferentiated epithelial cells. This should negatively affect transcriptional activity. Demethylation may release this repression in suprabasal cells and lead to increased transcription, as observed in situ. No evidence suggests a selective recognition of the viral DNA as part of a cellular defense mechanism.

1.4 Regulatory Effects of DNA Methylation via CpG Dinucleotides in E2 Binding Sites

The papillomavirus E2 gene encodes proteins that have the ability to bind the palindromic DNA sequence 5′-ACCGNNNNCGGT-3′, which occurs four times in the LCR of HPV16 and related HPV types. This sequence has two CpG methylation targets, and in vitro studies have shown that E2 proteins cannot bind the methylated target sequences (Thain et al. 1996). As expected, transfection experiments with unmethylated and methylated E2 site reporter genes and E2 factor expression vectors confirmed that methylation dramatically interferes with transcriptional transactivation (Kim et al. 2003).

This straightforward mechanism in vitro is much more complicated in vivo, on the one side due to the expression of different E2 proteins through differential splicing, some being transcriptional activators, some lacking the transcription activation domain, and on the other side due to multiple and opposing functions of E2 binding sites depending on the genomic context. E2 proteins can be (1) activators of transcription when their binding site is remote from a promoter, the binding sites functioning as E2 protein dependent enhancers. Alternatively, they can (2) repress transcription, when they bind target sites at the HPV E6 promoter, in part due to competition between E2 and the promoter factors SP1 and TFIID, whose binding sites overlap with E2 binding sites (Tan et al. 1994), and in part due to E2 complexes with histone modifying proteins (Smith et al. 2014). Lastly, (3) E2 also forms a complex with the replication factor E1 and increases its specificity and affinity to replication initiation sites, and (4) is involved in partitioning of papillomavirus genomes during mitosis (McBride 2013).

HPV methylation studies normally address only the second of these four functions. The reasoning goes as follows: For E2 protein to be expressed, the HPV genome must be continuous from the E6 promoter through the whole E2 gene, as E2 is translated from a polycistronic mRNA containing the E6, E7, E1 and E2 genes. This is the case when HPV genomes are episomal or exist as tandem repeats recombined with chromosomal DNA. In these two cases, the E2 protein serves a repressing feedback loop, binds to the E6 promoter, and decreases its activity. In this scenario, HPVs and their infected cells would have a growth advantage, if the E2 binding sites overlapping with the E6 promoter would be methylated, as E2 protein could not bind and could not lead to repression, increasing the amount of E6 and E7 oncoprotein production. No such advantage of host cells with HPV genomes with methylated E6 promoter sequences exists if no complete E2 transcript (and protein) can be delivered, which is the case when chromosomal recombination led to interruption of the E2 gene, a frequent scenario in cancer (see below).

There is agreement that these scenarios are regularly encountered, but different extents of this mechanism were reported in different studies (Schwarz et al. 1985; Kalantari et al. 2001; Peitsaro et al. 2002; Arias-Pulido et al. 2006; Bhattacharjee and Sengupta 2006; Brandsma et al. 2009; Snellenberg et al. 2012; Chaiwongkot et al. 2013; Mirabello et al. 2013; Bryant et al. 2014). Reasons for disagreement are technical limits to differentiate between integrated and episomal viral DNA, as, for example, integrated DNA often exists as large concatemers. A role for the E2 protein can be deduced from observations that, typically, the rate of CpG methylation through most of the LCR of HPV16 is by a factor of 2–3 lower than methylation of the four CpGs within the promoter-proximal E2 binding sites, suggesting that clones were selected that have eliminated the negative regulation of the E6 promoter by E2, as this repressor can now not bind anymore to its targets.

1.5 Effects of Papillomaviruses on the Cellular Epigenome

It is well established that extensive epigenomic changes are an intrinsic part of carcinogenesis of all tissues irrespective of their association with papillomaviruses (Sharma et al. 2010), and epigenetic changes contribute to carcinogenesis with a weight similar to that of mutations and aneuplodies. The same applies to cancer of the cervix (Wentzensen et al. 2009; Louvanto et al. 2015; Siegel et al. 2015), and those neoplasias, which have etiologies with and without HPVs such as anal and oral cancer (Hernandez et al. 2012; Jitesh et al. 2013). Although the cellular methylome of the same group of tumors may differ in the presence and the absence of HPVs (Sartor et al. 2011), there is no a priori need to assume that methylation may be affected by the functions of HPV gene products. Nevertheless, this may yet be the case, as the HPV-16 E7 oncoprotein was reported to associate in vitro and vivo with the DNA methyltransferase DNMT1 and to stimulate its activity (Burgers et al. 2007). This observation opens up the possibility that this epigenetic effect directly influences cellular proliferation pathways. Subsequent studies proposed as a consequence of this mechanism suppression of E-cadherin expression and reduced adhesion between squamous epithelial cells (Laurson et al. 2010; D'Costa et al. 2012) and extended the effect to interactions of both E6 and E7 protein with components of the histone modification machinery (Bodily et al. 2011; Hsu et al. 2012).

1.6 Differential Methylation of HPV Genomes in Malignant Lesions

It is known since the early days of HPV research in the 1980s that HPV genomes in cancer frequently exist in a form recombined with cellular DNA (Schwarz et al. 1985). It is now generally accepted that the transition from high-grade precursors (CIN III) to invasive carcinomas is accompanied by and possibly caused by this recombination (Mine et al. 2013), although it is still disputed whether all or only a subset of cancerous lesions contain HPV genomes in chromosomally recombined form (Kalantari et al. 2001; Peitsaro et al. 2002; Arias-Pulido et al. 2006;

Bhattacharjee and Sengupta 2006; Brandsma et al. 2009; Snellenberg et al. 2012; Chaiwongkot et al. 2013; Mirabello et al. 2013; Bryant et al. 2014). Recombination can result in interruption of the early polycistronic E6-E7-E1-E2 transcription unit. Failure to express E2 stimulates oncoprotein expression due to a lack of negative feedback repression of E2 on the E6 promoter. Beyond this, mechanisms for eliminating remaining episomal HPV genomes have recently been proposed as essential for cervical carcinogenesis (Mine et al. 2013).

In malignant and high-grade premalignant lesions, likely due to recombination with the cellular chromosomes, HPV genomes clearly undergo substantial methylation beyond the levels observed for episomal genomes (exceeding for some CpG residues 50%) as confirmed for HPV16 (Kalantari et al. 2004, 2014; Bhattacharjee and Sengupta 2006; Brandsma et al. 2009; Sun et al. 2011; Vinokurova and Knebel Doeberitz 2011; Xi et al. 2011; Clarke et al. 2012; Patel et al. 2012; Mirabello et al. 2013; Park et al. 2011; Verhoef et al. 2014; Frimer et al. 2015), HPV18 (Badal et al. 2004; Turan et al. 2006; Wentzensen et al. 2012; Kalantari et al. 2014; Vasiljevic et al. 2014), HPV31 (Wentzensen et al. 2012; Kalantari et al. 2014; Vasiljevic et al. 2014), HPV33 (Vasiljevic et al. 2014), HPV45 (Wentzensen et al. 2012; Kalantari et al. 2014), HPV52, and HPV58 (Murakami et al. 2013). For HPV16, this was reported not only for cervical but also vulval (Bryant et al. 2014), penile (Kalantari et al. 2008b), oral (Balderas-Loaeza et al. 2007), and anal cancer (Wiley et al. 2005; Hernandez et al. 2012). Methylation is relatively low in the LCR (which, together with the use of methylation sensitive restriction enzymes as opposed to bisulfite sequencing, led to an original misinterpretation of this mechanism, Badal et al. 2003), and is highest at certain CpGs in the late genes L2 and L1 (Brandsma et al. 2014; Mirabello et al. 2015).

Findings of increased methylation of HPV genomes correlating with the increasing severity of the lesion (from CIN I through CIN III to invasive cancer) were surprising and against intuition, as DNA methylation is normally seen as a transcription repression mechanism. The resolution of this contradiction came from two sources. Van Tine et al. (2004) reported in situ studies that cervical tumors typically contain numerous (i.e., up to a few hundred) HPV genome copies. All of these viral genomes are transcriptionally inactive, except one, which is the only source of E6 and E7 oncogene transcripts. In other words, some selective methylation mechanism targets these recombinant HPV genomes. Should all of them become methylated, HPV transcription would end, and such a clone would never grow into a detectable tumor. Only cells with one or few transcriptionally active HPV genomes grow into a detectable lesion.

This mechanism was further confirmed with the study of two cervical cancer cell lines, SiHa and CaSki. SiHa cells contain a single chromosomally recombined HPV16 genome, whose LCR is unmethylated and therefore transcriptionally active. CaSki cells contain about 500 HPV16 genomes, but generate a similar level of transcripts as SiHa cells. Not surprisingly, all HPV16 genomes in CaSki cells except one are methylated and transcriptionally inactive, oncogene transcripts being generated from the only unmethylated viral genome (Kalantari et al. 2004). However, it is not a necessary condition that most HPV genomes become methylated. The well-known cell line HeLa had been derived from a cervical adenocarcinoma and was shown to contain about 50 chromosomally recombined copies of HPV18 DNA. The analysis of its HPV18 genomes showed that the LCR and the E6 gene are generally not methylated and remain transcriptionally active (Johannsen and Lambert 2013), while parts of the genome that are upstream of the LCR, such as the L1 genes are heavily methylated (Turan et al. 2007).

It is unknown why chromosomally recombined HPV genomes become preferentially methylated. HPV DNA may be targeted by a methylation mechanism affecting all foreign DNA in mammalian cells (Dörfler et al. 2001). More recently a view emerged that the methylated state of DNA may be quite in general the default state of the hosts chromosomal DNA to lock genes in an off position (Edwards et al. 2010; Schuebeler 2015).

1.7 HPV Epigenomes and Cellular Epigenomic Properties of HPV Infected Cells as Cancer Biomarkers

Cancer of the cervix affects about 500,000 women every year, and about half of these die of this disease. It is the most prevalent cancer in women in many developing nations, but its incidence has been reduced in developed nations, to a large part through early diagnosis of precancerous lesions and surgical intervention. From the 1950s to the 1990s, diagnosis was mostly based on the Papanicolaou test (Pap test), which can be complemented with colposcopic observation of lesions. The Pap test is a staining test of a cervical smear obtained during a gynecological examination, which was developed without knowledge of the viral etiology of cervical cancer. The Pap test is a tremendous public health success, but it is less than satisfactory as it has a high rate of false negative diagnoses, as it misses many lesions. Since HPV infections are the sole underlying cause of precancerous cervical neoplasia, HPV DNA detection has become a valuable tool to amend or replace the Pap test. However, many women are carriers of HPV infections, which never progress toward malignancies. At this time, the best practice is to administer both a Pap test and an HPV DNA test on a patient, as well as interpreting the outcome in the context of the age and the previous diagnostic history of the patient (Saslow et al. 2012).

From these considerations, it is obvious that the triage of women with a positive Pap test or positive for HPV infection would benefit from the development of tests based on novel biomarkers. Detection of DNA methylation has the potential to be such a biomarker, whose detection can be technically standardized and made capable for high-throughput processing. This chapter has discussed that HPV DNA is either unmethylated or lowly methylated in asymptomatic infections and precancerous CIN I lesion while heavily methylated in cancer, an increase that begins in high-grade precursor lesions (CIN III). Methylation is particularly high at certain CpG dinucleotides in the late genes L1 and L2, identifying the best targets for HPV methylation analysis (Brandsma et al. 2009). A highly sensitive detection of these methylation changes may help to separate patients with malignantly progressing cervical lesions from those not undergoing such changes, as evaluated recently (Brandsma et al. 2014). In order to eliminate the time consuming DNA sequencing, HPV18 DNA methylation could be efficiently detected by PCR with methylation specific primers or with real-time PCR (Turan et al. 2007). As an alternative improvement toward clinical application, it has been shown that next-generation sequencing allows the establishment of the whole HPV16 methylome and eliminates laborious purification of PCR amplicons. Alternatively, pyrosequencing can target segments of HPV genomes of specific relevance for diagnosis. The same publication confirmed that high-grade precursors had a higher methylation than low-grade precursors, the decisive criterion for the usefulness to detect lesions likely to progress toward cancer (Mirabello et al. 2015). Beyond the analysis of the HPV genome, specific cellular genes such as DAPK and RARB are frequently methylated in cervical cancer (Wentzensen et al. 2009), and it may strengthen epigenomic testing to combine the measurement of HPV DNA methylation with that of the methylation status of such cellular genes (Sartor et al. 2011; Johannsen and Lambert 2013; Kalantari et al. 2014; Louvanto et al. 2015; Siegel et al. 2015). At this point, the utility of HPV methylation deserves to be further studied as a strategy to identify women at high risk for cervix cancer.

References

Arias-Pulido H, Peyton CL, Joste NE, Vargas H, Wheeler CM (2006) Human papillomavirus type 16 integration in cervical carcinoma in situ and in invasive cervical cancer. J Clin Microbiol 44:1755–1762

Badal V, Chuang LSH, Badal S, Tang E, Villa LL, Wheeler CM, Li BFL, Bernard HU (2003) CpG methylation of human papillomavirus-16 DNA in cervical cancer cell lines and in clinical specimens: genomic hypomethylation correlates with carcinogenic progression. J Virol 77:6227–6234

Badal S, Badal V, Calleja-Macias IE, Kalantari M, Chuang LSH, Li BFL, Bernard HU (2004) The human papillomavirus-18 genome is efficiently targeted by cellular DNA methylation. Virology 324:483–492

Balderas-Loaeza A, Anaya-Saavedra G, Ramirez-Amador VA, Guido-Jimenez MC, Kalantari M, Calleja-Macias IE, Bernard HU, Garcia-Carranca A (2007) Human papillomavirus-16 DNA methylation patterns support a causal association of the virus with oral squamous cell carcinomas. Int J Cancer 120:2165–2169

Bergvall M, Melendy T, Archambault J (2013) The E1 proteins. Virology 445:35–56

Bernard HU (2013) Regulatory elements in the papillomavirus genome. Virology 445:197–204

Bernard HU, Burk RD, Chen Z, van Doorslaer K, zur Hausen H, de Villiers EM (2010) Classification of papillomaviruses based on 189 PV types and proposal of taxonomic amendments. Virology 401:70–79

Bhattacharjee B, Sengupta S (2006) CpG methylation of HPV16 LCR at E2 binding site proximal to P97 is associated with cervical cancer in the presence of intact E2. Virology 354:280–285

Bodily JM, Mehta KP, Laimins LA (2011) Human papillomavirus E7 enhances hypoxia-inducible factor 1-mediated transcription by inhibiting binding of histone deacetylases. Cancer Res 71:1187–1195

Brandsma JL, Sun Y, Lizardi PM, Tuck DP, Zelterman D, Haines GK, Martel M, Harigopal M, Schofield K, Neapolitano M (2009) Distinct human papillomavirus type 16 methylomes in cervical cells at different stages of premalignancy. Virology 389:100–107

Brandsma JL, Harigopal M, Kiviat NB, Sun Y, Deng Y, Zelterman D, Lizardi PM, Shabanova VR, Levi A, Yaping T, Hu X, Feng Q (2014) Methylation of twelve CpGs in human papillomavirus type 16 (HPV16) as an informative biomarker for the triage of women positive for HPV16 infection. Cancer Prev Res 7:526–533

Bryant D, Onions T, Raybold R, Jones S, Tristram A, Hibbitts S, Fiander S, Powell N (2014) Increased methylation of human papillomavirus type 16 correlates with viral integration in vulval intraepithelial neoplasia. J Clin Virol 61:393–399

Buck CB, Day PM, Trus BL (2013) The papillomavirus major capsid protein L1. Virology 445:169–174

Burgers WA, Blanchon L, Pradhan S, de Launoit Y, Kouzarides T, Fuks F (2007) Viral oncoproteins target the DNA methyltransferases. Oncogene 26:1650–1655

Chaiwongkot A, Vinokurova S, Pientong C, Ekalaksananan T, Kongyingyoes B, Kleebkaow P, Chumworathayi B, Patarapadungkit N, Reuschenbach M, von Knebel Doeberitz M (2013) Differential methylation of E2 binding sites in episomal and integrated HPV16 genomes in preinvasive and invasive cervical lesions. Int J Cancer 132:2087–2094

Clarke MA, Wentzensen N, Mirabello L, Ghosh A, Wacholder S, Harari A, Lorincz A, Schiffman M, Burk RD (2012) Human papillomavirus DNA methylation as a potential biomarker for cervical cancer. Cancer Epidemiol Biomark Prev 21:2125–2137

D'Costa ZJ, Jolly C, Androphy EJ, Mercer A, Matthews CM, Hibma MH (2012) Transcriptional repression of E-cadherin by human papillomavirus type 16 E6. PLoS One 7:e48954

Dörfler W, Remus R, Müller K, Hohlweg U, Schubbert R (2001) The fate of foreign DNA in mammalian cells and organisms. Dev Biol 106:89–97

Edwards JR, O'Donnell AH, Rollins RA, Peckham HE, Lee C, Milekic MH, Chanrion B, Fu Y, Su T, Hibshoosh H, Gingrich JA, Haghighi F, Nutter R, Bestor TH (2010) Chromatin and sequence features that define the fine and gross structure of genomic methylation patterns. Genome Res 20:972–980

Frimer M, McAndrew T, Smith B, Harari A, Chen Z, Mirabello L, Wentzensen N, Goldberg GL, Rodriguez AZ, Schiffman M, Burk RD (2015) HPV16 CpG methyl-haplotypes are associated with cervix precancer and cancer in the Guanacaste natural history study. Gynecol Oncol 138:94–100

Hernandez JM, Siegel EM, Riggs B, Eschrich S, Elahi A, Qu X, Ajidahun A, Berglund A, Coppola D, Grady WM, Giuliano AR, Shibata D (2012) DNA methylation profiling across the spectrum of HPV-associated anal squamous neoplasia. PLoS One 7:e50533

Hsu CH, Peng KL, Jhang HC, Lin CH, Wu SY, Chiang CM, Lee SC, Yu WC, Juan LJ (2012) The HPV E6 oncoprotein targets histone methyltransferases for modulating specific gene transcription. Oncogene 31:2335–2349

Jitesh PV, Risk JM, Schache AG, Dhanda J, Lane B, Liloglou T, Shaw RJ (2013) The epigenetic landscape of oral squamous cell carcinoma. Br J Cancer 108:370–379

Johannsen E, Lambert PF (2013) Epigenetics of human papillomaviruses. Virology 445:205–212

Kalantari M, Blennow E, Hagmar B, Johannson B (2001) Physical state of HPV16 and chromosomal mapping of the integrated form in cervical carcinomas. Diagn Mol Pathol 10:46–54

Kalantari M, Calleja-Macias IE, Tewari K, Hagmar B, Barrera-Saldana HA, Wiley DJ, Bernard HU (2004) Conserved methylation patterns of human papillomavirus-16 DNA in asymptomatic infection and cervical neoplasia. J Virol 78:12762–12772

Kalantari M, Lee D, Calleja-Macias IE, Lambert PF, Bernard HU (2008a) Effects of cellular differentiation, chromosomal integration and treatment by 5′-deoxy-2′ azacytidine on human papillomavirus-16 DNA methylation in cultured cell lines. Virology 374:292–303

Kalantari M, Villa LL, Calleja-Macias IE, Bernard HU (2008b) Human papillomavirus-16 and 18 in penile carcinomas: DNA methylation, chromosomal recombination, and genomic variation. Int J Cancer 123:1832–1840

Kalantari M, Osann K, Calleja-Macias IE, Kim S, Yan B, Jordan S, Chase DM, Tewari KS, Bernard HU (2014) Methylation of human papillomavirus 16, 18, 31 and 45 L2 and L1 genes and the cellular DAPK gene: considerations fur use as biomarkers of the progression of cervical neoplasia. Virology 448:314–322

Kim K, Garner-Hamrick RP, Fisher C, Lee D, Lambert PF (2003) Methylation of patterns of papillomavirus DNA, its influence on E2 function, and implications in viral infection. J Virol 77:12450–12459

Laurson J, Khan S, Chung R, Cross K, Raj K (2010) Epigenetic repression of E-cadherin by human papillomavirus 17 E7 protein. Carcinogenesis 31:918–926

Louvanto K, Franco EL, Ramanakumar AV, Vasiljevic N, Scibior-Bentkowska D, Koushik A, Cusick J, Coutlee F, Lorincz AT, Biomarkers of Cervical Cancer Risk Study Team (2015) Methylation of viral and host genes and severity of cervical lesions associated with human papillomavirus type 16. Int J Cancer 136:E638–E645

McBride AA (2013) The papillomavirus E2 proteins. Virology 445:57–79

Mine KL, Shulzehenko N, Yambartsev A, Rochman M, Sanson GFO, Lando M, Varma S, Skinner J, Volfovsky N, Deng T, Brenna SMF, Carvalho CRN, Ribalta JCL, Bustin M, Matzinger P, Silva IDCG, Lyng H, Gerbase-DeLima M, Morgun A (2013) Gene network reconstruction reveals cell cycle and antiviral genes as major drivers of cervical cancer. Nat Commun 4:1806–1810

Mirabello L, Schiffman M, Ghosh A, Rodriguez AC, Vasilijevic N, Wentzensen N, Herrero R, Hildesheim A, Wacholder S, Scibior-Bentkowska D, Burk RD, Lorincz AT (2013) Elevated methylation of HPV16 DNA is associated with the development of high grade cervical intraepithelial neoplasia. Int J Cancer 132:1412–1422

Mirabello L, Frimer M, Harari A, McAndrew T, Smith B, Chen Z, Wentzensen N, Wacholder S, Castle PE, Raine-Bennett T, Schiffman M, Burk RD (2015) HPV16 methyl-haplotypes determined by a novel next-generation sequencing methods are associated with cervical precancer. Int J Cancer 136:E146–E153

Murakami I, Fujii T, Dan K, Saito M, Ohno A, Iwata T, Aoki D (2013) Methylation of human papillomavirus 52 and 58 is a candidate biomarker in cervical neoplasia. J Clin Virol 58:149–154

Park IS, Chang X, Loyo M, Wu G, Chuang A, Kim M, Chai YK, Lyford-Pike S, Westra WH, Saunders JR, Sidransky D, Pai S (2011) Characterization of the methylation patterns in human papillomavirus type 16 viral DNA in head and neck cancers. Cancer Prev Res 4:207–217

Patel DA, Rozek LS, Colacino JA, Zomeren-Dohm V, Ruffin M, Unger ER, Dolinoy DC, Swan DC, Onyekwuluje J, DeGraffinreid CR, Paskett ED (2012) Patterns of cellular and HPV16 methylation as biomarkers for cervical neoplasia. J Virol Methods 184:84–92

Peitsaro P, Johansson B, Syrjaenen S (2002) Integrated human papillomavirus type 16 is frequently found in cervical cancer precursors as demonstrated by a novel quantitative real-time PCR technique. J Clin Microbiol 40:886–891

Roman A, Munger K (2013) The papillomavirus E7 proteins. Virology 445:138–168

Rösl F, Arab A, Klevenz B, zur Hausen H (1993) The effect of DNA methylation on gene regulation of human papillomaviruses. J Gen Virol 74:791–801

Sartor MA, Dolinoy DC, Jones TR, Colacino JA, Prince ME, Carey TE, Rozek LS (2011) Genome-wide methylation and expression differences in HPV(+) and HPV(−) squamous cell carcinoma cell lines are consistent with different mechanisms of carcinogenesis. Epigenetics 6:777–787

Saslow D, Solomon D, Lawson HW et al (2012) American cancer society, American society for colposcopy and cervical pathology, and American society for clinical pathology screening guidelines for the prevention and early detection of cervical cancer. Am J Clin Pathol 137:516–542

Schuebeler D (2015) Function and information content of DNA methylation. Nature 517:321–326

Schwarz E, Freese UK, Gissmann L, Mayer W, Roggenbuck B, Stremlau A, zur Hausen H (1985) Structure and transcription of human papillomavirus sequences in cervical carcinoma cells. Nature 314:111–114

Sharma S, Kelly TK, Jones PA (2010) Epigenetics in cancer. Carcinogenesis 31:27–36

Siegel EM, Riggs BM, Delmas AL, Koch A, Hakam A, Brown KD (2015) Quantitative DNA methylation analysis of candidate genes in cervical cancer. PLoS One 10:e0122495

Smith JA, Haberstroh FS, White EA, Livingston DM, DeCaprio JA, Howley PM (2014) SMCX and components of the TIP60 complex contribute to E2 regulation of the HPV E6/E7 promoter. Virology 470:311–321

Snellenberg S, Schütze DM, Claassen-Cramer D, Meijer C, Snijders PJ, Steenbergen RD (2012) Methylation status of the E2 binding sites of HPV16 in cervical lesions determined with the Luminex XMAP system. Virology 422:357–365

Sugarawa K, Fujinaga K, Yamashita T, Ito Y (1983) Integration and methylation of shope papillomavirus DNA in the transplantable Vx2 and Vx7 rabbit carcinomas. Virology 130:88–99

Sun C, Reimers LL, Burk RD (2011) Methylation of HPV16 genome CpG sites is associated with cervix precancer and cancer. Gynecol Oncol 121:59–63

Tan SH, Leong LEC, Walker PA, Bernard HU (1994) The human papillomavirus type 16 transcription factor E2 binds with low cooperativity to two flanking binding sites and represses the E6 promoter through displacement of Sp1 and TFIID. J Virol 68:6411–6420

Thain A, Jenkins O, Clarke AR, Gaston K (1996) CpG methylation directly inhibits binding of the human papillomavirus type 16 E2 protein to specific DNA sequences. J Virol 70:7233–7235

Turan T, Kalantari M, Calleja-Macias IE, Villa LL, Cubie HA, Cuschieri K, Skomedal H, Barrera-Saldana HA, Bernard HU (2006) Methylation of the human papillomavirus-18 L1 gene: a biomarker of neoplastic progression? Virology 349:175–183

Turan T, Kalantari M, Cuschieri K, Cubie HA, Skomedal H, Bernard HU (2007) High-throughput detection of human papillomavirus-18 L1 gene methylation, a candidate biomarker for the progression of cervical neoplasia. Virology 361:185–193

Van Tine BA, Kappes JC, Banerjee NS, Knops J, Lai L, Steenbergen RD, Meijer CL, Snijders PJ, Chatis P, Broker TR, Moen PT Jr, Chow LT (2004) Clonal selection for transcriptionally active viral oncogenes during progression to cancer. J Virol 78:11172–11186

Vande Pol S, Klingelhutz AJ (2013) Papillomavirus E6 oncoproteins. Virology 445:115–137

Vasiljevic N, Scibior-Bentkowska D, Brentnall A, Cuzick J, Lorincz A (2014) A comparison of the methylation levels in HPV18, HPV31, and HPV33 genomes reveals similar associations with cervical precancers. J Clin Virol 59:161–166

Verhoef VM, Heideman DA, van Kemenade FJ, Rozendaal L, Bosgraaf RP, Hesselink AT, Bekkers RL, Massuger LF, Steenbergen RD, Snijders PJ, Berkhof J, Meijer CJ (2014) Methylation marker analysis and HPV16/18 genotyping in high-risk HPV positive self-sampled specimens to identify women with high grade CIN or cervical cancer. Gynecol Oncol 135:58–63

Vinokurova S, von Knebel Doeberitz M (2011) Differential methylation of the HPV16 upstream regulatory region during epithelial differentiation and neoplastic transformation. PLoS One 6: e24451

Wang JW, Roden RBS (2013) L2, the minor capsid protein of papillomaviruses. Virology 445:175–186

Wentzensen N, Sherman ME, Schiffman M, Wang SS (2009) Utility of methylation markers in cervical cancer early detection: appraisal of the state-of-the-science. Gynecol Oncol 112:293–299

Wentzensen N, Sun C, Ghosh A, Kinney W, Mirabello L, Wacholder S, Shaber R, LaMere B, Clarke M, Lorincz AT, Castle PE, Schiffman M, Burk RD (2012) Methylation of HPV18, HPV31, and HPV45 genomes and cervical intraepithelial neoplasia grade 3. J Natl Cancer Inst 104:1738–1749

Wettstein FO, Stevens JG (1983) Shope papillomavirus DNA is extensively methylated in non-virus-producing neoplasms. Virology 126:493–504

Wiley DJ, Huh J, Chang C, Kalantari M, Rao J, Goetz M, Msongsong E, Poulter M, Bernard HU (2005) Methylation of human papillomavirus DNA in samples of HIV-1 infected men screened for anal cancer. J Acqu Immunodef Syndr 39:143–151

Xi LF, Jiang M, Shen Z, Hulbert A, Zhou XH, Lin YY, Kiviat NB, Koutsky LA (2011) Inverse association between methylation of human papillomavirus type 16 DNA and risk of cervical intraepithelial neoplasia grades 2 or 3. PLoS One 6:e23897

Chapter 2
Epigenetic Regulation of Gammaherpesviruses: A Focus on Kaposi's Sarcoma-Associated Herpesvirus (KSHV/HHV-8)

Rosha Poudyal, Rolf Renne, and Michael P. Kladde

Abstract Gammaherpesviruses are ubiquitous in nature and infect a broad range of animal species. They have a biphasic life cycle that alternates between latent and lytic phases and are able to maintain a persistent infection for long periods. Both Epstein–Barr virus (EBV) and Kaposi's Sarcoma-associated Herpesvirus (KSHV) are oncogenic viruses that are known to cause several lymphoproliferative diseases in humans. EBV and KSHV persist as viral episomes that orchestrate very tightly controlled programs of gene expression, whereby a distinct subset of viral genes is expressed during the latent phase. Various stimuli can induce lytic reactivation of both viruses, which results in expression of lytic genes and is accompanied by changes in histone modification and DNA methylation. CTCF and cohesin binding provide segregation of chromatin loops as well as cross talk between different regions of the genome. Furthermore, noncoding RNAs (ncRNAs) provide an additional layer of epigenetic regulation for gammaherpesviruses. MAPit, a single-molecule footprinting assay, has revealed the occurrence of several subtypes of chromatin architecture at various KSHV promoters, suggesting the presence of heterogeneity within the population of KSHV viral episomes. In this chapter, we discuss the epigenetic regulation of gammaherpesviruses during latency and lytic reactivation, with a primary focus on KSHV.

Keywords Gammaherpesviruses • KSHV • EBV • Epigenetics • DNA methylation • Histone modifications • Epigenetic heterogeneity • CTCF • Cohesin • Non-coding RNA • microRNA • lncRNA • PAN RNA

R. Poudyal • M.P. Kladde (✉)
Department of Biochemistry and Molecular Biology, University of Florida, Gainesville, FL, USA
e-mail: kladde@ufl.edu

R. Renne
Department of Molecular Genetics and Microbiology, University of Florida, Gainesville, FL, USA

W. Doerfler, J. Casadesús (eds.), *Epigenetics of Infectious Diseases*, Epigenetics and Human Health, DOI 10.1007/978-3-319-55021-3_2

2.1 Introduction

Herpesviruses are ubiquitous, linear, double-stranded DNA viruses that infect a wide range of host animal species. More than 100 herpesviruses have been identified to date, which are classified into α, β, or γ subfamilies based on their genome organization and unique biological properties such as tissue tropism (Roizman 1982). Herpesviruses have a genome size ranging from 120 to 230 kilobases (kb), encoding 70–200 genes. Their genome is encapsulated in an icosahedral nucleocapsid that is approximately 100 nm in diameter (Liu and Zhou 2007). Herpesviruses establish a persistent infection and alternate between latent and lytic phases, exhibiting a gene expression pattern specific to each state. The γ-subfamily of herpesviruses is known to be oncogenic and is causally associated with several cancers, primarily in immunocompromised individuals. The γ-subfamily of herpesviruses is lymphotropic, but some have the capability to replicate in epithelial as well as endothelial cells. The two species of γ herpesviruses that infect humans are Epstein–Barr virus [EBV, also called Human Herpesvirus type 4 (HHV-4)] and Kaposi's Sarcoma-associated Herpesvirus (KSHV, also called HHV-8). EBV, discovered in 1964, infects approximately 90% of the adult world population and is largely associated with Hodgkin's lymphoma, Burkitt's lymphoma, and nasopharyngeal carcinoma (Epstein et al. 1964; Wei and Sham 2005; Maeda et al. 2009). KSHV, which was identified as the etiological agent of Kaposi's sarcoma (KS) in 1994, shows a diverse range of seroprevalence that varies among geographic regions (Table 2.1) (Cohen 2000; Chang et al. 1994; Chatlynne and Ablashi 1999; Wawer et al. 2001; Mohanna et al. 2005). KSHV has also been shown to cause two other neoplasms: Primary effusion lymphoma (PEL) and multicentric Castleman's disease (MCD) (Cesarman et al. 1995; Soulier et al. 1995).

Table 2.1 Worldwide seroprevalence of human γ-herpesviruses and their associated disorders

γ-herpesvirus	Worldwide seroprevalence	Associated disease
EBV (HHV-4)	>90% in adult population	Burkitt's lymphoma Central nervous system lymphomas Gastric carcinoma Hodgkin's lymphoma Nasopharyngeal carcinoma Non-Hodgkin's lymphoma Post-transplant lymphoproliferative disorders T-cell lymphoma
KSHV (HHV-8)	40% seroprevalence in sub-Saharan Africa 10% in Mediterranean countries 2–4% in northern Europe, Southeast Asia, and the Caribbean 5–20% in the United States	Classical, endemic, and AIDS-related Kaposi's Sarcoma Multicentric Castleman's disease Primary Effusion Lymphoma

2.1.1 Genome Organization and Circularization

The genomes of EBV (~172 kb) and KSHV (~140 kb) within their viral icosahedral capsids are linear, double-stranded DNA molecules flanked by terminal repeats (TRs) that are used for circularization of the virus. EBV, a *Lymphocryptovirus*, has a variable repeated 500 bp sequence TR and also contains multiple internal repeats interspersed within the unique regions (Young et al. 2007). KSHV, a *Rhadinovirus*, has a long unique region that is flanked by a variable number of 801 bp long TRs (Renne et al. 1996a). The TRs harbor sequences that serve as an origin of replication (Renne et al. 1996a; Zimmermann and Hammerschmidt 1995). In the capsid, both EBV and KSHV have no detectable levels of DNA methylation and core histones (Johannsen et al. 2004; Birdwell et al. 2014; Bechtel et al. 2005).

During primary infections, EBV viral particles bind to host cell surface receptor CD21, whereas KSHV binds to integrin and Ephrin A2, and are then internalized by endocytosis. The viral capsids, which house the viral genomes, are trafficked through the cytoplasm to the perinuclear region, where the viral genomes are ejected into the nucleus through nuclear pores (D'Addario et al. 2001; Akula et al. 2002; Chakraborty et al. 2012; Hahn et al. 2012). This mode of genome delivery prevents viral DNA from being degraded and exposed to DNA-dependent activators in the cytoplasm (Chandran 2010). Upon entering the host nucleus, the viral DNA circularizes by recombination within the TRs (Fig. 2.1). Genome circularization is an essential step for efficient viral infection as linear genomes are subject to exonucleolytic attack and can activate the host DNA damage response pathway (Weitzman et al. 2010; Deng et al. 2012). After circularization, the viral DNA of both EBV and KSHV become "chromatinized," acquiring histones and subsequently persisting in the nucleus of the host as multicopy, closed-circular, extrachromosomal episomes (Fig. 2.1) [reviewed in Knipe et al. (2013)]. These nuclear episomes have similar attributes to host cellular chromatin and are packaged into nucleosomes with a characteristic repeat length [reviewed in Knipe et al. (2013)].

The mechanism by which the viral DNA duplex establishes a successful non-reproductive, latent episomal state and is reactivated to a productive lytic phase is poorly understood. The genomes of both EBV and KSHV code for several proteins involved in immune evasion and cell cycle regulation, some of viral origin as well as copies of pirated cellular homologs (Ressing et al. 2015; Lee et al. 2012). In addition, both viruses encode microRNAs (miRNAs) and long noncoding RNAs (lncRNAs). The EBV genome contains about 80 protein coding genes, with two microRNA clusters within the *BART* and *BHRF1* genes [reviewed in Skalsky and Cullen (2015)]. KSHV has the coding potential for nearly 86 genes, 18 mature microRNAs from the KSHV latency-associated region (*KLAR*), and several non-coding RNAs [reviewed in Zhu et al. (2014)].

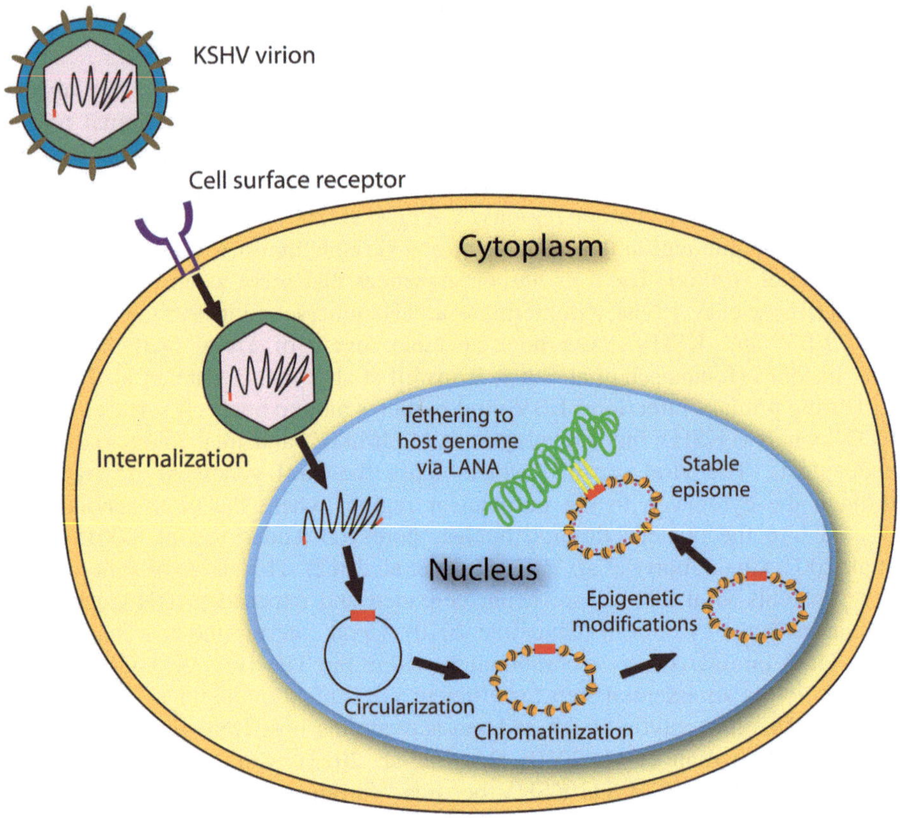

Fig. 2.1 Establishing a latent KSHV infection. During de novo infection, KSHV virions bind to the cell surface receptors, Integrin and Ephrin A2, and enter the cytoplasm by endocytosis. The viral capsid containing the linear, dsDNA viral genome docks with the nuclear pore and ejects the viral genome into the nucleus. Once inside the nucleus, the virus circularizes at its terminal repeats (TRs; *red rectangle*). During replication, the KSHV genome is assembled into nucleosomes, and cellular chromatin-modifying machinery is recruited to posttranslationally mark core histones and establish episomes that stably replicate as extrachromosomal minichromosomes. Methylation of DNA at inactive promoters and other regions is apparently established over the course of many months post infection. KSHV LANA protein (*yellow ellipses*) binds to the TRs and tethers the viral genome to host chromosomes. Viral episomes replicate synchronously with the host genome during S phase and are segregated to daughter cells during cell division, thereby enabling a persistent infection

2.1.2 Gene Expression During Latency

Gammaherpesviruses exhibit a biphasic life cycle with a persistent, but reversible, latent phase and a transient lytic reactivation phase. Both phases are characterized by distinct, tightly regulated gene expression profiles that are governed through the concerted action of histone modifications, DNA methylation, and noncoding RNAs (Tsurumi et al. 2005; Dourmishev et al. 2003). EBV displays more than one type of latency program, namely, Latency 0, Latency I, Latency II, and Latency III. Each of the latency programs is associated with the expression of a limited and distinct set of viral proteins and can vary in different cell types (Amon and Farrell 2005). Latency 0, observed in noncycling and resting B cells, is the most tightly regulated transcription program, where no viral genes are transcribed (Babcock et al. 2000). On the other hand, Latency III, observed in highly proliferating B cells, is the most transcriptionally permissive program, expressing all the gene products associated with latency, such as EBV nuclear antigen 1 (*EBNA1*), *EBNA2*, as well as latent membrane protein 1 (*LMP1*), *LMP2A*, and *LMP2B* [reviewed in Young and Rickinson (2004)].

KSHV may display less variation in latency types than EBV. Upon de novo infection, genes responsible for viral latency are expressed from the *KLAR* as a multicistronic transcript that encodes latency-associated nuclear antigen (LANA), virus-encoded Cyclin D homolog (vCyclin), viral Fas-associated death domain-like interleukin-1 beta-converting enzyme (FLICE)-inhibitory protein (vFLIP), and the Kaposin (K12) family of proteins (Zhong et al. 1996; Dittmer et al. 1998). No infectious viral particles are produced during latency. LANA, a functional ortholog of EBNA1 with respect to latent DNA replication and episome tethering, binds preferentially to the TRs of the viral genome and tethers it to the host chromosome. This ensures the faithful segregation of the viral genome to host daughter cells during cell division (Ballestas et al. 1999; Cotter and Robertson 1999). In addition, LANA is a multifunctional protein that is known to either activate or repress transcription of various cellular and viral genes (Renne et al. 2001; Garber et al. 2001; Fujimuro et al. 2003). The other *KLAR* gene products play key roles in host cell proliferation and survival (Chang et al. 1996; Thome et al. 1997; Ye et al. 2008). The chromatinization that leads to a condensed form of the episome provides protection to the viral genome from degradation while enabling tight regulation of gene expression [reviewed in Lieberman (2013)].

2.1.3 Gene Expression During Lytic Phase

For the completion of a full life cycle and maintenance of a persistent infection, gammaherpesviruses are required to undergo reactivation from latency, whereby new infectious viral particles are produced during lytic replication. Although the stimuli that promote the switch from latency to the lytic phase are not completely understood, various cellular phenomena, such as immune suppression, oxidative stress, and hypoxia, are known to trigger lytic reactivation [reviewed in Ye et al. (2011)]. In vitro, cells that are latently infected with EBV, KSHV, or both can be reactivated when treated with drugs that alter epigenetic modifications, e.g., sodium butyrate, a histone deacetylase inhibitor (HDACi), or 5-aza-2′-deoxycytidine, a DNA demethylating agent (Knipe et al. 2013; Shin et al. 2014; Shamay et al. 2006). During reactivation, the viral DNA is replicated by a viral polymerase along with a timely regulated cascade of gene expression, leading to the assembly and egress of mature infectious virions. Lytic viral genes are expressed in temporal fashion, activating three classes of lytic genes: immediate-early (IE), early (E), and late (L) genes (Renne et al. 1996b; Sun et al. 1999; Jenner et al. 2001).

The key viral IE proteins that are required for the transcriptional activation of other lytic genes are Zta in EBV and Replication and Transcription Activator (RTA/ORF50) in KSHV (Gl et al. 2007). Although a number of IE-lytic genes, such as *RTA*, *ORF45*, *K8.2*, *K4.2*, etc., are expressed upon reactivation in KSHV, it has been established that *RTA* functions as the master switch between latent and lytic gene expression (Sun et al. 1998). RTA is the only lytic viral protein that is both necessary and sufficient for the activation of several lytic promoters and replication of the viral genome (Wang et al. 2003; Guito and Lukac 2012). RTA is also known to auto-activate its own promoter through an RTA-responsive element and establish a positive feedback loop in the viral lytic gene expression (Deng et al. 2000). The expression of E genes is activated by IE gene products, which mostly include proteins that have enzymatic functions that are required for DNA replication (e.g., DNA polymerase I processivity factor ORF59) and for modulation of the immune system (MIR1/2) (Coscoy and Ganem 2001; Ishido et al. 2002; Majerciak et al. 2006). The L genes are expressed following the expression of E-lytic genes and are transcribed after lytic DNA replication (Honess and Roizman 1974). The L-lytic gene products consist of several viral structural proteins, including major capsid protein (MCP) encoded by ORF25, several membrane glycoproteins (K8.1), and a viral capsid antigen that facilitates the assembly and maturation of virions (Schulz and Yuan 2007). Viral tegument proteins that assist the virus during virion assembly, viral entry, and host immune evasion are also a part of the L-lytic gene expression period [reviewed in Sathish et al. (2012)].

2.2 Histone Modifications

Genomic DNA in the nucleus of all eukaryotic cells is associated with core histone proteins and nonhistone regulatory proteins to form chromatin. The fundamental repeating unit of chromatin, the nucleosome, comprises a histone octamer of two copies of each core histone protein (H2A, H2B, H3, and H4) wrapped by a left-handed superhelix of 147 bp of DNA, plus 20–80 bp of linker DNA (Li and Reinberg 2011; Kornberg and Lorch 1999; Luger 2003). Nucleosomes are organized in arrays and display higher levels of folding/condensation and chromatin organization.

The N-terminal tails of a histone octamer, which protrude from the globular domain of the nucleosome, can undergo at least eight distinct types of posttranslational covalent modification (Kouzarides 2007). An increasing number of amino acid residues (arginine, lysine, serine, and threonine) in the histone N-terminal tails are subject to posttranslational modifications, including acetylation, methylation, phosphorylation, ubiquitination, and ADP-ribosylation. Furthermore, the local concentration of differentially modified nucleosomes allows for the regulation of chromatin and largely determines the euchromatic (loosely packed) or heterochromatic (tightly packed) state of chromatin [reviewed in Ng and Cheung (2015)]. Histone acetylation, for example, is deposited locally following recruitment by site-specific DNA-binding proteins of histone acetyltransferases (HATs) that transfer an acetyl group from acetyl-CoA to various lysine residues in histones H3 and H4 (Takahashi et al. 2006; Wellen et al. 2009). High levels of histone acetyllysine have been causally linked to transcriptional activation (Grunstein 1997; Strahl and Allis 2000). Acetylation neutralizes the positive charge on histones, loosening their interaction with the negatively charged phosphodiester backbone of DNA (Widlund et al. 2000; Tse et al. 1998). Charge neutralization also weakens interactions between neighboring nucleosomes, leading to loss of higher-order compaction of the chromatin fiber, resulting in increased potential for gene transcription (Tse et al. 1998; Lee et al. 1993). In addition, interactions between histone acetyllysines and bromodomains lead to recruitment of histone acetyltransferases, providing a positive feedback loop of acetylation, and ATP-dependent remodelers to effect nucleosome disassembly or histone eviction (Chatterjee et al. 2011; Kingston and Narlikar 1999; Horn and Peterson 2001).

Multiple lysine (K) residues and arginine (R) residues in the histone H3 and H4 tails are also subject to methylation (Strahl et al. 1999). The lysine residues can accommodate mono-, di-, or trimethylated states, whereas the arginine residues can harbor mono- or dimethylated (symmetric or asymmetric) states. Using *S*-adenosyl-*L*-methionine (SAM) as a cofactor and methyl donor, histone methyltransferases (HMTs) catalyze the transfer of methyl groups to lysine or arginine residues on histone proteins (Struhl 1998). Promoter regions of active genes are marked with high levels of H3K4 methylation in addition to histone acetylation. Trimethylation of histone H3 lysine 4 (H3K4me3) is significantly enriched at transcription start sites (TSSs), and H3K36me3 is abundant in gene bodies and around the 3′ end of genes. Trimethylation of H3K9 and H3K27 is correlated with transcriptional repression. H3K27me3 levels are higher at silent promoters than at active promoters, and H3K9me3 is more prevalent in gene bodies and constitutive heterochromatin (Barski et al. 2007; Hahn et al. 2011). Previously, H3K4me3 and H3K27me3 were thought to have mutually exclusive localization, with strictly permissive or repressive functions, respectively. However, it has been found that there are many genes, particularly in embryonic stem cells, that possess both H3K4me3 and H3K27me3 modifications. Genes with such bivalent domains are frequently in a poised state that can be activated upon receiving particular stimuli (Bernstein et al. 2006). Similarly, bivalent histone modifications have also been observed in differentiated T cells (Roh et al. 2006).

Histone modifications regulate chromatin activity not only by altering histone–DNA interactions, but also by providing a landscape with differentially modified nucleosomes that can be recognized or "read" by other protein modules. Acetylated lysine residues on the histone tails create a docking site for bromodomains, which are found on various proteins and play a significant role in acetylation-dependent assembly of transcription regulator complexes (Yun et al. 2011). Methylated lysines, a very stable histone mark, are recognized by multiple proteins including

Tudor, chromodomain, PWWP, PHD, and BAH domains (Yun et al. 2011). Some chromodomain-containing proteins are also components of the RNA-induced transcriptional silencing complex, hence bridging distinct dimensions of gene regulation (Verdel et al. 2004).

Nucleosome positioning regulates accessibility of *trans*-acting factors to DNA and is a key regulator of DNA-templated processes, including transcription, DNA replication, and DNA repair (Ehrenhofer 2004; Richmond and Davey 2003; Owen-Hughes and Workman 1994). Consequently, regions of high nucleosome occupancy are associated with gene repression, whereas areas with extensive nucleosome remodeling correlate with promoters of active genes where RNA polymerase II (RNAPII) and other transcription factors (TFs) bind (Narlikar et al. 2002; Sun et al. 2009). Such dynamically remodeled promoters also have elevated levels of histone acetylation (Barski et al. 2007). A key characteristic of highly expressed genes is the presence of a nucleosome-free region (NFR) immediately upstream from or including the TSS (Schones et al. 2008; Jiang and Pugh 2009; Cairns 2009). Nucleosomes are also found to be positioned on the 5′ end of genes, creating a chromatin-organizing center that regulates access to TSSs for TF binding (Zhang et al. 2011). In biochemical studies targeting the ATP-dependent chromatin remodeler complex SWI/SNF to a dinucleosome by an activator, the activator-adjacent nucleosome is slid toward the more distal nucleosome from which one H2A/H2B dimer is first displaced, followed by displacement from the same nucleosome of the H3/H4 tetramer and remaining H2A/H2B dimer (Dechassa et al. 2010). This results in sliding of the activator-targeted mononucleosome and the complete displacement of the distal mononucleosome from the dinucleosome substrate. In brief, nucleosome positioning, histone modifications, and DNA methylation collectively work to generate chromatin states that regulate gene expression. Combinatorial contributions of epigenetic modifications add a dense layer of complexity to gene regulation.

2.2.1 Histone Modifications During Gammaherpesvirus Latency

Genome-wide studies have made it apparent that both EBV and KSHV have a complex pattern of histone modifications during latency (Knipe et al. 2013; Day et al. 2007; Minarovits 2006; Arvey et al. 2013; Günther and Grundhoff 2010; Toth et al. 2010; Hu et al. 2014). In EBV, the EBNA1 N-terminal domain interacts with the host cell protein EBNA1-binding protein (EBNA1BP2), which binds to AT-rich DNA (Sears et al. 2004). In KSHV, LANA has a DNA-binding and a chromatin-binding domain in its C- and N-terminal domains, respectively (Ballestas et al. 1999; Cotter and Robertson 1999; Garber et al. 2002). The LANA N-terminal domain interacts with host chromosomes by binding to the acidic patch on histone H2A-H2B dimers of nucleosomes (Barbera et al. 2006a, b). The C-terminal domain binds to multiple LANA binding sites (LBS) on the TRs, as well as interacts with cellular proteins that recognize modified histones, such as methyl CpG binding protein (MeCP2), and bromodomain proteins BRD2 and BRD4 (Garber et al. 2002; Hellert et al. 2015; Viejo-Borbolla et al. 2005; You et al. 2006). The tethering of the viral genomes to host chromosomes via EBNA1 or LANA enables the maintenance of a stable viral copy number over the course of many host cell divisions (Verma et al. 2013). Thus, both EBV and KSHV are tethered to host chromosomes via interactions with specific, virally encoded proteins.

The KSHV genome has been shown to be decorated with both activating histone marks, H3 acetylation (H3ac) and H3K4me3, and repressive marks, H3K9me3 and H3K27me3 (Günther and Grundhoff 2010; Toth et al. 2010). The localization of H3K4me3 and H3K27me3 is largely mutually exclusive in the latent KSHV genome (Günther and Grundhoff 2010; Toth et al. 2010). The *KLAR*, which expresses the latent genes, possesses activating histone marks, H3ac and H3K4me3, and colocalizes with transcriptionally active RNAPII (Günther and Grundhoff 2010; Toth et al. 2010). In contrast, IE and E-lytic genes in latent genomes possess either highly activating marks, H3ac and H3K4me3, or bivalent modification of H3K4me3 and H3K27me3 (Günther and Grundhoff 2010; Toth et al. 2010; Hu et al. 2014). As in embryonic stem cells, the bivalent modifications enable the rapid induction of genes in response to a triggering stimulus (Bernstein et al. 2006). The L-lytic genes in episomes are initially characterized by high levels of the repressive heterochromatic marks H3K9me3 and H3K27me3 (Günther and Grundhoff 2010; Toth et al. 2010).

The Polycomb repressive complex proteins compact chromatin with dense nucleosome occupancy, resulting in gene silencing (Schuettengruber and Cavalli 2009). The EZH2 catalytic subunit of Polycomb Repressive Complex 2 (PRC2) trimethylates H3K27 [reviewed in Riveraa et al. (2014)]. H3K27me3 recruits the PRC1 complex, because the modification serves as a docking site for a chromodomain-containing subunit of PRC1 (Cao et al. 2001). Recruitment of PRC1 prevents chromatin remodeling factors, such as SWI/SNF, from accessing nucleosomes and leads to the formation of a repressive chromatin state (Cao et al. 2001).

The repressive mark H3K27me3 and the PRC2 catalytic subunit EZH2 have been shown to colocalize and bind to the KSHV genome predominantly during latency (Günther and Grundhoff 2010; Toth et al. 2010). While H3K27me3 is widespread across the KSHV genome, H3K9me3 is restricted to two regions containing late genes (Günther and Grundhoff 2010; Toth et al. 2010). The histone demethylase JMJD2A (Jumonji domain 2A), which primarily removes H3K9me3, also binds to the KSHV genome and is thought to guard against the methylation of H3K9 (Chang et al. 2011). The binding of JMJD2A to chromatin prevents the formation of H3K9me3, which expedites the acetylation of H3K9 upon reactivation and enables robust induction of genes (Chang et al. 2011). Together, the PRC2 and JMJD2A protein complexes help maintain H3K27me3 and other silencing marks on lytic genes and repress lytic gene expression during latency (Knipe et al. 2013). LANA, which is continuously expressed during latency following de novo infection, interacts with several host transcriptional factors [MeCP2, DNA methyltransferases (DNMTs), etc.] and chromatin remodelers (MLL, CBP, etc.), leading to the epigenetic silencing of lytic genes and promotion of viral latency, respectively (Hu et al. 2014; Matsumura et al. 2010; Kim et al. 2013). Likewise, LANA interacts with hSET1 complex, a H3K4 methyltransferase, and may play a role in binding at latent promoters thereby protecting them from PRC2-mediated silencing (Hu et al. 2014).

2.2.2 Histone Modifications During Lytic Reactivation

Histone deacetylases (HDACs) remove acetyl groups from lysines and play a critical role in the regulation of gene expression. Gammaherpesvirus latency has been shown to be disrupted by HDAC inhibitors (Miller et al. 1996). Lytic reactivation mostly corresponds to the alteration in histone modifications of the viral genome by histone acetyltransferases (HATs). Chromatin remodeling post-acquisition of acetylation of histones H3 and H4 following the treatment with an HDAC inhibitor is observed both at *EBNA* and *RTA* promoters (Alazard et al. 2003; Lu et al. 2003). The *RTA* promoter in latent KSHV concurrently harbors both activating H3K4me3 and repressive H3K27me3 histone modifications, which keeps it poised for activation and rapid gene expression upon induction (Günther and Grundhoff 2010; Toth et al. 2010; Hu et al. 2014).

A wave of histone modifications occurs once lytic reactivation is induced. Demethylation of H3K27me3 by UTX (Ubiquitously transcribed tetratricopeptide repeat, X chromosome protein) or dissociation of EZH2 reverses PRC2-mediated repression of the *RTA* promoter (Günther and Grundhoff 2010; Toth et al. 2010). Reactivation by RTA leads to the dissociation of EZH2 from the genome, especially from IE and E-lytic genes, where decreasing levels of H3K27me3 are concomitant with increasing gene expression (Günther and Grundhoff 2010; Toth et al. 2010). RTA binds to its own promoter using cellular transcription factor CBF1, which recruits the CBP/p300 histone acetyltransferases, SWI/SNF chromatin remodelers, and the TRAP/mediator coactivators (Gwack et al. 2003). This feed-forward activation allows for efficient modification of the chromatin structure of the virus to a transcriptionally active state, which enables a complete viral reactivation cycle. Overall, histone modifications play an important role in the maintenance of latency as well as the lytic reactivation process in gammaherpesviruses. All these studies demonstrate that gammaherpesvirus genomes, like host chromosomes, show similar epigenetic modifications.

2.2.3 Intra- and Inter-locus Epigenetic Heterogeneity in KSHV Episomes

A major difference between viral and human genomes is that viral genomes often persist at a high copy number. Since ChIP-seq and ChIP-on-chip always measure the average state of overall genomes, it was formerly not clear if heterogeneity exists among the viral genomes. Heterogeneous gene expression has been reported in various models of KSHV-infected cells (Chang and Ganem 2013). Heterogeneous chromatin states at particular loci across the population of KSHV episomes could constitute an underlying reason for spontaneous activation of the virus in latently infected cells. The episomal chromatin state maintains the latent phase and represses the lytic phase (see histone modifications above and DNA methylation below). With the advent of single-cell and single-molecule techniques, epigenetic heterogeneity has been observed in cancer stems cells, pluripotent stem cells, and also the KSHV genome (Smallwood et al. 2014; Nabilsi et al. 2014; Boland et al. 2014; Darst et al. 2013).

NFRs, as discussed above, are essential for transcription factor binding and formation of the RNAPII transcription pre-initiation complex. Typically, accessibility to the nucleases DNase I and micrococcal nuclease (MNase) are used to map hypersensitivity and the positions of nucleosomes, respectively. Following MNase digestion of nuclear chromatin, increased accessibility at the *RTA* TSS due to remodeling of positioned nucleosomes was shown to be a regulatory step in the transition from the latent to lytic state in the KSHV viral life cycle (Lu et al. 2003). However, a population-averaged view of chromatin structure was obtained, obscuring detection of cell-to-cell epigenetic heterogeneity. To detect such heterogeneity, we used Methyltransferase Accessibility Protocol for individual templates (MAPit), a single-molecule, methylation-based footprinting technique, to examine KSHV chromatin structure. In MAPit, the native chromatin structures of nuclei are probed with the viral DNA methyltransferase (DNMT), M.CviPI, which modifies accessible GC dinucleotides, followed by bisulfite sequencing (Fig. 2.2) (Xu et al. 1998; Darst et al. 2012). As modification of GC sites can be unequivocally distinguished from endogenous CG methylation when analysis of methylated GCG sites is excluded, MAPit enables simultaneous detection of endogenous CG methylation, nucleosome positioning, and DNA-bound transcription factors at high resolution (Darst et al. 2010, 2012; Kilgore et al. 2007).

To visualize epigenetic heterogeneity across the populations of KSHV episomes and infected cells, three gene promoter regions (*LANA*, *RTA*, and *vIL6*) were investigated (Darst et al. 2013). MAPit was used to analyze the chromatin configurations present at the IE-lytic *RTA* promoter on KSHV episomes in TREx BCBL-1 RTA cells that contain a doxycycline-inducible *RTA* transgene (integrated into a host cell chromosome) (Darst et al. 2013; Nakamura et al. 2003). About 1–3% of TREx BCBL-1 RTA cells exhibit spontaneous lytic reactivation at any time. Many bisulfite sequencing reads, each representing the chromatin architecture of a specific promoter from one genome in the population, were obtained from cells that

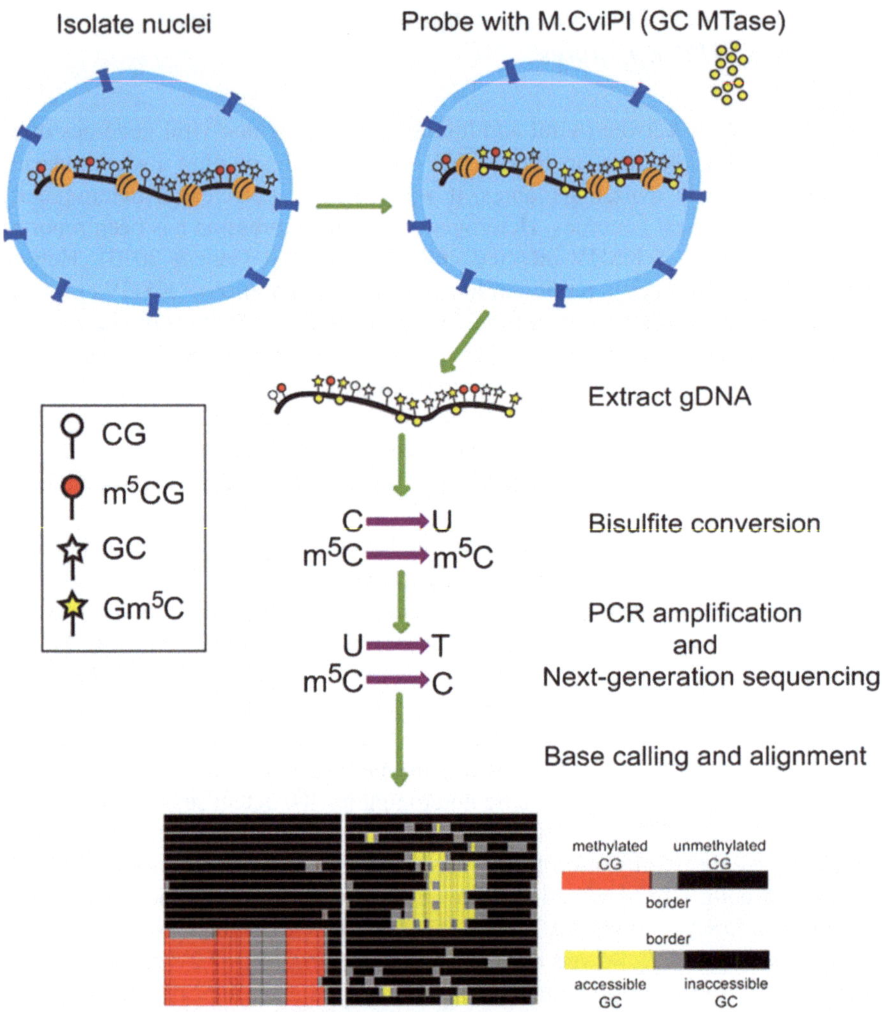

Fig. 2.2 Overview of Methyltransferase Accessibility Protocol for individual templates (MAPit) assay. Nuclei are isolated from cells and incubated with the GC methyltransferase, M.CviPI, which methylates cytosines in GC context. Following probing with M.CviPI, genomic DNA is extracted, bisulfite converted, and PCR amplified. The PCR product is either cloned for Sanger sequencing or prepared as a library for next-generation sequencing. The sequencing reads are aligned to the reference genome (with masked Cs), and methylation maps are generated. Each *horizontal line* represents a single DNA molecule/read (*bottom panel*). Any two consecutive endogenously methylated cytosines are shown in *red* patches, and exogenously methylated cytosines are shown in *yellow* patches. *Gray* represents the boundary between a methylated and unmethylated cytosine patch. For CG map, *red* represents hypermethylation whereas *black* represents hypomethylation. For GC map, *yellow* represents chromatin accessibility, and *black* represents either a protein footprint or a closed, inaccessible chromatin structure. *Red-filled circle*, endogenous CG methylation; *unfilled circle*, unmethylated cytosine; *yellow-filled star*, M.CviPI-accessible and methylated GC; *unfilled star*, M.CviPI-inaccessible and unmethylated GC

were latently infected or in which viral reactivation had been induced. Bisulfite sequencing reads from uninduced and lytically reactivated TREx BCBL1-RTA cells were pooled and populated five different clusters following unsupervised hierarchical clustering (Fig. 2.3). These chromatin structures varied from completely inaccessible, most likely transcriptionally inactive (Fig. 2.3, cluster i), to almost completely accessible at every GC site, presumably corresponding to lytically replicating viral genomes that were stripped of core histones (Fig. 2.3, cluster v). Strikingly, about 40% of the episomes in the latently infected cells from all three analyzed promoters were completely inaccessible to the viral GC DNMT, even at high concentrations of the enzyme, suggesting genome-wide compaction of a subpopulation of KSHV episomes (Fig. 2.3, cluster i). Interestingly, lytic reactivation significantly depleted the percentage of copies of the *RTA* promoter in a cluster displaying a short ~75 bp NFR in favor of another cluster with a 150–175 bp NFR (Fig. 2.3, clusters ii and iii), consistent with the chromatin remodeling observed previously by MNase digestion (Lu et al. 2003). However, the ~40% of *RTA* promoters that were inaccessible in uninduced cells, were not significantly depleted by lytic reactivation, suggesting that the pool of highly compacted episomes was less responsive to lytic reactivation stimuli and did not show a shift in chromatin architecture (Darst et al. 2013).

MAPit was also used to analyze the chromatin configurations present at the latent *LANA* promoter. Bisulfite sequencing reads from the *LANA* promoter populated nine different hierarchical clusters, i.e., nine different overall chromatin configurations. Each cluster exhibited varying degrees of chromatin accessibility, in particular, displaying different NFR sizes at the *LANA* latent TSS (Darst et al. 2013). To our knowledge, these data constitute the first observation of episome-to-episome variation in chromatin structure. In contrast, CTCF (discussed in detail below) exhibited strong footprints at its three binding sites in the *LANA* promoter in a significant fraction of promoter copies in each identified subpopulation, demonstrating uniform organization at specific regions.

Another independent method called Formaldehyde-Assisted Isolation of Regulatory Elements (FAIRE), which identifies regulatory regions on a population average basis, has showed that only around 8% of the latent KSHV genome has an open chromatin structure, and that most of the viral episomes are covered in nucleosomes (Nagy and Price 2009; Hilton et al. 2013). The study also showed the presence of two broad subpopulations based on the regions of open chromatin within the KSHV episomes: ones that are bound to CTCF and map to transcriptionally inactive loci and ones that are in the open chromatin region but not bound by CTCF and are presumably actively transcribing or poised for transcription (Hilton et al. 2013). In sum, considering our single-molecule studies along with the FAIRE-seq analysis, it seems plausible that at any given time in latency and during lytic reactivation, specific loci in KSHV episomes adopt diverse epigenetic configurations that likely vary in their potential to be transcriptionally active.

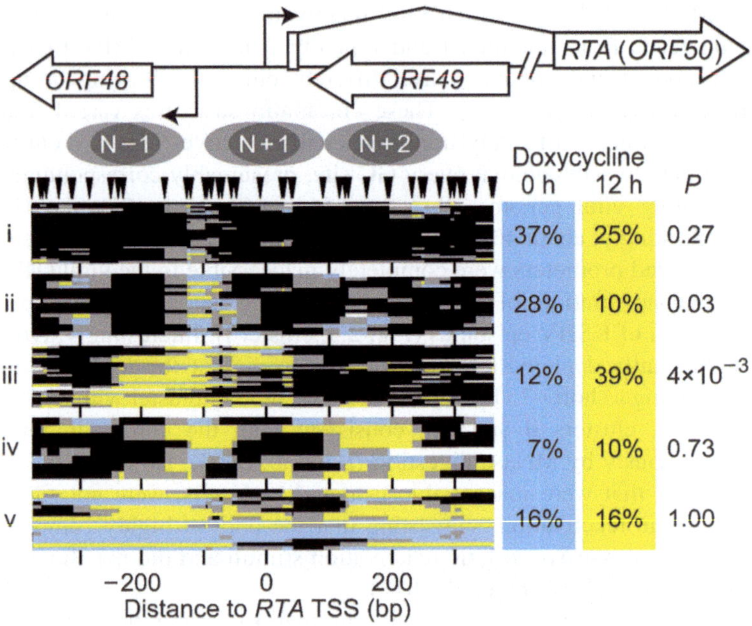

Fig. 2.3 Various classes of chromatin architecture at the *RTA* promoter in KSHV in TREx BCBL1-RTA reactivated cells. MAPit assay was performed on nuclei isolated from TREx BCBL1-RTA cells in which the *RTA* transgene was either uninduced (0 h doxycycline) or lytically induced (12 h doxycycline). The *top panel* depicts the organization of the amplicon from the KSHV genome that was analyzed by bisulfite sequencing. Each *horizontal line* represents one sequenced read or molecule from one episome in the population. Regions accessible to the viral GC DNMT M.CviPI were colored *light blue* and *yellow* in reads obtained from uninduced and induced cells, respectively. Regions inaccessible were colored *black*. *White* indicates areas of nonalignment to the reference sequence. Reads from both uninduced and induced cells were pooled and subjected to unsupervised hierarchical clustering. The coloring scheme and clustering facilitate pattern recognition. The GC methylation plot is centered on the *RTA* TSS. Each *black triangle* represents a GC site. Regions of protection against GC DNMT methylation corresponding to nucleosomes $N-1, N+1$, and $N+2$ are indicated. Five distinct clusters, ranging from closed (cluster i, *top*) to open (cluster v, *bottom*), were observed, demonstrating epigenetic heterogeneity across the population of analyzed KSHV episomes. Values on the *right* are the expectation *P* values for the proportion of reads from each sample. Cluster (iii), displaying an approximately 150 bp NFR, is presumably the transcriptionally active subset of episomes. Upon lytic induction of the virus, a statistically significant fraction of episomes in cluster (ii) (with ~75 bp NFR) is depleted (i.e., low number of reads with accessibility colored *yellow*), and a highly significant increase in the fraction of induced episomes in cluster (iii) (~150 bp NFR; most reads with accessibility are colored *yellow*). This indicates active chromatin remodeling at the *RTA* TSS during lytic gene reactivation in response to *RTA* activation. Figure reproduced from Darst RP, Haecker I, Pardo CE, Renne R, Kladde MP (2013) Epigenetic diversity of Kaposi's sarcoma-associated herpesvirus. Nucleic Acids Res 41:2993–3009 under the CCA license

2.3 DNA Methylation

DNA methylation predominantly occurs at carbon 5 (C-5) on cytosine (m^5C) in the context of CG dinucleotides in vertebrates (Bird 1986; Deaton and Bird 2011). Mammalian DNMT3A and DNMT3B establish initial patterns of de novo CG methylation during embryogenesis [reviewed in Feng et al. (2010)]. This activity is enhanced by interaction of the catalytic domains of either DNMT3A or DNMT3B with the carboxyl-terminal domain of DNMT3-like (DNMT3L), which has no known catalytic activity of its own (Bestor 2000; Cheng and Blumenthal 2008; Suetake et al. 2004; Jia et al. 2007). Interaction of the DNMT3A-DNMT3L complex with nucleosomes is inhibited by trimethylation of histone H3K4 in transcriptionally active regions of chromatin (Ooi et al. 2007). DNMT1 is preferentially recruited to hemi-methylated CG sites in replication forks by ubiquitin-like with PHD and RING finger domains 1 (UHRF1) and therefore, in the classical view, is the primary activity for maintenance of DNA methylation patterns in vertebrates (Sharif et al. 2007). However, biochemical studies using purified DNMT1 have demonstrated that the enzyme can also perform de novo methylation (Pradhan et al. 1999; Bacolla et al. 1999, 2001). Additionally, siRNA depletion of DNMT3A, DNMT3B, and DNMT3L, either alone or in all combinations, revealed that DNMT1 can de novo methylate CG sites in cells (Tiedemann et al. 2014). Aberrant hypermethylation of CG islands (GC-rich regions of ~1 kb with a high content of unmethylated CG dinucleotides) in promoters, which are predominantly unmethylated in non-diseased somatic cells, is a characteristic phenotype of many cancers (Robertson and Wolffe 2000).

The methylation status of CG dinucleotides is recognized by a family of nuclear proteins called methyl-CG binding proteins. In humans, members of the methyl-CG binding family of proteins are related to each other by the presence of a methyl-CpG-binding domain (MBD) that often recognizes m^5CG (Hendrich and Bird 1998). DNA methylation can either directly repress local transcription by inhibiting the binding of transcription factors or indirectly by recruiting MBDs and remodeling chromatin into a repressive conformation. MeCP2 recruits and binds to a multiprotein corepressor complex containing Sin3A-histone deacetylases (HDAC1/2), which leads to chromatin reassembly and increased nucleosome occupancy (Robertson and Wolffe 2000; Theisen et al. 2013; Jones et al. 1998).

DNA methylation plays an important role in the establishment of viral latency, although more so in EBV than in KSHV. Treating cells that are latently infected with EBV or KSHV with 5-azacytidine, a DNA demethylating agent, induces stochastic viral lytic reactivation, indicating that DNA methylation plays a vital role in controlling the lytic phase (Masucci et al. 1989; Pantry and Medveczky 2009). This suggests that direct methylation of lytic promoters or indirect regulation by methylation of other genes is essential for repression of lytic genes during latency (Chen et al. 2001). However, while much work has focused on deciphering methylation regulatory mechanisms in EBV and KSHV, the role of various DNMTs in the establishment and maintenance of the viral genomes has not been rigorously elucidated.

DNA methylation is crucial in propagating the EBV life cycle as reactivation of some lytic viral promoters is methylation dependent (Bhende et al. 2004). Zta, the lytic activator, preferentially binds to its cognate target elements when CG motifs within them are methylated; therefore, some lytic viral genes require accumulation of DNA methylation for transcriptional activation by Zta (Bhende et al. 2004; Karlsson et al. 2008; Kalla et al. 2011; Woellmer and Hammerschmidt 2013; Hong et al. 2017). EBV latency stages (I–III) are defined by the methylation state of a distinct set of gene promoters; however, DNA methylation is usually not observed at transcriptionally active latent promoters (Tao and Robertson 2003; Fejer et al. 2009). Lytic replication of the EBV genome resets its genome to an epigenetically naïve state as the newly synthesized viral DNA molecules lack detectable epigenetic modifications (Woellmer and Hammerschmidt 2013).

DNA methylation appears to exert less epigenetic control in KSHV. During latency, methylation is neither detectable at the *LANA* latency-associated promoter nor at proximal promoters such as *K9*, *ORF45*, *K7*, and *ORF8*. In contrast, DNA methylation is present at transcriptionally silent regions and at many lytic gene promoters (Günther and Grundhoff 2010; Toth et al. 2010). The KSHV genome is initially unmethylated in the viral capsid. In contrast to EBV, after de novo infection in model cell systems, KSHV episomes are not prone to rapid accumulation of DNA methylation. Instead, DNA methylation steadily accumulates over the course of many months, primarily over transcriptionally inactive promoters (Günther et al. 2014). This gradual accumulation of DNA methylation on KSHV, as compared to rapid association with and modifications of histones, strongly suggests that chromatinization and posttranslational histone modifications play primary roles in the establishment and maintenance of latency upon infection of the host by the virus. DNA methylation might reinforce the inhibition of expression of lytic cycle gene products at later times after productive infection. DNA methylation is undetectable at the *RTA* promoter in several KSHV-infected cell lines, suggesting that the repression of this gene promoter, and perhaps several others, during latency is achieved by an alternative mechanism (Günther et al. 2014). Furthermore, methylation of histone H3K4 and DNA are mutually exclusive, so regions with active or bivalent histone modifications usually lack DNA methylation (Rose and Klose 2014). Therefore, nucleosome positioning and histone modifications rather than DNA methylation appear to be the key players in establishment of latency in KSHV.

2.4 Chromatin Organization by CTCF and the Cohesin Complex

CTCF (CCCTC-binding factor) is a nuclear transcription factor that is involved in the three-dimensional organization of eukaryotic genomes [reviewed in Ong and Corces (2014)]. CTCF contains 11 zinc fingers with varying target recognition specificity that bind in different combinations to the consensus sequence CCGCGNGGNGGCAG, producing a 50–60 bp footprint (Ohlsson et al. 2001, 2010; Kim et al. 2007). CTCF blocks enhancer function when bound between enhancers and promoters, thus serving as a chromatin insulator (Ohlsson et al. 2001; Herold et al. 2012).

CG methylation blocks CTCF binding to many genomic sites and thus regulates enhancer blocking activity, as exemplified by allele-specific, parent-of-origin imprinting of the mammalian *H19*/*IGF2* locus (Szabo et al. 1998, 2004). CTCF binds to the unmethylated imprinting control region (ICR) on the maternally inherited and hypomethylated allele, preventing an enhancer from communicating with and activating the distally located *IGF2* promoter. As a consequence, this enhancer is redirected to activate transcription from the more proximal *H19* promoter. By contrast, on the paternally inherited, hypermethylated allele, methylation of a CG dinucleotide in the ICR blocks CTCF binding, alleviating its enhancer blocking activity. This allows the enhancer to communicate with and induce *IGF2* expression at the expense of *H19* activation (Szabo et al. 2004).

At least some of these regulatory properties are likely governed by CTCF's ability to organize chromatin structure. Sites bound by CTCF in the human genome are strongly depleted of nucleosomes and DNA methylation (Kelly et al. 2012). By contrast, arrays of well-positioned nucleosomes reside adjacent to CTCF binding sites (Fu et al. 2008). At CTCF-bound sites located at least 10 kb from promoters, a sufficient distance to ensure strong depletion of histone H3K4 trimethylation, DNA methylation often accumulates and is confined to linker DNA, i.e., excluded from nucleosomal DNA wrapped around the core histone octamer (Kelly et al. 2012).

In EBV, CTCF has been found to bind at key regulatory regions, namely, the Cp, the Qp, and the latent *LMP1*/*2* promoters. It can serve the function of a chromatin boundary element or affect the transcription level of surrounding genes. Deletion of the *EBNA2* CTCF binding site has been shown to increase the level of *EBNA2* transcription, whereas overexpression of CTCF can repress *EBNA2* expression (Chau et al. 2006). CTCF binds upstream of the Qp and acts as an insulator, impeding accumulation of DNA methylation at the promoter. Chromatin-bound CTCF also prevents the spread of the repressive histone modification H3K9me3. When binding of the factor is abolished, the repressive chromatin mark spreads to the Qp and silences transcription. Thus, CTCF plays an essential role in preventing the promiscuous transcription of genes during latent EBV infection and also blocks the epigenetic silencing of the Qp (Tempera et al. 2010).

CTCF and the cohesin complex associate at several regions on the KSHV genome, with the strongest colocalization upstream of the major latency-associated transcript region (Stedman et al. 2008). Deletion of the CTCF-binding site from the viral genome disrupts cohesin binding, resulting in deregulated and elevated expression of neighboring lytic genes (Stedman et al. 2008). Chromatin conformation capture (3C) assay has shown that the KSHV *KLAR* is physically juxtaposed to the promoter region of *RTA (ORF50)* (Kang et al. 2011). CTCF and cohesin mediate the formation of two major chromatin loops in the KSHV episome: (1) a short 10 kb loop between the cluster of three CTCF sites upstream of the *LANA* ORF and the 3′ end of the *K12* gene, which appears to organize the entire *KLAR* region into a topological domain and (2) a larger loop (>50 kb) between the CTCF binding site cluster and the promoter for *RTA* (*ORF50*) (Knipe et al. 2013; Kang et al. 2011). Disruption of CTCF-cohesin binding by genetic mutation of the viral genome leads to the elimination of the chromatin loops, decreased intra- and extracellular KSHV copy number, reduced expression of IE genes (*RTA* and *ORF69*), and increased LANA expression (Kang et al. 2011). Furthermore, knockdown of cohesin results in a dramatic 100-fold increase in viral yield, indicating the importance of CTCF and cohesin-mediated control of latent and lytic viral gene expression (Li et al. 2014).

In addition to looping and partitioning the viral genome into domains, CTCF serves as a boundary element that facilitates the organization of nucleosomes. At the KSHV *LANA* promoter, CTCF occupies a cluster of three sites spanning ~150 bp, including intervening sequences. These CTCF sites in *LANA* reside between the open reading frame (*ORF73*) and lytic TSS, which is located downstream of a second latent TSS used during latency and lytic reactivation (Sun et al. 1998). An array of positioned nucleosomes is arranged downstream of the CTCF sites and CG methylation is apparently restricted to the linkers (Darst et al. 2013), as was observed in human chromosomes (Kelly et al. 2012). Ablation of CTCF binding at *LANA* by genetic mutation of all three binding sites disrupts nucleosome positioning, reduces the efficiency of de novo infection, leads to alterations in RNAPII binding, accumulation of the unspliced form of the multicistronic transcript, and changes in histone modifications in the latency control region (Kang et al. 2013). Thus, CTCF along with the cohesin complex serves as an organizer of higher-order chromatin structure in the coordination of latent and lytic gene expression and viral replication, as blocks of latent and lytic genes are separated within the virus by CTCF.

2.5 Noncoding RNAs and Epigenetic Control

A significant fraction of the eukaryotic genome has been reported to transcribe functional RNA molecules that do not code for proteins, i.e., noncoding RNAs (ncRNAs) [reviewed in Palazzo and Lee (2015)]. These ncRNAs regulate gene expression at both the transcriptional and posttranscriptional level. There are reports of regulatory RNAs that are longer than 200 nt, long noncoding RNAs (lncRNAs), and shorter RNAs (sRNAs) that are less than 200 nt in length (Bartel 2004; Kapranov et al. 2007; Mercer et al. 2009). In the context of this chapter, we focus on ncRNA targets that contribute to the epigenetic regulation of latent and lytic replication.

MicroRNAs (miRNAs) are 20–22 nt sRNAs that regulate gene expression by targeting mRNAs for degradation or translational repression and regulate cellular processes such as apoptosis and tumorigenesis (Bartel 2004; Mercer et al. 2009; Ponting et al. 2009). MiRNAs are synthesized by RNAPII as a long primary miRNA (pri-miRNA) transcript, which is then serially processed first by the nuclear enzyme Drosha, in conjunction with its cofactor DGCR8, and subsequently by Dicer activity to yield short functional miRNAs (Hutvagner et al. 2001; Chendrimada et al. 2005). Many viruses, including herpesviruses, also encode miRNAs; more than 225 viral miRNAs have been identified to date [reviewed in Zhu et al. (2014) and Grundhoff and Sullivan (2011)].

Gammaherpesviruses encode numerous miRNAs. The EBV genome harbors two differentially expressed miRNA clusters, one within the *BART* gene and a second within *BHRF1*. The *BHRF1* cluster codes for three pre-miRNAs, whereas at least 20 miRNAs are transcribed from the *BART* pri-miRNA cluster (Cai et al. 2006). MiRNAs generated from the miR-*BART* cluster in infected cells have been shown to target anti-apoptotic cellular and tumor suppressor genes like *CASZ1a*, *OCT1*, *PAK2*, and *NDRG1* and enhance carcinogenesis (Kang et al. 2015; Kanda et al. 2015). One of the targets of the EBV-miR-BHRF1-2 from the *BHRF1* cluster is the PR domain zinc finger protein 1 (*PRDM1*) transcript, which encodes a repressor in the β-interferon pathway. Repression of *PRDM1* promotes viral survival and lymphomagenesis (Ma et al. 2015).

In KSHV-infected cells, several viral miRNAs have been demonstrated to regulate both host and viral gene expression during latency as well as the lytic phase (Samols et al. 2005; Pfeffer et al. 2005). KSHV miRNAs have also been shown to induce metabolic transformation, repress inhibitors of oncogenic transformation, and promote survival and proliferation of infected cells, hence facilitating the pathogenesis of KSHV (Yogev et al. 2014; Forte et al. 2015; Yang et al. 2014; Kieffer-Kwon et al. 2015). The KSHV genome encodes 12 pre-miRNAs from two different locations in the genome, which yield 25 mature miRNAs [reviewed in Zhu et al. (2014)]. At the molecular level, KSHV miRNAs, particularly miR-K12-3 and miR-K12-11, target host transcription factors, such as Myc, Ets-1, and C/EBPα, which are known to activate transcription of *RTA* (Plaisance-Bonstaff et al. 2014). Thus, KSHV miRNAs directly facilitate and provide tight regulation of viral latency maintenance by downregulating many host transcription factors which, in turn, leads to downregulation of the expression of *RTA* (Plaisance-Bonstaff et al. 2014). Besides mimicking cellular oncogenic miRNAs and targeting host transcription factors, KSHV miRNAs have also been implicated in regulating genes coding for cellular transport and membrane proteins such as *KCNS1*, *PRAM1*, *IPO5*, and *EDA* (Gottwein et al. 2011; Quan et al. 2015).

PAR-CLIP and HITS-CLIP techniques have identified more than 2000 targets for KSHV miRNAs (Gottwein et al. 2011; Haecker et al. 2012). For instance, KSHV miR-K12-4-5p decreases the expression of host retinoblastoma-like protein 2 (RBL2), which is a known repressor of *DNMT3A* and *DNTM3B* gene transcription (Lu et al. 2010). Therefore, by downregulating RBL2 expression via miR-K12-4-5p, cellular de novo DNMT expression is upregulated. The resulting global increases in DNA methylation of the viral and host genomes attenuates the gene expression program of the viral lytic cycle (Lu et al. 2010).

Besides miRNA, both EBV and KSHV transcribe highly abundant ncRNAs that can bind to host factors and enhance viral pathogenesis. EBV expresses two non-polyadenylated ncRNAs, *EBER1* and *EBER2* (EBV-encoded RNA 1 and 2), which are 166 and 172 nt, respectively, that form double-stranded RNA structures with hairpin loops (Lerner et al. 1981; Arrand and Rymo 1982). *EBER* expression is seen in nasopharyngeal carcinoma, Burkitt's lymphoma biopsies, and Burkitt's lymphoma and lymphoblastoid cell lines (Jat and Arrand 1982; Minarovits et al. 1992). In latently infected cells, *EBER* expression can exceed 5×10^6 copies per cell (Lerner et al. 1981; Clemens 1993). EBER has been proposed to enhance B cell survival by facilitating specific gene expression and preventing apoptosis, some of which is achieved via epigenetic regulation. For instance, *EBER2* facilitates the recruitment of Paired box protein 5 (PAX5), a host DNA-binding transcription factor and important regulator of B-cell lineage differentiation, to the TRs on the viral episome through interaction with nascent TR transcripts (Lee et al. 2015). Binding of PAX5 to the TRs promotes viral latency by regulating the transcription levels of latency-associated factors *EBNA1*, *EBNA2*, and lytic activator *Zta* (Arvey et al. 2012).

KSHV transcribes a ncRNA that acts as a major regulator of latent and lytic replication. During lytic infection a highly abundant, 1.1 kb viral lncRNA, termed polyadenylated nuclear RNA (PAN RNA), is highly expressed. PAN RNA, which is retained in the nucleus, plays key roles in immune modulation as well as viral gene expression and replication (Rossetto and Pari 2011; Borah et al. 2011; Rossetto et al. 2013). The PAN RNA is transcribed from the KSHV genome within a region between *K6* and *ORF16*, slightly overlapping the 5′ of the RNA with the 3′ end of *ORF K7* [reviewed in Rossetto and Pari (2014)]. PAN RNA is an E-gene product that accounts for up to 80% of the polyadenylated RNA transcripts during lytic phase. Although detection of PAN RNA in latency has been reported, this has been attributed mostly to transcription in cells undergoing spontaneous lytic reactivation (Arias et al. 2014). PAN RNA expression during lytic reactivation is regulated by RTA. RTA binds with high affinity to a *cis*-acting RTA-response element located in the promoter of PAN RNA transcription (Song et al. 2001; Massimelli et al. 2013).

During the lytic phase of KSHV infection, polyadenylate-binding protein cytoplasmic 1 (PABPC1) is relocalized from the cytoplasm to the nucleus as a consequence of expression of viral shutoff exonuclease (SOX) protein (Borah et al. 2011, 2012). PABPC1 complexes with nuclear ORF57 protein, allowing the protein to interact with a 9 nucleotide core element at the 5′ end of PAN RNA, consequently increasing the stability of PAN RNA (Massimelli et al. 2013). During lytic reactivation, PAN RNA serves as a molecular scaffold for the chromatin-modifying enzymes JMJD3 and UTX, an H3K27 demethylase complex, and MLL2, an H3K4me3 methyltransferase. Thus, PAN RNA enables transcriptional activation

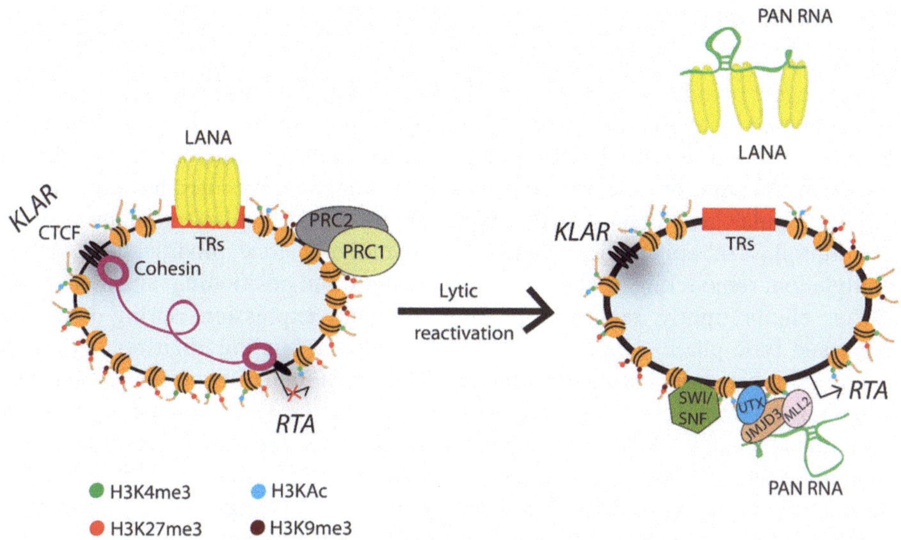

Fig. 2.4 Epigenetic regulation of KSHV episomes during latency and lytic reactivation. During latency, only genes in the KSHV latency-associated region (*KLAR*) possess activating histone marks (H3ac/H3K4me3), whereas bivalent genes (e.g., *RTA*) also possess the repressive histone mark H3K27me3. CTCF (*black* ellipses) and cohesin (*purple* rings) stabilize latency by forming a long-range, looped interaction (*purple* tether of two cohesin rings) between the IE gene *RTA* and the *KLAR*. The Polycomb Repressive Complex (PRC), comprised of PRC1 and PRC2, maintains silencing and/or poising of lytic genes as well as a highly condensed state of most non-transcribed chromatin. Upon viral reactivation in the lytic phase, histone demethylases UTX and JMJD3 and the histone H3K4 methyltransferase MLL2, among other activities, are recruited to transcriptionally inactive or poised promoters via PAN RNA. SWI/SNF, a complex that disassembles nucleosomes in an ATP-dependent manner, is also recruited to evict nucleosomes to create nucleosome-free regions for assembly of the RNAPII pre-initiation complex. The combined recruitment of these chromatin-associated activities reverses Polycomb-mediated repression, repositions nucleosomes, and allows for transcriptional activation of lytic genes

of KSHV lytic genes by relieving their repression by removal of the repressive H3K27me3 mark (Fig. 2.4) (Rossetto and Pari 2012). PAN RNA can also associate with the *RTA* promoter and recruit JMJD3, UTX, and MLL2, thus promoting a positive feedback regulatory mechanism that perpetuates RTA expression during reactivation. However, given that *RTA* is an IE gene and PAN RNA is an E gene, the initial derepression of PRC2-mediated silencing of the *RTA* promoter is likely achieved by another mechanism (Rossetto and Pari 2012). PAN RNA acts as an RNA aptamer by directly interacting with LANA and promotes the dissociation of LANA from the viral episome to allow for efficient viral reactivation (Fig. 2.4) (Campbell et al. 2014). Thus, PAN RNA is a multifunctional lncRNA that interacts with multiple chromatin-modifying enzymes and factors to modulate KSHV gene expression programs.

2.6 Conclusions

The life cycle of gammaherpesviruses fluctuates between latent and lytic states and is coordinated by an intricate series of regulatory mechanisms. The viral genome is quickly circularized and chromatinized once it enters the nucleus to establish a persistent infection of the host. To do so, the gammaherpesviruses have mimicked various cellular gene homologs that facilitate evasion of the host immune system. The interplay of epigenetic mechanisms, namely, histone modifications, DNA methylation, remodeling of nucleosome occupancy and positioning, and noncoding RNAs, elicits tightly regulated patterns of gene expression during both the latent and lytic phases (Fig. 2.4). CTCF and cohesin binding elicit regulation by means of higher-order chromatin structure. The CTCF-cohesin complex propagates formation of DNA loops that keep various parts of the viral genome segregated. Viral miRNA and PAN RNA have been identified as further key players that affect the critical balance between viral latency and lytic reactivation as well as modify the expression status of host genes. Much progress has been made; however, a detailed molecular understanding of the viral and cellular epigenetic mechanisms that enable the gammaherpesviruses to successfully establish latency and ultimately transition to viral lytic reactivation remains elusive. Using a single-molecule footprinting assay, we have recently shown that epigenetic diversity exists within the viral episomal population and not all episomes might be transcriptionally active at once (Fig. 2.3). This adds another dimension to understanding the molecular switch between latency and lytic reactivation.

Acknowledgement We thank all those who have made contributions to the field of gammaherpesvirus research, especially KSHV, and apologize to authors whose work could not be cited in this chapter due to space limitations. We thank Peter Turner for making suggestions and carefully revising the manuscript. This work was supported by NIH grants R01 CA088763, CA119917, and R21 DE024703 to Rolf Renne, and R01 CA155390 to Michael P. Kladde.

Conflicts of Interest The authors declare no conflict of interest.

References

Akula SM, Pramod NP, Wang FZ, Chandran B (2002) Integrin α3β1 (CD 49c/29) is a cellular receptor for Kaposi's sarcoma-associated herpesvirus (KSHV/HHV-8) entry into the target cells. Cell 108:407–419

Alazard N, Gruffat H, Hiriart E, Sergeant A, Manet E (2003) Differential hyperacetylation of histones H3 and H4 upon promoter-specific recruitment of EBNA2 in Epstein-Barr virus chromatin. J Virol 77:8166–8172

Amon W, Farrell PJ (2005) Reactivation of Epstein-Barr virus from latency. Rev Med Virol 15: 149–156

Arias C, Weisburd B, Stern-Ginossar N, Mercier A, Madrid AS, Bellare P et al (2014) KSHV 2.0: a comprehensive annotation of the Kaposi's sarcoma-associated herpesvirus genome using next-generation sequencing reveals novel genomic and functional features. PLoS Pathog 10: e1003847

Arrand JR, Rymo L (1982) Characterization of the major Epstein-Barr virus-specific RNA in Burkitt lymphoma-derived cells. J Virol 41:376–389

Arvey A, Tempera I, Tsai K, Chen H, Tikhmyanova N, Klichinsky M et al (2012) An atlas of the Epstein-Barr virus transcriptome and epigenome reveals host-virus regulatory interactions. Cell Host Microbe 12:233–245

Arvey A, Tempera I, Lieberman PM (2013) Interpreting the Epstein-Barr virus (EBV) epigenome using high-throughput data. Viruses 5:1042–1054

Babcock GJ, Hochberg D, Thorley-Lawson AD (2000) The expression pattern of Epstein-Barr virus latent genes in vivo is dependent upon the differentiation stage of the infected B cell. Immunity 13:497–506

Bacolla A, Pradhan S, Roberts RJ, Wells RD (1999) Recombinant human DNA (cytosine-5) methyltransferase. II Steady-state kinetics reveal allosteric activation by methylated DNA. J Biol Chem 274:33011–33019

Bacolla A, Pradhan S, Larson JE, Roberts RJ, Wells RD (2001) Recombinant human DNA (Cytosine-5) methyltransferase. III. Allosteric control, reaction order, and influence of plasmid topology and triplet repeat length on methylation of the fragile X CGG CCG sequence. J Biol Chem 276:18605–18613

Ballestas ME, Chatis PA, Kaye KM (1999) Efficient persistence of extrachromosomal KSHV DNA mediated by latency-associated nuclear antigen. Science 284:641–644

Barbera AJ, Chodaparambil JV, Kelley-Clarke B, Joukov V, Walter JC, Luger K, Kaye KM (2006a) The nucleosomal surface as a docking station for Kaposi's sarcoma herpesvirus LANA. Science 311:856–861

Barbera AJ, Chodaparambil JV, Kelley-Clarke B, Luger K, Kaye KM (2006b) Kaposi's sarcoma-associated herpesvirus LANA hitches a ride on the chromosome. Cell Cycle 5:1048–1052

Barski A, Cuddapah S, Cui K, Roh TY, Schones DE, Wang Z et al (2007) High-resolution profiling of histone methylations in the human genome. Cell 129:823–837

Bartel DP (2004) MicroRNAs: genomics, biogenesis, mechanism, and function. Cell 116:281–297

Bechtel JT, Winant RC, Ganem D (2005) Host and viral proteins in the virion of Kaposi's sarcoma-associated herpesvirus. J Virol 79:4952–4964

Bernstein BE, Mikkelsen TS, Xie X, Kamal M, Huebert DJ, Cuff J et al (2006) A bivalent chromatin structure marks key developmental genes in embryonic stem cells. Cell 125:315–326

Bestor TH (2000) The DNA methyltransferases of mammals. Hum Mol Genet 9:2395–2402

Bhende P, Seaman W, Delecluse H, Kenney S (2004) The EBV lytic switch protein, Z, preferentially binds to and activates the methylated viral genome. Nat Genet 36:1099–1104

Bird AP (1986) CpG-rich islands and the function of DNA methylation. Nature 321:209–213

Birdwell CE, Queen KJ, Kilgore PC, Rollyson P, Trutschl M, Cvek U, Scott RS (2014) Genome-wide DNA methylation as an epigenetic consequence of Epstein-Barr virus infection of immortalized keratinocytes. J Virol 88:11442–11458

Boland MJ, Nazor KL, Loring LF (2014) Epigenetic regulation of pluripotency and differentiation. Circ Res 115:311–324

Borah S, Darricarrere N, Darnell A, Myoung J, Steitz JA (2011) A viral nuclear noncoding RNA binds re-localized poly(A) binding protein and is required for late KSHV gene expression. PLoS Pathog 7:e1002300

Borah S, Nichols LA, Hassman LM, Kedes DH, Steitz JA (2012) Tracking expression and sub-cellular localization of RNA and protein species using high-throughput single cell imaging flow cytometry. RNA 18:1573–1579

Cai X, Schäfer A, Lu S, Bilello JP, Desrosiers RC, Edwards R et al (2006) Epstein-Barr virus microRNAs are evolutionarily conserved and differentially expressed. PLoS Pathog 2: e0020023

Cairns BR (2009) The logic of chromatin architecture and remodelling at promoters. Nature 461: 193–198

Campbell M, Kim KY, Chang PC, Huerta S, Shevchenko B, Wang DH et al (2014) A lytic viral long noncoding RNA modulates the function of a latent protein. J Virol 88:1843–1848

Cao R, Wang L, Wang H, Xia L, Erdjument-Bromage H, Tempst P et al (2001) Role of histone H3 lysine 27 methylation in Polycomb-group silencing. Science 298:1039–1043

Cesarman E, Chang Y, Moore PS, Said JW, Knowles DM (1995) Kaposi's sarcoma-associated herpesvirus-like DNA sequences in AIDS-related body-cavity-based lymphomas. N Engl J Med 332:1186–1191

Chakraborty S, Veettil MV, Bottero V, Chandran B (2012) Kaposi's sarcoma-associated herpesvirus interacts with EphrinA2 receptor to amplify signaling essential for productive infection. Proc Natl Acad Sci USA 109:E1163–E1172

Chandran B (2010) Early events in Kaposi's sarcoma-associated herpesvirus infection of target cells. J Virol 84:2188–2199

Chang HH, Ganem DA (2013) Unique herpesviral transcriptional program in KSHV-infected lymphatic endothelial cells leads to mTORC1 activation and rapamycin sensitivity. Cell Host Microbe 13:429–440

Chang Y, Cesarman E, Pessin MS, Lee F, Culpepper J, Knowles DM, Moore PS (1994) Identification of herpesvirus-like DNA sequences in AIDS-associated Kaposi's sarcoma. Science 266: 1865–1869

Chang Y, Moore PS, Talbot SJ, Boshoff CH, Zarkowska T, Godden-Kent D et al (1996) Cyclin encoded by KS herpesvirus. Nature 382:410

Chang PC, Fitzgerald LD, Hsia DA, Izumiya Y, Wu CY, Hsieh WP et al (2011) Histone demethylase JMJD2A regulates Kaposi's sarcoma-associated herpesvirus replication and is targeted by a viral transcriptional factor. J Virol 85:3283–3293

Chatlynne LG, Ablashi DV (1999) Seroepidemiology of Kaposi's sarcoma-associated herpesvirus (KSHV). Semin Cancer Biol 9:175–185

Chatterjee N, Sinha D, Lemma-Dechassa M, Tan S, Shogren-Knaak M, Bartholomew B (2011) Histone H3 tail acetylation modulates ATP-dependent remodeling through multiple mechanisms. Nucleic Acids Res 39:8378–8391

Chau CM, Zhang XY, McMahon SB, Lieberman PM (2006) Regulation of Epstein-Barr virus latency type by the chromatin boundary factor CTCF. J Virol 80:5723–5732

Chen J, Ueda K, Sakakibara S, Okuno T, Parravicini C, Corbellino M, Yamanishi K (2001) Activation of latent Kaposi's sarcoma-associated herpesvirus by demethylation of the promoter of the lytic transactivator. Proc Natl Acad Sci USA 98:4119–4124

Chendrimada TP, Gregory RI, Kumaraswamy E, Norman J, Cooch N, Nishikura K, Shiekhattar R (2005) TRBP recruits the Dicer complex to Ago2 for microRNA processing and gene silencing. Nature 436:740–744

Cheng X, Blumenthal RM (2008) Mammalian DNA methyltransferases: a structural perspective. Structure 16:341–350

Clemens MJ (1993) The small RNAs of Epstein-Barr virus. Mol Biol Rep 17(2):81–92

Cohen JI (2000) Epstein-Barr virus infection. N Engl J Med 17:481–492

Coscoy L, Ganem D (2001) A viral protein that selectively downregulates ICAM-1 and B7-2 and modulates T cell costimulation. J Clin Invest 107:1599–1606

Cotter MA, Robertson ES (1999) The latency-associated nuclear antigen tethers the Kaposi's sarcoma-associated herpesvirus genome to host chromosomes in body cavity-based lymphoma cells. Virology 264:254–264

D'Addario M, Libermann TA, Xu J, Ahmad A, Menezes J (2001) Epstein-Barr virus and its glycoprotein-350 upregulate IL-6 in human B-lymphocytes via CD21, involving activation of NF-κB and different signaling pathways. J Mol Biol 308:501–514

Darst RP, Pardo CE, Ai L, Brown KD, Kladde MP (2010) Bisulfite sequencing of DNA. Curr Protoc Mol Biol Chapter 7:Unit 7.9.1–7.9.16

Darst RP, Pardo CE, Pondugula S, Gangaraju VK, Nabilsi NH, Bartholomew B, Kladde MP (2012) Simultaneous single-molecule detection of endogenous C-5 DNA methylation and chromatin accessibility using MAPit. Methods Mol Biol 833:125–141

Darst RP, Haecker I, Pardo CE, Renne R, Kladde MP (2013) Epigenetic diversity of Kaposi's sarcoma-associated herpesvirus. Nucleic Acids Res 41:2993–3009

Day L, Chau CM, Nebozhyn M, Rennekamp AJ, Showe M, Lieberman PM (2007) Chromatin profiling of Epstein-Barr virus latency control region. J Virol 81:6389–6401

Deaton AM, Bird AP (2011) CpG islands and the regulation of transcription. Genes Dev 25: 1010–1022

Dechassa ML, Sabri A, Pondugula S, Kassabov SR, Chatterjee N, Kladde MP, Bartholomew B (2010) SWI/SNF has intrinsic nucleosome disassembly activity that is dependent on adjacent nucleosomes. Mol Cell 38:590–602

Deng H, Young A, Sun R (2000) Auto-activation of the *RTA* gene of human herpesvirus-8/ Kaposi's sarcoma-associated herpesvirus. J Gen Virol 81:3043–3048

Deng Z, Wang Z, Lieberman PM (2012) Telomeres and viruses: common themes of genome maintenance. Front Oncol 2. doi:10.3389/fonc.2012.00201

Dittmer D, Lagunoff M, Renne R, Staskus K, Haase A, Ganem D (1998) A cluster of latently expressed genes in Kaposi's sarcoma-associated herpesvirus. J Virol 72:8309–8315

Dourmishev LA, Dourmishev AL, Palmeri D, Schwartz RA, Lukac DM (2003) Molecular genetics of Kaposi's sarcoma-associated herpesvirus (human herpesvirus-8) epidemiology and pathogenesis. Microbiol Mol Biol Rev 67:175–212

Ehrenhofer MAE (2004) Chromatin dynamics at DNA replication, transcription and repair. Eur J Biochem 271:2335–2349

Epstein MA, Cantab MD, Achong BG, Barr YM (1964) Virus particles in cultured lymphoblasts from Burkitt's lymphoma. Lancet 283:702–703

Fejer G, Koroknai A, Banati F, Gyory I, Salamon D, Hans Wolf H et al (2009) Latency type-specific distribution of epigenetic marks at the alternative promoters Cp and Qp of Epstein-Barr virus. J Gen Virol 89:1364–1370

Feng S, Jacobsen SE, Reik W (2010) Epigenetic reprogramming in plant and animal development. Science 330:622–627

Forte E, Raja AN, Shamulailatpam P, Manzano M, Schipma MJ, Casey JL, Gottwein E (2015) MicroRNA-mediated transformation by the Kaposi's sarcoma-associated herpesvirus Kaposin locus. J Virol 89:2333–2341

Fu YT, Sinha M, Peterson CL, Weng ZP (2008) The insulator binding protein CTCF positions 20 nucleosomes around its binding sites across the human genome. PLoS Genet 4:e1000138

Fujimuro M, Wu FY, ApRhys C, Kajumbula H, Young DB, Hayward GS, Hayward SD (2003) A novel viral mechanism for dysregulation of beta-catenin in Kaposi's sarcoma-associated herpesvirus latency. Nat Med 9:300–306

Garber AC, Shu MA, Hu J, Renne R (2001) DNA binding and modulation of gene expression by the latency-associated nuclear antigen of Kaposi's sarcoma-associated herpesvirus. J Virol 75: 7882–7892

Garber AC, Hu J, Renne R (2002) Latency-associated nuclear antigen (LANA) cooperatively binds to two sites within the terminal repeat, and both sites contribute to the ability of LANA to suppress transcription and to facilitate DNA replication. J Biol Chem 277:27401–27411

Gl M, El-Guindy A, Countryman J, Ye J, Gradoville L (2007) Lytic cycle switches of oncogenic human gammaherpesviruses. Adv Cancer Res 97:81–109

Gottwein E, Corcoran DL, Mukherjee N, Skalsky RL, Hafner M, Nusbaum JD et al (2011) Viral microRNA targetome of KSHV-infected primary effusion lymphoma cell lines. Cell Host Microbe 10:515–526

Grundhoff A, Sullivan CS (2011) Virus-encoded microRNAs. Virology 411:325–343

Grunstein M (1997) Histone acetylation in chromatin structure and transcription. Nature 389: 349–352

Guito J, Lukac DM (2012) KSHV *RTA* promoter specification and viral reactivation. Front Microbiol 3:1–21
Günther T, Grundhoff A (2010) The epigenetic landscape of latent Kaposi sarcoma-associated herpesvirus genomes. PLoS Pathog 6:e1000935
Günther T, Schreiner S, Dobner T, Tessmer U, Grundhoff A (2014) Influence of ND10 components on epigenetic determinants of early KSHV latency establishment. PLoS Pathog 10: e1004274
Gwack Y, Baek HJ, Nakamura H, Lee SH, Meisterernst M, Roeder RG, Jung JU (2003) Principal role of TRAP/mediator and SWI/SNF complexes in Kaposi's sarcoma-associated herpesvirus RTA-mediated lytic reactivation. Mol Cell Biol 23:2055–2067
Haecker I, Gay LA, Yang Y, Hu J, Morse AM, McIntyre LM, Renne R (2012) Ago HITS-CLIP expands understanding of Kaposi's sarcoma-associated herpesvirus miRNA function in primary effusion lymphomas. PLoS Pathog 8:e1002884
Hahn MA, Wu X, Li AX, Hahn T, Pfeifer GP (2011) Relationship between gene body DNA methylation and intragenic H3K9me3 and H3K36me3 chromatin marks. PLoS One 6:e18844
Hahn AS, Kaufmann JK, Wies E, Naschberger E, Jea P-I (2012) The ephrin receptor tyrosine kinase A2 is a cellular receptor for Kaposi's sarcoma-associated herpesvirus. Nat Med 18: 961–966
Hellert J, Weidner-Glunde M, Krausze J, Lunsdorf H, Ritter C, Schulz TF, Luhrs T (2015) The 3D structure of KSHV LANA C-terminal domain bound to DNA. Proc Natl Acad Sci USA 112: 6694–6699
Hendrich B, Bird A (1998) Identification and characterization of a family of mammalian methyl-CpG binding proteins. Mol Cell Biol 18:6538–6547
Herold M, Bartkuhn M, Renkawitz R (2012) CTCF: insights into insulator function during development. Development 139:1045–1057
Hilton IB, Simon JM, Lieb JD, Davis IJ, Damania B, Dittmer DP (2013) The open chromatin landscape of Kaposi's sarcoma-associated herpesvirus. J Virol 87:11832–11842
Honess RW, Roizman B (1974) Regulation of herpesvirus macromolecular synthesis. I Cascade regulation of the synthesis of three groups of viral proteins. J Virol 14:8–19
Hong S, Wang D, Horton JR, Zhang X, Speck SH, Blumenthal RM, Cheng X (2017) Methyl-dependent and spatial-specific DNA recognition by the orthologous transcription factors human AP-1 and Epstein-Barr virus Zta. Nucleic Acids Res. doi:10.1093/nar/gkx057
Horn PJ, Peterson CL (2001) The bromodomain: a regulator of ATP-dependent chromatin remodeling? Front Biosci 6:D1019–D1023
Hu J, Yang Y, Turner PC, Jain V, McIntyre LM, Renne R (2014) LANA binds to multiple active viral and cellular promoters and associates with the H3K4methyltransferase hSET1 complex. PLoS Pathog 10:e1004240
Hutvagner G, McLachlan J, Pasquinelli AE, Balint E, Tuschl T, Zamore PD (2001) Transcription and processing of human microRNA precursors. Mol Cell 16:861–865
Ishido S, Wang C, Lee BS, Cohen GB, Jung JU (2002) Downregulation of major histocompatibility complex class I molecules by Kaposi's sarcoma-associated herpesvirus K3 and K5 proteins. J Virol 74:5300–5309
Jat P, Arrand JR (1982) In vitro transcription of two Epstein-Barr virus specified small RNA molecules. Nucleic Acids Res 10:3407–3425
Jenner RG, Alba MM, Boshoff C, Kellam P (2001) Kaposi's sarcoma-associated herpesvirus latent and lytic gene expression as revealed by DNA arrays. J Virol 75:891–902
Jia D, Jurkowska RZ, Zhang X, Jeltsch A, Cheng X (2007) Structure of Dnmt3a bound to Dnmt3L suggests a model for de novo DNA methylation. Nature 449:248–251
Jiang C, Pugh BF (2009) Nucleosome positioning and gene regulation: advances through genomics. Nat Rev Genet 10:161–172
Johannsen E, Luftig M, Chase MR, Weicksel S, Cahir-McFarland E, Illanes D et al (2004) Proteins of purified Epstein-Barr virus. Proc Natl Acad Sci USA 101:16286–16291
Jones PL, Veenstra GJ, Wade PA, Vermaak D, Kass SU, Landsberger N et al (1998) Methylated DNA and MeCP2 recruit histone deacetylase to repress transcription. Nat Genet 19:187–191

Kalla M, Gobel C, Hammerschmidt W (2011) The lytic phase of Epstein-Barr virus requires a viral genome with 5-methylcytosine residues in CpG sites. J Virol 86:447–458

Kanda T, Miyata M, Kano M, Kondo S, Yoshizaki T, Iizasa H (2015) Clustered microRNAs of the Epstein-Barr virus cooperatively downregulate an epithelial cell-specific metastasis suppressor. J Virol 89:2684–2697

Kang H, Wiedmer A, Yuan Y, Robertson E, Lieberman PM (2011) Coordination of KSHV latent and lytic gene control by CTCF-cohesin mediated chromosome conformation. PLoS Pathog 7: e1002140

Kang H, Cho H, Sung GH, Lieberman PM (2013) CTCF regulates Kaposi's sarcoma-associated herpesvirus latency transcription by nucleosome displacement and RNA polymerase programming. J Virol 87:1789–1799

Kang D, Skalsky RL, Cullen BR (2015) EBV *BART* microRNAs target multiple pro-apoptotic cellular genes to promote epithelial cell survival. PLoS Pathog 11:e1004979

Kapranov P, Cheng J, Dike S, Nix DA, Duttagupta R, Willingham AT et al (2007) RNA maps reveal new RNA classes and a possible function for pervasive transcription. Science 5830: 1484–1488

Karlsson QH, Schelcher C, Verrall E, Petosa C, Sinclair AJ (2008) The reversal of epigenetic silencing of the EBV genome is regulated by viral bZIP protein. Biochem Soc Trans 36: 637–639

Kelly TK, Liu Y, Lay FD, Liang G, Berman BP, Jones PA (2012) Genome-wide mapping of nucleosome positioning and DNA methylation within individual DNA molecules. Genome Res 22:2497–2506

Kieffer-Kwon P, Happel C, Uldrick TS, Ramalingam D, Ziegelbauer JM (2015) KSHV microRNAs repress tropomyosin 1 and increase anchorage-independent growth and endothelial tube formation. PLoS One 11:e0135560

Kilgore JA, Hoose SA, Gustafson TL, Porter W, Kladde MP (2007) Single-molecule and population probing of chromatin structure using DNA methyltransferases. Methods 41:320–332

Kim TH, Abdullaev ZK, Smith AD, Ching KA, Loukinov DI, Green RD et al (2007) Analysis of the vertebrate insulator protein CTCF binding sites in the human genome. Cell 128:1231–1245

Kim KY, Huerta SB, Izumiya C, Wang DH, Martinez A, Shevchenko B et al (2013) Kaposi's sarcoma-associated herpesvirus (KSHV) latency-associated nuclear antigen regulates the KSHV epigenome by association with the histone demethylase KDM3A. J Virol 87: 6782–6793

Kingston RE, Narlikar GJ (1999) ATP-dependent remodeling and acetylation as regulators of chromatin fluidity. Genes Dev 13:2339–2352

Knipe DM, Lieberman PM, Jung JU, McBride AA, Morris KV, Ott M et al (2013) Snapshots: chromatin control of viral infection. Virology 435:141–156

Kornberg RD, Lorch Y (1999) Twenty-five years of the nucleosome, fundamental particle of the eukaryote chromosome. Cell 98:285–294

Kouzarides T (2007) Chromatin modifications and their function. Cell 128:693–705

Lee DY, Hayes JJ, Pruss D, Wolffe AP (1993) A positive role for histone acetylation in transcription factor access to nucleosomal DNA. Cell 72:73–84

Lee HR, Brulois K, Wong L, Jung JU (2012) Modulation of immune system by Kaposi's sarcoma-associated herpesvirus: lessons from viral evasion strategies. Front Microbiol 3. doi:10.3389/fmicb.2012.00044

Lee N, Moss WN, Yario TA, Steitz JA (2015) EBV noncoding RNA binds nascent RNA to drive host PAX5 to viral DNA. Cell 160:607–618

Lerner MR, Andrews NC, Miller G, Steitz JA (1981) Two small RNAs encoded by Epstein-Barr virus and complexed with protein are precipitated by antibodies from patients with systemic lupus erythematosus. Proc Natl Acad Sci USA 78:805–809

Li G, Reinberg D (2011) Chromatin higher-order structures and gene regulation. Curr Opin Genet Dev 21:175–186

Li DJ, Verma D, Mosbruger T, Swaminathan S (2014) CTCF and Rad21 act as host cell restriction factors for Kaposi's sarcoma-associated herpesvirus (KSHV) lytic replication by modulating viral gene transcription. PLoS Pathog 10:e1003880

Lieberman PM (2013) Keeping it quiet: chromatin control of gammaherpesvirus latency. Nat Rev Microbiol 11:863–875
Liu F, Zhou ZH (2007) Comparative virion structures of human herpesviruses. In: Arvin A, Campadelli-Fiume G, Edward M et al (eds) Human herpesviruses: biology, therapy, and immunoprophylaxis. Cambridge University Press, Cambridge
Lu F, Zhou J, Wiedmer A, Madden K, Yuan Y, Lieberman PM (2003) Chromatin remodeling of the Kaposi's sarcoma-associated herpesvirus ORF50 promoter correlates with reactivation from latency. J Virol 77:11425–11435
Lu F, Stedman W, Yousef M, Renne R, Lieberman PM (2010) Epigenetic regulation of Kaposi's sarcoma-associated herpesvirus latency by virus-encoded microRNAs that target *RTA* and the cellular Rbl2-DNMT pathway. J Virol 84:2697–2706
Luger K (2003) Structure and dynamic behavior of nucleosomes. Curr Opin Genet Dev 13: 127–135
Ma J, Nie K, Redmond D, Liu Y, Elemento O, Knowles DM, Tam W (2015) EBV-miR-BHRF1-2 targets *PRDM1/Blimp1*: potential role in EBV lymphomagenesis. Leukemia. doi:10.1038/leu. 2015.285
Maeda E, Akahane M, Kiryu S, Kato N, Yoshikawa T, Hayashi N et al (2009) Spectrum of Epstein-Barr virus-related diseases: a pictorial review. Jpn J Radiol 27:4–19
Majerciak V, Yamanegi K, Zheng ZM (2006) Gene structure and expression of Kaposi's sarcoma-associated herpesvirus ORF56, ORF57, ORF58, and ORF59. J Virol 80:11968–11981
Massimelli MJ, Majerciak V, Kruhlak M, Zheng ZM (2013) Interplay between polyadenylate-binding protein 1 and Kaposi's sarcoma-associated herpesvirus ORF57 in accumulation of polyadenylated nuclear RNA, a viral long noncoding RNA. J Virol 87:243–256
Masucci MG, Contreras-Salazar B, Ragnar E, Falk K, Minarovits J, Ernberg I, Klein G (1989) 5-Azacytidine up regulates the expression of Epstein-Barr virus nuclear antigen 2 (EBNA-2) through EBNA-6 and latent membrane protein in the Burkitt's lymphoma line rael. J Virol 63: 3135–3141
Matsumura S, Persson LM, Wong L, Wilson AC (2010) The latency-associated nuclear antigen interacts with MeCP2 and nucleosomes through separate domains. J Virol 84:2318–2330
Mercer TR, Dinger ME, Mattick JS (2009) Long non-coding RNAs: insights into functions. Nat Rev Genet 10:155–159
Miller G, Rigsby MO, Heston L, Grogan E, Sun R, Metroka C et al (1996) Antibodies to butyrate-inducible antigens of Kaposi's sarcoma-associated herpesvirus in patients with HIV-1 infection. N Engl J Med 334:1292–1297
Minarovits J (2006) Epigenotypes of latent herpesvirus genomes. Curr Top Microbiol Immunol 310:61–80
Minarovits J, Hu LF, Marcsek Z, Minarovits-Kormuta S, Klein G, Ernberg I (1992) RNA polymerase III-transcribed EBER 1 and 2 transcription units are expressed and hypomethylated in the major Epstein-Barr virus-carrying cell types. J Gen Virol 73:1687–1692
Mohanna S, Maco V, Bravo F, Gotuzzo E (2005) Epidemiology and clinical characteristics of classic Kaposi's sarcoma, seroprevalence, and variants of human herpesvirus 8 in South America: a critical review of an old disease. Int J Infect Dis 9:239–250
Nabilsi NH, Deleyrolle LP, Darst RP, Riva A, Reynolds BA, Kladde MP (2014) Multiplex mapping of chromatin accessibility and DNA methylation within targeted single molecules identifies epigenetic heterogeneity in neural stem cells and glioblastoma. Genome Res 24:329–339
Nagy PL, Price DH (2009) Formaldehyde-assisted isolation of regulatory elements. Wiley Interdiscip Rev Syst Biol Med 1:400–406
Nakamura H, Lu M, Gwack Y, Souvlis J, Zeichner S, Jung J (2003) Global changes in Kaposi's sarcoma-associated virus gene expression patterns following expression of a tetracycline-inducible *RTA* transactivator. J Virol 77:4205–4220
Narlikar GJ, Fan H, Kingston RE (2002) Cooperation between complexes that regulate chromatin structure and transcription. Cell 108:475–487
Ng MK, Cheung P (2015) A brief histone in time: understanding the combinatorial functions of histone PTMs in the nucleosome context. Biochem Cell Biol 94:1–10

Ohlsson R, Renkawitz R, Lobanenkov V (2001) CTCF is a uniquely versatile transcription regulator linked to epigenetics and disease. Trends Genet 17:520–527

Ohlsson R, Lobanenkov V, Klenova E (2010) Does CTCF mediate between nuclear organization and gene expression? Bioessays 32:37–50

Ong C, Corces VG (2014) CTCF: an architectural protein bridging genome topology and function. Nat Rev Genet 15:234–246

Ooi SK, Qiu C, Bernstein E, Li K, Jia D, Yang Z et al (2007) DNMT3L connects unmethylated lysine 4 of histone H3 to de novo methylation of DNA. Nature 448:714–717

Owen-Hughes T, Workman JL (1994) Experimental analysis of chromatin function in transcription control. Crit Rev Eukaryot Gene Expr 4:403–441

Palazzo AF, Lee ES (2015) Non-coding RNA: what is functional and what is junk? Front Genet 6:2. doi:10.3389/fgene.2015.00002

Pantry SN, Medveczky PG (2009) Epigenetic regulation of Kaposi's sarcoma-associated herpesvirus replication. Semin Cancer Biol 19:153–157

Pfeffer S, Sewer A, Lagos-Quintana M, Sheridan R, Sander C, Grässer FA et al (2005) Identification of microRNAs of the herpesvirus family. Nat Methods 2:269–276

Plaisance-Bonstaff K, Choi HS, Beals T, Krueger BJ, Boss IW, Gay LA et al (2014) KSHV miRNAs decrease expression of lytic genes in latently infected PEL and endothelial cells by targeting host transcription factors. Viruses 6:4005–4023

Ponting CP, Oliver PL, Reik W (2009) Evolution and functions of long noncoding RNAs. Cell 136:629–641

Pradhan S, Bacolla A, Wells RD, Roberts RJ (1999) Recombinant human DNA (cytosine-5) methyltransferase. I. Expression, purification, and comparison of de novo and maintenance methylation. J Biol Chem 274:33002–33010

Quan L, Qiu T, Liang J, Li M, Zhang Y, Tao K (2015) Identification of target genes regulated by KSHV miRNAs in KSHV-infected lymphoma cells. Pathol Oncol Res 21:875–880

Renne R, Lagunoff M, Zhong W, Ganem D (1996a) The size and conformation of Kaposi's sarcoma-associated herpesvirus (human herpesvirus 8) DNA in infected cells and virions. J Virol 70: 8151–8154

Renne R, Zhong W, Herndier B, Mcgrath M, Abbey N, Kedes D, Ganem D (1996b) Lytic growth of Kaposi's sarcoma-associated herpesvirus (human herpesvirus 8) in culture. Nat Med 2:342–346

Renne R, Barry C, Dittmer D, Compitello N, Brown PO, Ganem D (2001) Modulation of cellular and viral gene expression by the latency-associated nuclear antigen of Kaposi's sarcoma-associated herpesvirus. J Virol 75:458–468

Ressing ME, van Gent M, Gram AM, Hooykaas MJ, Piersma SJ, Wiertz EJ (2015) Immune evasion by Epstein-Barr virus. Curr Top Microbiol Immunol 391:355–381

Richmond TJ, Davey CA (2003) The structure of DNA in the nucleosome core. Nature 423: 145–150

Riveraa C, Gurard-Levinb ZA, Almounzib G, Loyolaa A (2014) Histone lysine methylation and chromatin replication. Biochim Biophys Acta 1839:1433–1439

Robertson KD, Wolffe AP (2000) DNA methylation in health and disease. Nat Rev Genet 1:11–19

Roh T, Cuddapah S, Cui K, Zhao K (2006) The genomic landscape of histone modifications in human T cells. Proc Natl Acad Sci USA 103:15782–15787

Roizman B (1982) The family Herpesviridae: general description, taxonomy, and classification. In: Roizman B (ed) The viruses. Springer, New York, pp 1–23

Rose NR, Klose RJ (2014) Understanding the relationship between DNA methylation and histone lysine methylation. Biochim Biophys Acta 1839:1362–1372

Rossetto CC, Pari GS (2011) Kaposi's sarcoma-associated herpesvirus noncoding polyadenylated nuclear RNA interacts with virus- and host cell-encoded proteins and suppresses expression of genes involved in immune modulation. J Virol 85:13290–13297

Rossetto CC, Pari G (2012) KSHV PAN RNA associates with demethylases UTX and JMJD3 to activate lytic replication through a physical interaction with the virus genome. PLoS Pathog 8: e1002680

Rossetto CC, Pari GS (2014) PAN's labyrinth: Molecular biology of Kaposi's sarcoma-associated herpesvirus (KSHV) PAN RNA, a multifunctional long noncoding RNA. Viruses 6:4212–4226

Rossetto CC, Tarrant-Elorza M, Verma S, Purushothaman P, Pari GS (2013) Regulation of viral and cellular gene expression by Kaposi's sarcoma-associated herpesvirus polyadenylated nuclear RNA. J Virol 87:5540–5553

Samols MA, Hu J, Skalsky RL, Renne R (2005) Cloning and identification of a microRNA cluster within the latency-associated region of Kaposi's sarcoma-associated herpesvirus. J Virol 79: 9301–9305

Sathish N, Wang X, Yuan Y (2012) Tegument proteins of Kaposi's sarcoma-associated herpesvirus and related gamma-herpesviruses. Front Microbiol 3:1–13

Schones DE, Cui K, Cuddapah S, Roh TY, Barski A, Wang Z et al (2008) Dynamic regulation of nucleosome positioning in the human genome. Cell 132:887–898

Schuettengruber B, Cavalli G (2009) Recruitment of polycomb group complexes and their role in the dynamic regulation of cell fate choice. Development 136:3531–3542

Schulz TF, Yuan C (2007) KSHV gene expression and regulation. In: Arvin A, Campadelli-Fiume G, Edward M et al (eds) Human herpesviruses: biology, therapy, and immunoprophylaxis. Cambridge University Press, Cambridge

Sears J, Ujihara M, Wong S, Ott C, Middeldorp J, Aiyar A (2004) The amino terminus of Epstein-Barr virus (EBV) nuclear antigen 1 contains AT hooks that facilitate the replication and partitioning of latent EBV genomes by tethering them to cellular chromosomes. J Virol 78: 11487–11505

Shamay M, Krithivas A, Zhang J, Hayward SD (2006) Recruitment of the de novo DNA methyltransferase Dnmt3a by Kaposi's sarcoma-associated herpesvirus LANA. Proc Natl Acad Sci USA 103:14554–14559

Sharif J, Muto M, Takebayashi S, Suetake I, Iwamatsu A, Endo TA et al (2007) The SRA protein Np95 mediates epigenetic inheritance by recruiting Dnmt1 to methylated DNA. Nature 450: 908–912

Shin HJ, DeCotiis J, Giron M, Palmeri D, Lukac DM (2014) Histone deacetylase classes I and II regulate Kaposi's sarcoma-associated herpesvirus reactivation. J Virol 88:1281–1292

Skalsky RL, Cullen BR (2015) EBV noncoding RNAs. Curr Top Microbiol Immunol 391: 181–217

Smallwood SA, Lee HJ, Angermueller C, Krueger F, Saadeh H, Peat J et al (2014) Single-cell genome-wide bisulfite sequencing for assessing epigenetic heterogeneity. Nat Methods 11: 817–820

Song MJ, Brown HJ, Wu TT, Sun R (2001) Transcription activation of polyadenylated nuclear RNA by RTA in human herpesvirus 8/Kaposi's sarcoma-associated herpesvirus. J Virol 75: 3129–3140

Soulier J, Grollet L, Oksenhendler E, Cacoub P, Cazalshatem MD, Babinet P et al (1995) Kaposi's sarcoma-associated herpesvirus-like DNA sequences in multicentric Castleman's disease. Blood 86:1276–1280

Stedman W, Kang H, Lin S, Kissil JL, Bartolomei MS, Lieberman PM (2008) Cohesins localize with CTCF at the KSHV latency control region and at cellular *c-myc* and *H19/IGF2* insulators. EMBO J 27:654–666

Strahl BD, Allis CD (2000) The language of covalent histone modifications. Nature 403:41–45

Strahl BD, Ohba R, Cook RG, Allis CD (1999) Methylation of histone H3 at lysine 4 is highly conserved and correlates with transcriptionally active nuclei in *Tetrahymena*. Proc Natl Acad Sci USA 96:14967–14972

Struhl K (1998) Histone acetylation and transcriptional regulatory mechanisms. Genes Dev 12: 599–606

Suetake I, Shinozaki F, Miyagawa J, Takeshima H, Tajima S (2004) DNMT3L stimulates the DNA methylation activity of Dnmt3a and Dnmt3b through a direct interaction. J Biol Chem 279:27816–27823

Sun R, Lin SF, Gradoville L, Yuan Y, Zhu F, Miller G (1998) A viral gene that activates lytic cycle expression of Kaposi's sarcoma-associated herpesvirus. Proc Natl Acad Sci USA 95: 10866–10871

Sun R, Lin SF, Staskus K, Gradoville L, Grogan E, Haase A, Miller G (1999) Kinetics of Kaposi's sarcoma-associated herpesvirus gene expression. J Virol 73:2232–2242

Sun W, Xie W, Xu F, Grunstein M, Li K (2009) Dissecting nucleosome free regions by a segmental semi-Markov model. PLoS One 4:e4721

Szabo PE, Pfeifer GP, Mann JR (1998) Characterization of novel parent-specific epigenetic modifications upstream of the imprinted mouse *H19* gene. Mol Cell Biol 18:6767–6776

Szabo PE, Tang SHE, Silva FJ, Tsark WMK, Mann JR (2004) Role of CTCF binding sites in the *Igf2/H19* imprinting control region. Mol Cell Biol 24:4791–4800

Takahashi H, McCaffery JM, Irizarry RA, Boeke JD (2006) Nucleocytosolic acetyl-coenzyme A synthetase is required for histone acetylation and global transcription. Mol Cell 23:207–217

Tao Q, Robertson K (2003) Stealth technology: how Epstein-Barr virus utilizes DNA methylation to cloak itself from immune detection. Clin Immunol 109:53–63

Tempera I, Wiedmer A, Dheekollu J, Lieberman PM (2010) CTCF prevents the epigenetic drift of EBV latency promoter Qp. PLoS Pathog 6:e1001048

Theisen JW, Gucwa JS, Yusufzai T, Khuong MT, Kadonaga JT (2013) Biochemical analysis of histone deacetylase-independent transcriptional repression by MeCP2. J Biol Chem 288: 7096–7104

Thome M, Schneider P, Hofmann K, Fickenscher H, Meinl E, Neipel F et al (1997) Viral FLICE-inhibitory proteins (FLIPs) prevent apoptosis induced by death receptors. Nature 286:517–521

Tiedemann RL, Putiri EL, Lee J, Hlady RA, Kashiwagi K, Ordog T et al (2014) Acute depletion redefines the division of labor among DNA methyltransferases in methylating the human genome. Cell Rep 9:1554–1556

Toth Z, Maglinte DT, Lee SH, Lee HR, Wong LY, Brulois KF et al (2010) Epigenetic analysis of KSHV latent and lytic genomes. PLoS Pathog 6:e1001013

Tse C, Sera T, Wolffe AP, Hansen JC (1998) Disruption of higher-order folding by core histone acetylation dramatically enhances transcription of nucleosomal arrays by RNA polymerase III. Mol Cell Biol 18:4629–4638

Tsurumi T, Fujita M, Kudoh A (2005) Latent and lytic EBV replication strategies. Rev Med Virol 15:3–15

Verdel A, Jia S, Gerber S, Sugiyama T, Gygi S, Grewal SIS, Moazed D (2004) RNAi-mediated targeting of heterochromatin by the RITS complex. Science 303:672–676

Verma SC, Cai Q, Kreider E, Lu J, Robertson ES (2013) Comprehensive analysis of LANA interacting proteins essential for viral genome tethering and persistence. PLoS One 8:e74662

Viejo-Borbolla A, Ottinger M, Brüning E, Bürge A, König R, Kati E et al (2005) Brd2/RING3 interacts with a chromatin-binding domain in the Kaposi's sarcoma-associated herpesvirus latency-associated nuclear antigen 1 (LANA-1) that is required for multiple functions of LANA-1. J Virol 79:13618–13629

Wang SE, Wu FY, Yu Y, Hayward GS (2003) CCAAT/enhancer-binding protein-α is induced during the early stages of Kaposi's sarcoma-associated herpesvirus (KSHV) lytic cycle reactivation and together with the KSHV replication and transcription activator (RTA) cooperatively stimulates the viral *RTA*, *MTA*, and *PAN* promoters. J Virol 77:9590–9612

Wawer MJ, Eng SM, Serwadda D, Sewankambo NK, Kiwanuka N, Li C, Gray RH (2001) Prevalence of Kaposi sarcoma-associated herpesvirus compared with selected sexually transmitted diseases in adolescents and young adults in rural Rakai district, Uganda. Sex Transm Dis 28:77–81

Wei WI, Sham JS (2005) Nasopharyngeal carcinoma. Lancet 365:2041–2045

Weitzman MD, Lilley CE, Chaurushiya MS (2010) Genomes in conflict: maintaining genome integrity during virus infection. Annu Rev Microbiol 64:61–81

Wellen KE, Hatzivassiliou G, Sachdeva UM, Bui TV, Cross JR, Thompson CB (2009) ATP-citrate lyase links cellular metabolism to histone acetylation. Science 324:1076–1080

Widlund HR, Vitolo JM, Thiriet C, Hayes JJ (2000) DNA sequence-dependent contributions of core histone tails to nucleosome stability: differential effects of acetylation and proteolytic tail removal. Biochemistry 39:3835–3841

Woellmer A, Hammerschmidt W (2013) Epstein-Barr virus and host cell methylation: regulation of latency, replication and virus reactivation. Curr Opin Virol 3:260–265

Xu M, Kladde MP, Van Etten JL, Simpson RT (1998) Cloning, characterization and expression of the gene coding for cytosine-5-DNA methyltransferase recognizing GpC. Nucleic Acids Res 26:3961–3966

Yang Y, Boss IW, McIntyre LM, Renne R (2014) A systems biology approach identified different regulatory networks targeted by KSHV miR-K12-11 in B cells and endothelial cells. BMC Genomics 15. doi:10.1186/1471-2164-15-668

Ye FC, Zhou FC, Xie JP, Kang T, Greene W, Kuhne K et al (2008) Kaposi's sarcoma-associated herpesvirus latent gene *vFLIP* inhibits viral lytic replication through NF-κB-mediated suppression of the AP-1 pathway: a novel mechanism of virus control of latency. J Virol 82: 4235–4249

Ye F, Lei X, Gao SJ (2011) Mechanisms of Kaposi's sarcoma-associated herpesvirus latency and reactivation. Adv Virol. doi:10.1155/2011/193860

Yogev O, Lagos D, Enver T, Boshoff C (2014) Kaposi's sarcoma herpesvirus microRNAs induce metabolic transformation of infected cells. PLoS Pathog 10:e1004400

You J, Srinivasan V, Denis GV, Harrington WJ, Ballestas ME, Kaye KM, Howley ML (2006) Kaposi's sarcoma-associated herpesvirus latency-associated nuclear antigen interacts with bromodomain protein Brd4 on host mitotic chromosomes. J Virol 80:8909–8919

Young LS, Rickinson AB (2004) Epstein-Barr virus: 40 years on. Nat Rev Cancer 4:757–768

Young LS, Arrand JR, Murray PG (2007) EBV gene expression and regulation. In: Arvin A, Campadelli-Fiume G, Edward M et al (eds) Human herpesviruses: biology, therapy, and immunoprophylaxis. Cambridge University Press, Cambridge

Yun M, Wu J, Workman JL, Li B (2011) Readers of histone modifications. Cell Res 21:564–578

Zhang Z, Wippo CJ, Wal M, Ward E, Korber P, Pugh BF (2011) A packing mechanism for nucleosome organization reconstituted across a eukaryotic genome. Science 332:977–980

Zhong W, Wang H, Herndier B, Ganem D (1996) Restricted expression of Kaposi sarcoma-associated herpesvirus (human herpesvirus 8) genes in Kaposi sarcoma. Proc Natl Acad Sci USA 93:6641–6646

Zhu Y, Haecker I, Yang Y, Gao SJ, Renne R (2014) γ-Herpesvirus-encoded miRNAs and their roles in viral biology and pathogenesis. Curr Opin Virol 3:266–275

Zimmermann J, Hammerschmidt W (1995) Structure and role of the terminal repeats of Epstein-Barr virus in processing and packaging of virion DNA. J Virol 69:3147–3155

Chapter 3
Discoveries in Molecular Genetics with the Adenovirus 12 System: Integration of Viral DNA and Epigenetic Consequences

Walter Doerfler

Abstract Starting in the 1960s, the human adenovirus type 12 (Ad12) system has been used in my laboratory to investigate basic mechanisms in molecular biology and viral oncology. Ad12 replicates in human cells but undergoes a completely abortive cycle in Syrian hamster cells. Ad12 induces neuro-ectodermal tumors in newborn hamsters (*Mesocricetus auratus*). Each tumor cell or Ad12-transformed hamster cell carries multiple copies of integrated Ad12 DNA. Ad12 DNA usually integrates at one chromosomal site which is not specific since Ad12 DNA can integrate at many different locations in the hamster genome. Epigenetic research occupies a prominent role in tumor biology. We have been using the human Ad12 Syrian hamster cell system for the analysis of epigenetic alterations in Ad12-infected cells and in Ad12-induced hamster tumors. Virion or free intracellular Ad12 DNA remains unmethylated at CpG sites, whereas the integrated viral genomes become de novo methylated in specific patterns. Inverse correlations between promoter methylation and activity were described for the first time in this system and initiated active research in the field of DNA methylation and epigenetics. Today, promoter methylation has been recognized as an important factor in long-term genome silencing. We have also discovered that the insertion of foreign (Ad12, bacteriophage lambda, plasmid) DNA into mammalian genomes can lead to genome-wide alterations in methylation and transcription patterns in the recipient genomes. This concept has been verified recently in a pilot study with human cells which had been rendered transgenomic for a 5.6 kbp bacterial plasmid. Currently, we study epigenetic effects on cellular methylation and transcription patterns in Ad12-infected cells and in Ad12-induced hamster tumor cells. These epigenetic alterations are considered crucial elements in (viral) oncogenesis.

Keywords Abortive infection of hamster cells with Ad12 • Adenovirus type 12 (Ad12) • Ad12-induced hamster tumors • Ad12 viral oncogenesis • A model

---

W. Doerfler (✉)
Institute for Clinical and Molecular Virology, University Erlangen-Nürnberg Medical School, Schlossgarten 4, 91054 Erlangen, Germany

Institute of Genetics, University of Cologne, Zülpicherstraße 47, 50674 Köln, Germany
e-mail: walter.doerfler@viro.med.uni-erlangen.de

W. Doerfler, J. Casadesús (eds.), *Epigenetics of Infectious Diseases*, Epigenetics and Human Health, DOI 10.1007/978-3-319-55021-3_3

for Ad12-induced oncogenesis • De novo methylation of integrated foreign DNA • Epigenetics • Epigenetic consequences of foreign DNA insertions • Epstein–Barr Virus immortalization and epigenetic changes • Foreign DNA in the environment • Genome-wide epigenetic alterations • Hit-and-run mechanism of viral oncogenesis • Integration of foreign (viral) DNA into mammalian genomes • Promoter methylation and long-term silencing • Stability of induced epigenetic changes in the genome • Viral oncogenesis

3.1 Background on Interest in DNA Methylation and Epigenetic Effects

In the history of research on viruses and their interactions with host cells, unexpected observations have frequently drawn the investigator to the study of mechanisms far beyond the realms of virology. For several decades, molecular virology has thus succeeded in occupying a pioneering role in molecular genetics. In the late 1960s and the 1970s, my laboratory at Rockefeller University in New York City, NY, and later at the Institute of Genetics in Köln has been investigating the integration of human adenovirus type 12 (Ad12) DNA into the genome of hamster cells (Doerfler 1968, 1970; Groneberg et al. 1977; Sutter et al. 1978; Sutter and Doerfler 1980; Stabel et al. 1980; Hochstein et al. 2007). Ad12 had been shown to induce tumors at the site of virus inoculation into newborn hamsters (Trentin et al. 1962). This discovery had attracted my interest and led to detailed studies on the molecular biology of the DNA virus Ad12. In a 1970 publication in the Journal of Virology, I had raised the question of whether the integration of the Ad12 genome into the hamster genome was the decisive precondition (*conditio sine qua non*) for the oncogenic consequences of Ad12 infection (Doerfler 1970).

At the same time, we analyzed details of the structure and sequence of the adenovirus DNA molecule (Doerfler 1969; Sprengel et al. 1994), established the denaturation maps of Ad2 and Ad12 DNA (Doerfler and Kleinschmidt 1970; Doerfler et al. 1983), and demonstrated that the DNAs of Ad2 and Ad12 did not contain 5-methyl-deoxycytidine (5-mC) (Günthert et al. 1976; Wienhues and Doerfler 1985). One of these studies (Günthert et al. 1976) also demonstrated that the DNA of hamster cells transformed by Ad12 contains 3.11 and 3.14% 5-methylcytosine (cell lines HA12/7 and T637 cells, respectively), whereas the DNA from untransformed hamster cells (BHK21 cells) contains only 2.22% 5-methylcytosine. These results suggested early on that levels of DNA methylation in transformed or tumor cells were fundamentally altered (Günthert et al. 1976).

In the late 1970s and early 1980s, we used restriction enzyme analyses of cellular DNA with integrated Ad12 genomes in Ad12-transformed cells to elucidate the viral DNA integration patterns and to localize more precisely the site of foreign DNA insertion. In the course of this work, we recognized that the restriction endonuclease HpaII, which cuts unmethylated virion Ad12 DNA frequently to small fragments, did not cleave the integrated Ad12 DNA efficiently. In contrast, the restriction endonuclease MspI, known to be methylation insensitive (Waalwijk and Flavell 1978), cut the integrated viral genome to small fragments (Sutter et al. 1978; Sutter and Doerfler 1980). HpaII and MspI are isoschizomers, i.e., restriction endonucleases which recognize the same 5′-CCGG-3′ sites, where HpaII is blocked by a 5-mC residue in the 3′-position, whereas MspI is refractory to this inhibition. These data documented that the integrated Ad12 genomes had become extensively de novo methylated. These thus roughly determined profiles of Ad12 DNA methylation were more precisely mapped by using the bisulfite sequencing technique in a later publication (Hochstein et al. 2007). We went on to show that there was an inverse correlation between the levels of methylation in the different regions (early versus late) of the integrated adenovirus genomes and their genetic activities (Sutter and Doerfler 1980; Vardimon et al. 1980). These data were the first to functionally relate genetic activity and promoter methylation in eukaryotic and viral genomes and became the basis for our interest in problems of DNA methylation which have been continuously pursued in different systems until today.

Here, I will refrain from describing all of our research on DNA methylation of the past 30 years, but instead refer the reader to recently published reviews on these topics (Doerfler 2011, 2012, 2016). Based on these earlier studies, we have continued to investigate the biological meaning of DNA methylation and the fifth nucleotide, 5-methyldeoxycytidine. Here are some of the key references to our work on DNA methylation: Sutter and Doerfler (1980), Vardimon et al. (1980), Doerfler (1983, 2011), Toth et al. (1989), Kochanek et al. (1990), Kochanek et al. (1993), Orend et al. (1995), Heller et al. (1995), Zeschnigk et al. (1997), Naumann et al. (2009), and Weber et al. (2015).

3.1.1 Introduction to the Adenovirus System

Epigenetic research has gradually occupied a prominent role also in virology. Ad12 offers the opportunity to study virus infections in a productive (human cells) and a completely abortive system (Syrian hamster cells). In addition, Ad12-induced hamster tumors allow epigenetic analyses in an efficient viral oncogenesis model. Ad12 tumorigenesis in hamsters was originally discovered by Trentin, Yabe, and Taylor in 1962. Ad12 induces undifferentiated neuro-ectodermal tumors in >90% of the inoculated and surviving newborn hamsters (*Mesocricetus aureatus*) within 3–6 weeks after inoculation (Hohlweg et al. 2003). A combination of important parameters, like high incidence and short latency between infection and tumorigenesis, coupled with Ad12 DNA integration in all tumor cells (Knoblauch et al. 1996; Hilger-Eversheim and Doerfler 1997) facilitates the molecular analysis of tumor induction by this DNA virus. My laboratory has been working on the molecular biology, genetics, and epigenetics of adenoviruses since 1966 (Doerfler 1968, 1969, 1970; Doerfler and Kleinschmidt 1970): molecular strategies of adenovirus types 2 (Ad2) and Ad12 (199 citations in PubMed); studies on foreign DNA integration (61 citations); and investigations on the role of DNA methylation in epigenetic regulatory mechanisms in adenovirus infection and transformation (117 citations) as well as in human genetics (>70 citations). We were probably the first laboratory to initiate epigenetic investigations of an (adeno-) viral system (Günthert et al. 1976; Sutter and Doerfler 1980; Vardimon et al. 1980). A more detailed account of previous work has been summarized in Weber et al. (2016b) and will be briefly summarized in the following sections.

3.1.1.1 Ad12-Syrian Hamster Cells: The Abortive System

When viruses transcend their natural host range, which has been developing over evolutionary timescales, their impact on the noncanonical host can be catastrophic. Human Ad12 infecting hamster cells and leading to oncogenic transformation is a case in point and only one of several examples. The retrovirus HIV of originally simian origin or the coronavirus SARS of canine origin upon their adaptation to the human organism are examples of extraordinary medical importance. Hence the study of the interaction of human Ad12 with cells of the Syrian hamster has been considered of importance to understand this abortive interaction of oncogenic relevance at the molecular level.

One of the characteristics of the Ad12-hamster cell system is a strictly abortive infection cycle (review Hösel et al. 2003). The block of Ad12 replication lies before viral DNA replication (Doerfler 1969) and late gene transcription which cannot be detected (Ortin et al. 1976). Ad12 adsorption, cellular uptake, and transport of the viral DNA to the nucleus were less efficient in the nonpermissive hamster cells than in permissive human cells. However, many of the early functions of the Ad12 genome were expressed in BHK21 cells, though at a low level. In the downstream region of the major late promoter (MLP) of Ad12 DNA, a mitigator element of 33 nucleotide pairs in length was identified which contributed to the inactivity of the MLP in hamster cells and its markedly decreased activity in human cells (Zock and Doerfler 1990). The E1 functions of Ad2 or Ad5 were capable of partly complementing these Ad12 deficiencies in hamster cells in that Ad12 viral DNA replication and late gene transcription could proceed, e.g., in a BHK hamster cell line, BHK297-C131, which carried in an integrated form and constitutively expressed the E1 region of Ad5 DNA (Klimkait and Doerfler 1985). Nevertheless, the late Ad12 mRNAs, which were synthesized in this system and carried the authentic Ad12 nucleotide sequence, failed to be translated to structural viral proteins (Schiedner et al. 1994). Hence, infectious virions were not produced even in this partly complemented system. There appears to exist an additional translational block for late Ad12 mRNAs in hamster cells. We have further shown that the overexpression of the Ad12 preterminal protein (pTP) or of E1A genes facilitated the synthesis of full-length, authentic Ad12 DNA in Ad12-infected BHK21 hamster cells. Apparently, the Ad12 pTP had a hitherto unknown function in eliciting full cycles of Ad12 DNA replication even in nonpermissive BHK21 cells when sufficient levels of Ad12 pTP were produced (Hösel et al. 2001). The amounts of Ad12 DNA in the nuclei or cytoplasm of the complemented hamster cells were about 2 orders of magnitude [2 h postinfection (p.i.)] and 4–5 orders of magnitude (48 h p. i.) lower than in permissive human cells. Cell line BHK21-hCAR is transgenic for and expresses the human coxsackie and adenovirus receptor (hCAR) gene. Nuclear uptake of Ad12 DNA in BHK21-hCAR cells was markedly increased compared to that in naïve BHK21 cells. Ad12 elicited a cytopathic effect in BHK21-hCAR cells but not in BHK21 cells. Quantitative PCR or [^{3}H]-thymidine labeling followed by zone velocity sedimentation however failed to detect Ad12 DNA replication in BHK21 or BHK21-hCAR cells. Newly assembled Ad12 virions could not be detected. Thus, the block in Ad12 DNA replication in hamster cells was not released by the hCAR-enhanced nuclear import of Ad12 DNA (Hochstein et al. 2008).

We pursue the possibility that the completely abortive infection cycle of Ad12 in hamster cells ensures the survival of Ad12-induced hamster tumor cells which all carry multiple copies of genomically integrated Ad12 DNA. In this way, the viral genomes are immortalized and expanded into a huge number of tumor cells. Moreover, the totally abortive cycle of Ad12 in Syrian hamster cells is one of the decisive preconditions for the ability of Ad12 to induce tumor cells and tumors in Syrian hamsters, since Ad12-infected cells are capable of surviving virus infection, in contrast to the Ad12-human cell system in which all infected cells are killed (Doerfler 1991).

3.1.2 Foreign DNA in the Environment

DNA, free or embedded in tissue remnants, abounds all over the living world. These DNA molecules might still reside in complexes with specific proteins and RNAs attached. Their stability is variable and dependent on numerous, unidentified environmental factors. DNA fragments with free termini and DNA protein complexes belong to the most stable and aggressive molecules in nature and represent versatile recombination partners which can target the genomes of recipient cells. The pathways of DNA from cellular remnants in the environment to the nucleus of successfully entered cells and their genomes are poorly understood. DNA carried by viruses is well equipped for entry into specialized cells. In an earlier study on the fate of adenovirus type 2 (Ad2) DNA added directly to a culture of actively replicating human KB cells, between 3 and 9% of this DNA became associated with the cells in a DNase-resistant form (Groneberg et al. 1975). At 24 hr after the addition of DNA to the culture medium, 70% of the cell-associated DNA was found in the nucleus, and this DNA had been endonucleolytically fragmented. Direct penetration of the cytoplasmic membrane, perhaps during cell division, appeared the most likely mechanism for DNA uptake in these experiments. I suspect that there are less haphazardly acting mechanisms via DNA-sensitive receptors on the cell's surface.

3.1.3 Biology of Ad12-Induced Hamster Tumors

Histological and Immuno-histochemical Properties (Hohlweg et al. 2003) The following is a quote from this article in which the histopathological analysis was performed by Reinhardt Büttner, now in Cologne.

> Independent of location and size, the histology of all Ad12-induced tumors revealed small, rounded, uniformly stained cells with a large hyperchromatic nucleus and granular chromatin, numerous mitotic cells, and Homer–Wright rosette structures characteristic of primitive neuro-ectodermal tumors (Fig. 3.1). Immuno-histochemical assays for the expression of tissue-specific markers identified vimentin, synaptophysin, and neuronal-specific enolase in all tumors. The presence of the latter two proteins was compatible with the neuronal origin of the Ad12-induced tumors, whereas vimentin was typical for their mesenchymal derivation. Tests for additional tissue markers proved marginally positive (S-100, glia fiber protein) or negative (Table 3.1).

Figure 3.1 and Table 3.1 were taken from Hohlweg et al. (2003).

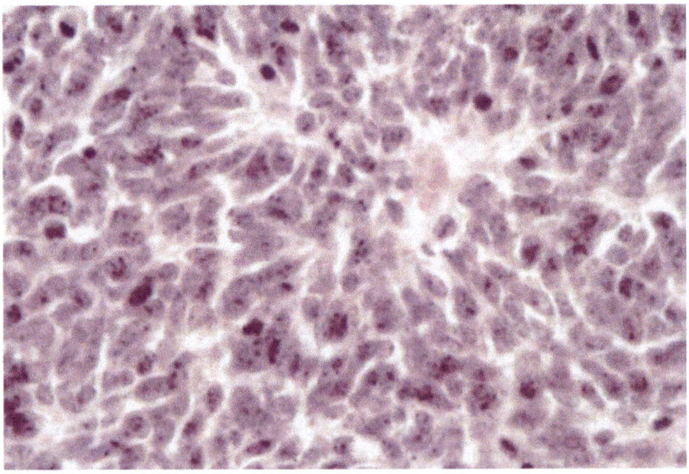

Fig. 3.1 Histological section of an Ad12-induced hamster tumor stained with hematoxylin and eosin

Table 3.1 Immuno-histological properties of Ad12-induced hamster tumors

Immuno-histological staining
Cytokeratin –
Vimentin +
Neuronal-specific enolase (NSE) +
Synaptophysin (+)
S-100a (+)
Gliafiber protein (+)
Ewing sarcoma marker: MIC2/CD99 –
Chromogranin A –

3.1.4 Characteristics of Chromosomally Integrated Adenovirus DNA

- Ad12-transformed cells or Ad12-induced hamster tumor cells carry up to >30 copies of viral genomes chromosomally integrated by covalent bonds between viral and cellular DNAs (Stabel et al. 1980; Hilger-Eversheim and Doerfler 1997; Knoblauch et al. 1996; Hohlweg et al. 2003; Hochstein et al. 2007).
- Although multiple copies of viral DNA are integrated, there is most frequently only one site of Ad12 DNA insertion on the chromosomes. In one study, 59/60 Ad12-induced tumors showed only one chromosomal site of Ad12 DNA integration as detected by fluorescent in situ hybridization (FISH) (Hilger-Eversheim and Doerfler 1997). The integratiion site was different in each tumor.
- In different Ad12-transformed cells or Ad12-induced tumor cells, viral DNA integration occurred at different sites in the cellular genome. There is no evidence for a specific cellular site of viral DNA insertion (Deuring et al. 1981b; Doerfler et al. 1983; Knoblauch et al. 1996; Hilger-Eversheim and Doerfler 1997).
- At the sites of recombination between the Ad12 and the cellular genomes, short or patchy nucleotide sequence homologies were frequently observed (Gahlmann et al. 1982; Stabel and Doerfler 1982). Moreover, the cellular pre-integration sites of viral DNA insertion often showed transcriptional activity and presumably an open chromatin structure (Schulz et al. 1987; Hochstein et al. 2007).
- Most of the integrated Ad12 genomes appeared to be intact, but fragmented genomes were also observed (Stabel et al. 1980; Hochstein et al. 2007). At the site of foreign DNA integration, cellular DNA sequences have been found to be completely conserved (Gahlmann and Doerfler 1983). In other instances, some of the abutting cellular DNA sequences were deleted.
- Recombination of Ad12 DNA with a cloned hamster cell DNA fragment, which had previously been identified as an integration site of Ad12 DNA, could be elicited also in a cell-free system by using high-salt nuclear extracts from hamster cells (Jessberger et al. 1989; Tatzelt et al. 1993).
- There is evidence that adenovirus DNA can recombine with cellular DNA also in human cells productively infected with adenoviruses (Burger and Doerfler 1974; Schick et al. 1976). An integrated state of viral DNA is difficult to prove since human cells productively infected with adenoviruses do not survive the infection. However, the discovery of a symmetric recombinant between Ad12 DNA and human cellular DNA has documented that Ad12 DNA does in fact recombine with cellular DNA even in productively infected cells (Deuring et al. 1981a; Deuring and Doerfler 1983).
- All of the aforementioned characteristics of Ad12 DNA integration in hamster cells have been confirmed in studies on the integration of replication-deficient adenovirus vector genomes in mouse cells (Stephen et al. 2010). In this system, adenoviral infection is compatible with cell survival due to the replication deficiency of the adenoviral vector genomes used. The results described in this system confirmed all our earlier findings in the Ad12 hamster cell system and in human cells productively infected with Ad12.

3.1.5 The Consequences of Inserting Foreign DNA into Established Mammalian Genomes

We have studied the consequences of foreign DNA insertions into the mammalian genome, namely (i) the de novo methylation of the transgenomes and (ii) alterations in the epigenetic stability of the recipient genomes. Several independently investigated mammalian systems with integrates of adenovirus DNA, bacteriophage lambda DNA, plasmid DNA, EBV DNA, or the telomerase gene as well as expansions of a CGG repeat were investigated (Heller et al. 1995; Remus et al. 1999; Müller et al. 2001; Naumann et al. 2014; Weber et al. 2015, 2016a).

3.1.5.1 Hypermethylation of Integrated Ad12 DNA, the Transgenome

As mentioned above, the virion genomes of Ad12 and Ad2 lack 5-methyldeoxycytidine nucleotides (Günthert et al. 1976). Free intracellular adenovirus DNA also remains unmethylated (Wienhues and Doerfler 1985; Kämmer and Doerfler 1995). In contrast, the integrated form of Ad12 DNA in Ad12-transformed hamster cells or in Ad12-induced hamster tumor cells becomes hyper-methylated in specific patterns (Sutter et al. 1978; Sutter and Doerfler 1980; Orend et al. 1995; Hochstein et al. 2007). De novo methylation appeared to be initiated at certain regional sites and extend from there by spreading (Toth et al. 1989; Orend et al. 1995).

3.1.5.2 Promoter CpG Methylation and Promoter Silencing

In 1979/1980, we documented an inverse correlation between adenoviral gene expression and CpG methylation of integrated adenovirus DNA in Ad12- and Ad2-transformed hamster cells (Sutter and Doerfler 1980; Vardimon et al. 1980; Doerfler 1983). These data were extended to activity studies of promoter-indicator gene constructs by using adenoviral promoters in order to document that pre-methylated promoters led to gene silencing, whereas unmethylated promoters allowed gene transcription (Vardimon et al. 1982; Kruczek and Doerfler 1983; Langner et al. 1984, 1986; Knebel and Doerfler 1986; Munnes et al. 1998). The silencing effect of promoter methylation could be partly or completely reversed by the expression of a viral trans-activator/ enhancer sequence, e.g., of the E1 proteins of Ad2 (Langner et al. 1986; Weisshaar et al. 1988), or by the close vicinity of a strong viral promoter/enhancer element, e.g., that of the human cytomegalovirus (Knebel-Mörsdorf et al. 1988). Today, it has been generally appreciated that many factors collaborate in regulating eukaryotic promoter function. Nevertheless, promoter methylation remains one of the dominant and experimentally convenient factors when studying the long-term silencing of gene activities.

3.1.5.3 Foreign DNA Integration into Mammalian Genomes Leads to Alterations in Methylation and Transcription Patterns

Genome-Wide Increases in DNA Methylation in Ad12-Transformed Cells—Stability of Changes Even After the Loss of All Viral Genomes: A "Hit-and-Run" Mechanism

The Ad12-transformed cell line T637 originated from BHK21 hamster cells following the infection of these cells with Ad12 and the selection of cells which expressed early Ad12 gene products (Strohl et al. 1970). In comparison to the levels of DNA methylation in the ~900 copies of the retrotransposon intracisternal A particle (IAP) genomes in the parent BHK21 cells, methylation of the IAP sequences in T637 cells, which are transgenomic for Ad12 DNA, was very markedly increased as detected by Southern blot hybridization (Heller et al. 1995). The extent of methylation augmentation in the T637 cell genome suggested that alterations in CpG methylation involved the entire cellular genome, since IAP sequences are distributed over many hamster chromosomes, frequently on their short arms (Heller et al. 1995; Meyer zu Altenschildesche et al. 1996). Methylation in other parts of the T637 genome—including single copy genes—was also enhanced (Heller et al. 1995). These hypermethylation patterns persisted in TR3 cells, a revertant of T637 cells (Groneberg et al. 1978; Groneberg and Doerfler 1979) which had lost all Ad12 DNA sequences.[1] Hence, the effects of foreign DNA insertions on cellular CpG methylation patterns were not dependent on the continued presence of the originally causative insertion of foreign (Ad12) genomes. This mechanism has the characteristics of a "hit-and-run" event. Since we consider the genome-wide methylation effects of Ad12 integration as crucial to the transformation and its oncogenic consequences (Doerfler 1995, 2000, 2011, 2012), the much debated possibility of a "hit-and-run" mechanism in viral oncogenesis is again raised by these results and has to be considered highly relevant.

Alterations of Cellular DNA Methylation and Transcription Patterns Are Also Elicited in Bacteriophage Lambda or Bacterial Plasmid Transgenomic Cells

Alterations of CpG DNA methylation patterns were also observed in BHK21 cells transgenomic for bacteriophage lambda or bacterial plasmid DNA (Heller et al. 1995). Alterations of DNA methylation in the lambda DNA trans-genomic cells were documented in more detail by bisulfite sequencing of a subsegment of the IAP transposon DNA in these cells (Remus et al. 1999). Bisulfite sequencing (Frommer

[1] By using the very sensitive PCR technique, which was not available in 1979, the revertant cell line TR3 of the Ad12-transformed hamster cell line T637 (Groneberg et al. 1978; Groneberg and Doerfler 1979) has recently been shown to be completely devoid of any Ad12 genome segments (S. Weber and W. Doerfler, unpublished studies).

et al. 1992; Clark et al. 1994) provides a positive display of all 5-mC residues in a DNA sequence. The inserted bacteriophage lambda DNA became also de novo methylated. There was no evidence that the increase in IAP DNA methylation might have preexisted in some of the non-transgenomic BHK21 cell clones (Remus et al. 1999).

A wide array of cellular DNA segments and genes was analyzed in hamster cells transgenomic for Ad12 or bacteriophage lambda DNA (Müller et al. 2001) for alterations in their transcriptional profiles as well by using the techniques of methylation-sensitive representational difference analysis (MS-RDA) (Ushijima et al. 1997) and suppressive subtractive hybridization. The data demonstrated that the insertion of foreign (Ad12 or bacteriophage lambda) DNA into an established mammalian genome can lead to extensive alterations also in cellular DNA transcription patterns (Müller et al. 2001).

Alterations of CpG Methylation Patterns Way Upstream of the FMR1 Boundary in Human Cells Immortalized by Epstein–Barr Virus (EBV) or by Transformation with the Telomerase Gene

We had previously investigated CpG methylation patterns in the human FMR1 segment (Genç et al. 2000; Gray et al. 2007) and in the human Prader–Willi region on chromosome 15q11-13 (Zeschnigk et al. 1997; Schumacher et al. 1998). The genome segment upstream of the FMR1 (fragile X mental retardation 1) gene on chromosome Xq27.3 contains several genetic signals (Naumann et al. 2009). Among them, we have described a DNA methylation boundary which is located 65–70 CpGs upstream of the CGG repeat in the gene's untranslated first exon and has been detected in any human (or mouse) cell type investigated (Naumann et al. 2009). In patients with the fragile X syndrome (FXS) (OMIM 300624), the methylation boundary is lost and, as a consequence, de novo methylation spreads downstream into the FMR1 promoter region and subsequently leads to promoter inactivation (Naumann et al. 2009). Loss of the FMR1 gene product is the cause for the FXS (for review O'Donnell and Warren 2002). This stable methylation boundary appears to help protect the promoter against the spreading of de novo methylation (Naumann et al. 2009, 2010). In cells transgenomic for EBV DNA or for the telomerase gene, the large number of normally methylated CpGs in the far-upstream region of the boundary is decreased about fourfold (Naumann et al. 2014). We have interpreted this marked decrease of DNA CpG methylation in a well-studied part of the human genome as a consequence of the introduction of foreign genomes (EBV or telomerase gene) into human cells (Naumann et al. 2014).

A Model System to Study the Epigenomic Destabilization in Human Cells Transgenomic for a 5.6 kbp Bacterial Plasmid

We have recently described a model system to study the effects of foreign DNA insertion into human cells in culture (Weber et al. 2015). Human cells from cell line HCT116 were rendered transgenomic for a 5.6 kbp bacterial plasmid. Transgenomic cell clones were selected with foreign plasmids stably integrated, most likely at different genomic sites. In five non-transgenomic HCT116 control clones, transcription and methylation patterns proved very similar among individual cell clones. These control data facilitated comparisons of these patterns between non-transgenomic and transgenomic clones. In 4.7% of the 28.869 gene segments analyzed, the transcriptional activities were upregulated (907 genes) or downregulated (436 genes) (Weber et al. 2015) (Fig. 3.2a). Upregulations were frequently found in small nucleolar RNA genes which regulate RNA metabolism and in genes involved in signaling pathways. Genome-wide methylation profiling was performed for 361,983 CpG sites. In comparisons of methylation levels in five transgenomic versus four non-transgenomic cell clones, 3791 CpGs were differentially methylated, 1504 CpGs were hyper-methylated, and 2287 were hypomethylated (Fig. 3.2b). These differential values, both for transcriptional activities and methylation patterns, were statistically corrected for minute differences in some of the 28.869 genome segments in non-transgenomic cell clones. The importance of transgenome size, CG or gene content, copy number, and the mechanism (s) responsible for the observed epigenetic alterations have not yet been investigated.

As a corollary to this earlier study (Weber et al. 2015), we have investigated whether the alterations in transcriptional and methylation profiles had extended also to repetitive genome elements like the HERV and LINE-1.2 sequences in the same transgenomic HCT116 cell clones which had exhibited epigenetic alterations in the above studied parts of the human genome. Such differences were not found. Apparently in the cell clones selected for this investigation, the HERV and LINE elements had not responded to foreign DNA insertions (Weber et al 2016a). In addition, this work provided a survey of the CpG modifications in the human endogenous viral sequences HERV-K, HERV-W, and HERV-E and in LINE-1.2 whose methylation levels ranged between 60 and 98%. At least some of these elements were transcribed into RNA as determined by reverse transcription and PCR. Obviously, there are enough unmethylated control sequences to facilitate transcription of at least some of the tested elements into RNA.

3.1.6 Résumé

Based on the study of human Ad12 as an oncogenic DNA virus, the fate of foreign DNA in mammalian systems and the epigenetic consequences of foreign DNA insertions in general have been a long-term interest in my laboratory (Doerfler et al.

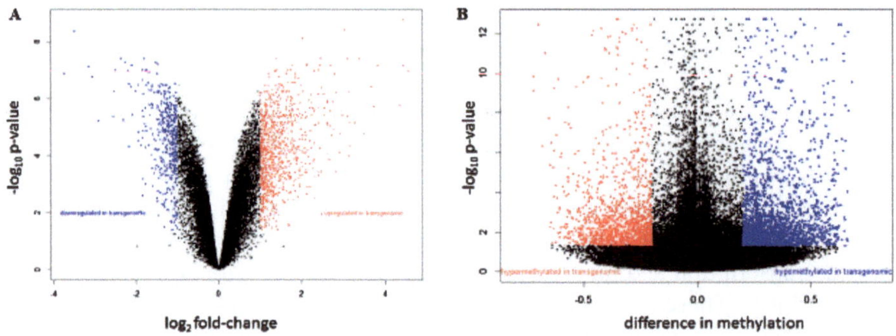

Fig. 3.2 Alterations in patterns of transcription (**a**) and methylation (**b**) in pC1-5.6 transgenomic HCT116 cell clones as compared to non-transgenomic cells. (**a**) Volcano plot displays non-standardized signals (log2 fold-change) on the x-axis against standardized signals ($-$log10 FDR-adjusted p-value) on the y-axis for the comparison of five non-transgenomic against seven transgenomic cell clones of all 28,869 genes analyzed. Upregulated genes in transgenomic cell clones were displayed in *red* and downregulated genes in *blue* (FC $\pm$ 2, adjusted p-values <0.05; $n = 1343$ genes). (**b**) Volcano plot displays differences in methylation on the x-axis against standardized methylation ($-$log10 FDR-adjusted p-value) on the y-axis for the comparison of four non-transgenomic against five pC1-5.6 transgenomic cell clones of all 361,983 CpGs interrogated. Hyper-methylated CpGs in transgenomic cell clones were displayed in *red* and hypomethylated CpGs in blue ($\Delta\beta$ value ≥ 0.2, adjusted p-value <0.05; $n = 3791$ CpGs). This Figure and its legends were taken with permission from Weber et al. (2015)

1983; Weber et al. 2016b). Foreign DNA which emanates from a panoply of sources is ubiquitous and abundant in our environment. Research about the fate of this very stable and biologically potent molecule in the environment is a medically highly relevant topic. How can DNA interact with and be taken up by living cells, how frequently is it integrated into the invaded cell's genome, and what are the consequences of these interactions for cell survival and genetic integrity—oncogenicity?

In studies on the integrated state of Ad12 DNA in Ad12-transformed hamster cells, we discovered that the CpG methylation profiles in some of their endogenous retrotransposon sequences and in several cellular genes were increased. This augmented methylation persisted in revertants of the transformed cells that had lost all Ad12 genomes ("hit-and-run" mechanism). Moreover, alterations of DNA methylation and transcription profiles were documented in Ad12 DNA- and in bacteriophage λ DNA-transgenomic cells.

- I have previously hypothesized that epigenetic effects in mammalian genomes due to the insertion of foreign DNA are a general phenomenon (Doerfler 2012). These alterations might play a role in (viral) oncogenesis and are possibly instrumental during evolution as a consequence of multiple retroviral DNA insertions into ancient genomes. Over evolutionary times, these alterations of transcription profiles might have led to novel phenotypes that were then selected for or against depending on environmental conditions during evolution (Doerfler 2016).

- To examine the general significance of these observations, we designed a model system for proof-of-principle assessment. Human cells from cell line HCT116 were rendered transgenomic by transfecting a 5.6 kbp bacterial plasmid and selecting cell clones with foreign plasmids stably integrated, most likely at different genomic sites.
- In five non-transgenomic HCT116 control clones without the plasmid, transcription and methylation patterns proved similar, if not identical, among five individual cell clones. This finding opened the possibility for comparisons of these patterns between non-transgenomic and transgenomic clones.
- In 4.7% of the 28,869 gene segments analyzed, the transcriptional activities were upregulated (907 genes) or downregulated (436 genes) in plasmid-transgenomic cell clones in comparison to control clones. A significant gene set enrichment was found in 43 canonical pathways. Frequent upregulations were noted in small nucleolar RNA genes that regulate RNA metabolism and in genes involved in signaling pathways.
- Genome-wide methylation profiling was performed for 361,983 CpG sites. In comparisons of methylation levels in five transgenomic versus four non-transgenomic cell clones, 3791 CpGs were differentially methylated, 1504 CpGs were hyper-methylated, and 2287 were hypo-methylated.
- Thus, the epigenetic effects in the wake of foreign DNA integration events can be considered a very significant effect also in human cells. We still lack insights into the role of transgenome size, gene or CG content, or copy number of the transgenome. The mechanism(s) underlying the observed epigenetic alterations are unknown. Extent and location of alterations in genome activities and CpG methylation might depend on the site(s) of foreign DNA insertion.
- In the same cell clones studied as described above, differences in methylation and transcription profiles in some of the HERV and LINE-1.2 repetitive elements were not observed.
- We note that genome manipulations in general—work with transgenomic or knocked cells and organisms—have assumed a major role in molecular biology and medicine. The consequences of cellular genome manipulations for epigenetic stability have so far received unwarrantedly limited attention. Before drawing far-reaching conclusions from work with cells or organisms with manipulated genomes, critical considerations for and careful analyses of their epigenomic stability will prove prudent.

With previous and current research described here, we have barely scratched the surface of the problem but are now poised to ask more precise questions. The Ad12 system has been a very reliable guide to this approach which has in due course been extended also to other types of foreign DNA molecules. We will now pursue more far-reaching questions and again use the Ad12 system as a versatile model organism and guide.

Acknowledgements Research summarized in this chapter has been supported between 1972 and 2002 at the Institute of Genetics in Köln by the DFG (SFB 74 and 274) and by the Center for

Molecular Medicine Cologne (CMMC—TP13). In Erlangen, we received funding from the DFG (DO 165/28), the Fritz Thyssen Foundation in Köln (Az. 10.07.2.138), and the Staedtler Foundation in Nürnberg (WW/eh 01/15) at different times between 2002 and the present.

References

Burger H, Doerfler W (1974) Intracellular forms of adenovirus DNA. III. Integration of the DNA of adenovirus type 2 into host DNA in productively infected cells. J Virol 13:975–992

Clark SJ, Harrison J, Paul CL, Frommer M (1994) High sensitivity mapping of methylated cytosines. Nucleic Acids Res 22:2990–2999

Deuring R, Doerfler W (1983) Proof of recombination between viral and cellular genomes in human KB cells productively infected by adenovirus type 12: structure of the junction site in a symmetric recombinant (SYREC). Gene 26:283–289

Deuring R, Klotz G, Doerfler W (1981a) An unusual symmetric recombinant between adenovirus type 12 DNA and human cell DNA. Proc Natl Acad Sci USA 78:3142–3146

Deuring R, Winterhoff U, Tamanoi F, Stabel S, Doerfler W (1981b) Site of linkage between adenovirus type 12 and cell DNAs in hamster tumour line CLAC3. Nature 293:81–84

Doerfler W (1968) The fate of the DNA of adenovirus type 12 in baby hamster kidney cells. Proc Natl Acad Sci USA 60:636–643

Doerfler W (1969) Nonproductive infection of baby hamster kidney cells (BHK21) with adenovirus type 12. Virology 38:587–606

Doerfler W (1970) Integration of the deoxyribonucleic acid of adenovirus type 12 into the deoxyribo-nucleic acid of baby hamster kidney cells. J Virol 6:652–666

Doerfler W (1983) DNA methylation and gene activity. Annu Rev Biochem 52:93–124

Doerfler W (1991) Abortive infection and malignant transformation by adenoviruses: integration of viral DNA and control of viral gene expression by specific patterns of DNA methylation. Adv Virus Res 39:89–128

Doerfler W (1995) The insertion of foreign DNA into mammalian genomes and its consequences: a concept in oncogenesis. Adv Cancer Res 66:313–344

Doerfler W (2000) Foreign DNA in mammalian systems. Wiley-VCH, Weinheim

Doerfler W (2011) Epigenetic consequences of foreign DNA integration: global alterations of methylation and transcription patterns in recipient genomes. Rev Med Virol 21:336–346

Doerfler W (2012) The impact of foreign DNA integration on tumor biology and evolution via epigenetic alterations. Epigenomics 4:41–49

Doerfler W, Kleinschmidt AK (1970) Denaturation pattern of the DNA of adenovirus type 2 as determined by electron microscopy. J Mol Biol 50:579–593

Doerfler W (2016) Beware of manipulations on the genome: epigenetic destabilization through (foreign) DNA insertions. Epigenomics 8:587–591

Doerfler W, Gahlmann R, Stabel S, Deuring R, Lichtenberg U, Schulz M, Eick D, Leisten R (1983) On the mechanism of recombination between adenoviral and cellular DNAs: the structure of junction sites. Curr Top Microbiol Immunol 109:193–228

Frommer M, McDonald LE, Millar DS, Collis CM, Watt F, Grigg GW, Molloy PL, Paul CL (1992) A genomic sequencing protocol that yields a positive display of 5-methylcytosine residues in individual DNA strands. Proc Natl Acad Sci USA 89:1827–1831

Gahlmann R, Doerfler W (1983) Integration of viral DNA into the genome of the adenovirus type 2-transformed hamster cell line HE5 without loss or alteration of cellular nucleotides. Nucleic Acids Res 11:7347–7361

Gahlmann R, Leisten R, Vardimon L, Doerfler W (1982) Patch homologies and the integration of adenovirus DNA in mammalian cells. EMBO J 1:1101–1104

Genç B, Müller-Hartmann H, Zeschnigk M, Deissler H, Schmitz B, Majewski F, von Gontard A, Doerfler W (2000) Methylation mosaicism of 5′-(CGG)$_n$-3′ repeats in fragile X, premutation and healthy individuals. Nucleic Acids Res 28:2141–2152

Gray S, Gerhardt J, Doerfler W, Small LE, Fanning E (2007) An origin of DNA replication in the promoter region of the human fragile X mental retardation (FMR1) gene. Mol Cell Biol 27:426–437

Groneberg J, Doerfler W (1979) Revertants of adenovirus type-12-transformed hamster cells have lost part of the viral genomes. Int J Cancer 24:67–74

Groneberg J, Brown DT, Doerfler W (1975) Uptake and fate of the DNA of adenovirus type 2 in KB cells. Virology 64:115–131

Groneberg J, Chardonnet Y, Doerfler W (1977) Integrated viral sequences in adenovirus type 12-transformed hamster cells. Cell 10:101–111

Groneberg J, Sutter D, Soboll H, Doerfler W (1978) Morphological revertants of adenovirus type 12-transformed hamster cells. J Gen Virol 40:635–645

Günthert U, Schweiger M, Stupp M, Doerfler W (1976) DNA methylation in adenovirus, adenovirus-transformed cells, and host cells. Proc Natl Acad Sci USA 73:3923–3927

Heller H, Kämmer C, Wilgenbus P, Doerfler W (1995) Chromosomal insertion of foreign (adenovirus type 12, plasmid, or bacteriophage lambda) DNA is associated with enhanced methylation of cellular DNA segments. Proc Natl Acad Sci USA 92:5515–5519

Hilger-Eversheim K, Doerfler W (1997) Clonal origin of adenovirus type 12-induced hamster tumors: nonspecific chromosomal integration sites of viral DNA. Cancer Res 57:3001–3009

Hochstein N, Muiznieks I, Mangel L, Brondke H, Doerfler W (2007) The epigenetic status of an adenovirus transgenome upon long-term cultivation in hamster cells. J Virol 81:5349–5361

Hochstein N, Webb D, Hösel M, Seidel W, Auerochs S, Doerfler W (2008) Human CAR gene expression in non-permissive hamster cells boosts entry of type 12 adenovirions and nuclear import of viral DNA. J Virol 82:4159–4163

Hohlweg U, Hösel M, Dorn A, Webb D, Hilger-Eversheim K, Remus R, Schmitz B, Mende Y, Büttner R, Schramme A, Corzilius L, Niemann A, Doerfler W (2003) Intraperitoneal dissemination of Ad12-induced undifferentiated neuro-ectodermal hamster tumors: de novo methylation and transcription patterns of integrated viral and of cellular genes. Virus Res 98:45–56

Hösel M, Webb D, Schröer J, Schmitz B, Doerfler W (2001) The over-expression of the adenovirus type 12 pTP or E1A gene facilitates Ad12 DNA replication in non-permissive BHK21 hamster cells. J Virol 75:16041–16053

Hösel M, Webb D, Schröer J, Doerfler W (2003) The abortive infection of Syrian hamster cells with human adenovirus type 12. Curr Top Microbiol Immunol 272:415–440

Jessberger R, Heuss D, Doerfler W (1989) Recombination in hamster cell nuclear extracts between adenovirus type 12 DNA and two hamster preinsertion sequences. EMBO J 8:869–878

Kämmer C, Doerfler W (1995) Genomic sequencing reveals absence of DNA methylation in the major late promoter of adenovirus type 2 DNA in the virion and in productively infected cells. FEBS Lett 362:301–305

Klimkait T, Doerfler W (1985) Adenovirus types 2 and 5 functions elicit replication and late expression of adenovirus type 12 DNA in hamster cells. J Virol 55:466–474

Knebel D, Doerfler W (1986) N^6-methyldeoxyadenosine residues at specific sites decrease the activity of the E1A promoter of adenovirus type 12 DNA. J Mol Biol 189:371–375

Knebel-Mörsdorf D, Achten S, Langner KD, Rüger R, Fleckenstein B, Doerfler W (1988) Reactivation of the methylation-inhibited late E2A promoter of adenovirus type 2 by a strong enhancer of human cytomegalovirus. Virology 166:166–174

Knoblauch M, Schröer J, Schmitz B, Doerfler W (1996) The structure of adenovirus type 12 DNA integration sites in the hamster cell genome. J Virol 70:3788–3796

Kochanek S, Toth M, Dehmel A, Renz D, Doerfler W (1990) Interindividual concordance of methylation profiles in human genes for tumor necrosis factors α and β. Proc Natl Acad Sci USA 87:8830–8834

Kochanek S, Renz D, Doerfler W (1993) DNA methylation in the Alu sequences of diploid and haploid primary human cells. EMBO J 12:1141–1151

Kruczek I, Doerfler W (1983) Expression of the chloramphenicol acetyltransferase gene in mammalian cells under the control of adenovirus type 12 promoters: effect of promoter methylation on gene expression. Proc Natl Acad Sci USA 80:7586–7590

Langner KD, Vardimon L, Renz D, Doerfler W (1984) DNA methylation of three 5′ C-C-G-G 3′ sites in the promoter and 5′ region inactivates the E2a gene of adenovirus type 2. Proc Natl Acad Sci USA 81:2950–2954

Langner KD, Weyer U, Doerfler W (1986) Trans-effect of the E1 region of adenoviruses on the expression of a prokaryotic gene in mammalian cells: resistance to 5′-CCGG-3′ methylation. Proc Natl Acad Sci USA 83:1598–1602

Meyer zu Altenschildesche G, Heller H, Wilgenbus P, Tjia ST, Doerfler W (1996) Chromosomal distribution of the hamster intracisternal A-particle (IAP) retrotransposons. Chromosoma 104:341–344

Müller K, Heller H, Doerfler W (2001) Foreign DNA integration. Genome-wide perturbations of methylation and transcription in the recipient genomes. J Biol Chem 276:14271–14278

Munnes M, Patrone G, Schmitz B, Romeo G, Doerfler W (1998) A 5′-CG-3′-rich region in the promoter of the transcriptionally frequently silenced RET proto-oncogene lacks methylated cytidine residues. Oncogene 17:2573–2584

Naumann A, Hochstein N, Weber S, Fanning E, Doerfler W (2009) A distinct DNA-methylation boundary in the 5′-upstream sequence of the FMR1 promoter binds nuclear proteins and is lost in fragile X syndrome. Am J Hum Genet 85:606–616

Naumann A, Weber S, Hochstein N, Fanning E, Doerfler W (2010) Safeguarding the human FMR1 promoter: the methylation barrier in its 5′-upstream region. In: Abstract, Annual meeting of the American Society of Human Genetics in Washington, DC, 2010

Naumann A, Kraus C, Hoogeveen A, Ramirez CM, Doerfler W (2014) Stable DNA methylation boundaries and expanded trinucleotide repeats: role of DNA insertions. J Mol Biol 426:2554–2566

O'Donnell WT, Warren ST (2002) A decade of molecular studies of fragile X syndrome. Annu Rev Neurosci 25:315–338

Orend G, Knoblauch M, Kämmer C, Tjia ST, Schmitz B, Linkwitz A, Meyer zu Altenschildesche G, Maas J, Doerfler W (1995) The initiation of de novo methylation of foreign DNA integrated into a mammalian genome is not exclusively targeted by nucleotide sequence. J Virol 69:1226–1242

Ortin J, Scheidtmann K-H, Greenberg R, Westphal M, Doerfler W (1976) Transcription of the genome of adenovirus type 12. III. Maps of stable RNA from productively infected human cells and abortively and transformed hamster cells. J Virol 20:355–372

Remus R, Kämmer C, Heller H, Schmitz B, Schell G, Doerfler W (1999) Insertion of foreign DNA into an established mammalian genome can alter the methylation of cellular DNA sequences. J Virol 73:1010–1022

Schick J, Baczko K, Fanning F, Groneberg J, Burger H, Doerfler W (1976) Intracellular forms of adenovirus DNA: integrated form of adenovirus DNA appears early in productive infection. Proc Natl Acad Sci USA 73:1043–1047

Schiedner G, Schmitz B, Doerfler W (1994) Late transcripts of adenovirus type 12 DNA are not translated in hamster cells expressing the E1 region of adenovirus type 5. J Virol 68:5476–5482

Schulz M, Freisem-Rabien U, Jessberger R, Doerfler W (1987) Transcriptional activities of mammalian genomes at sites of recombination with foreign DNA. J Virol 61:344–353

Schumacher A, Buiting K, Zeschnigk M, Doerfler W, Horsthemke B (1998) Methylation analysis of the PWS/AS region does not support an enhancer competition model of genomic imprinting on human chromosome 15. Nat Genet 19:324–325

Sprengel J, Schmitz B, Heuss-Neitzel D, Zock C, Doerfler W (1994) Nucleotide sequence of human adenovirus type 12 DNA: comparative functional analysis. J. Virol. 68:379–389

Stabel S, Doerfler W (1982) Nucleotide sequence at the site of junction between adenovirus type 12 DNA and repetitive hamster cell DNA in transformed cell line CLAC1. Nucleic Acids Res 10:8007–8023

Stabel S, Doerfler W, Friis RR (1980) Integration sites of adenovirus type 12 DNA in transformed hamster cells and hamster tumor cells. J Virol 36:22–40

Stephen SL, Montini E, Sivanandam VG, Al-Dhalimy M, Kestler HA, Finegold M, Grompe M, Kochanek S (2010) Chromosomal integration of adenoviral vector DNA in vivo. J Virol 84:9987–9994

Strohl WA, Rouse H, Teets K, Schlesinger RW (1970) The response of BHK21 cells to infection with type 12 adenovirus. 3. Transformation and restricted replication of superinfecting type 2 adenovirus. Arch Gesamte Virusforsch 31:93–112

Sutter D, Doerfler W (1980) Methylation of integrated adenovirus type 12 DNA sequences in transformed cells is inversely correlated with viral gene expression. Proc Natl Acad Sci USA 77:253–256

Sutter D, Westphal M, Doerfler W (1978) Patterns of integration of viral DNA sequences in the genomes of adenovirus type 12-transformed hamster cells. Cell 14:569–585

Tatzelt J, Fechteler K, Langenbach P, Doerfler W (1993) Fractionated nuclear extracts from hamster cells catalyze cell-free recombination at selective sequences between adenovirus DNA and a hamster preinsertion site. Proc Natl Acad Sci USA 90:7356–7360

Toth M, Lichtenberg U, Doerfler W (1989) Genomic sequencing reveals a 5-methylcytosine-free domain in active promoters and the spreading of preimposed methylation patterns. Proc Natl Acad Sci USA 86:3728–3732

Trentin JJ, Yabe Y, Taylor G (1962) The quest for human cancer viruses. Science 137:835–841

Ushijima T, Morimura K, Hosoya Y, Okonogi H, Tatematsu M, Sugimura T, Nagao M (1997) Establishment of methylation-sensitive-representational difference analysis and isolation of hypo- and hypermethylated genomic fragments in mouse liver tumors. Proc Natl Acad Sci USA 94:2284–2289

Vardimon L, Neumann R, Kuhlmann I, Sutter D, Doerfler W (1980) DNA methylation and viral gene expression in adenovirus-transformed and -infected cells. Nucleic Acids Res 8:2461–2473

Vardimon L, Kressmann A, Cedar H, Maechler M, Doerfler W (1982) Expression of a cloned adenovirus gene is inhibited by *in vitro* methylation. Proc Natl Acad Sci USA 79:1073–1077

Waalwijk C, Flavell RA (1978) MspI, an isochizomer of HpaII which cleaves both unmethylated and methylated HpaII sites. Nucleic Acids Res 5:3231–3236

Weber S, Hofmann A, Herms S, Hoffmann P, Doerfler W (2015) Destabilization of the human epigenome: consequences of foreign DNA insertions. Epigenomics 7:745–755

Weber S, Jung S, Doerfler W (2016a) DNA Methylation in HERV (K, W, E) and LINE sequences and their transcription remain unchanged by foreign DNA insertions. Epigenomics 8:157–165

Weber S, Hofmann A, Naumann A, Hoffmann P, Doerfler W (2016b) Epigenetic alterations upon the insertion of foreign DNA into mammalian genomes: oncogenesis and evolution. In: Doerfler W, Boehm P (eds) Epigenetics – a different way of looking at genetics. Springer, Berlin, pp 123–143

Weisshaar B, Langner KD, Jüttermann R, Müller U, Zock C, Klimkait T, Doerfler W (1988) Reactivation of the methylation-inactivated late E2A promoter of adenovirus type 2 by E1A (13S) functions. J Mol Biol 202:255–270

Wienhues U, Doerfler W (1985) Lack of evidence for methylation of parental and newly synthesized adenovirus type 2 DNA in productive infections. J Virol 56:320–324

Zeschnigk M, Schmitz B, Dittrich B, Buiting K, Horsthemke B, Doerfler W (1997) Imprinted segments in the human genome: different DNA methylation patterns in the Prader-Willi/ Angelman syndrome region as determined by the genomic sequencing method. Hum Mol Genet 6:387–395

Zock C, Doerfler W (1990) A mitigator sequence in the downstream region of the major late promoter of adenovirus type 12 DNA. EMBO J 9:1615–1623

Chapter 4
Epigenetic Consequences of Epstein–Barr Virus Infection

Christine E. Birdwell, Krista Queen, and Rona S. Scott

Abstract The biphasic nature of the Epstein–Barr virus (EBV) life cycle is tightly regulated by epigenetic modifications. Silencing of viral gene expression associated with latency is accompanied by repressive chromatin modifications and DNA methylation. Latency is integral to the lifelong persistence of EBV, yet persistence also relies on viral replication and virus production for infection of naïve hosts. Thus, EBV can overcome and disrupt the repressive epigenetic environment of the latent viral genomes. Viral modulators of the host epigenetic machinery are not only involved in establishing the latent and lytic viral epigenetic states but also reprogram the host epigenome in ways that are likely beneficial to the virus, but can carry long term consequences to the host. Here, we will review epigenetic aspects of EBV life cycle control and consequences to the host cell.

Keywords Epstein–Barr virus • EBV • Latency • Reactivation • Lymphoma • Carcinoma • DNA methylation • Chromatin

---

C.E. Birdwell
Department of Microbiology and Immunology, Louisiana State University Health Sciences Center—Shreveport, Shreveport, LA, USA

Center for Molecular and Tumor Virology, Louisiana State University Health Sciences Center—Shreveport, Shreveport, LA, USA

K. Queen
Department of Microbiology and Immunology, Louisiana State University Health Sciences Center—Shreveport, Shreveport, LA, USA

Division of Viral Diseases, National Center for Immunizations and Respiratory Diseases, Centers for Disease Control and Prevention, Atlanta, GA, USA

R.S. Scott (✉)
Department of Microbiology and Immunology, Louisiana State University Health Sciences Center—Shreveport, Shreveport, LA, USA

Center for Molecular and Tumor Virology, Louisiana State University Health Sciences Center—Shreveport, Shreveport, LA, USA

Feist-Weiller Cancer Center, Louisiana State University Health Sciences Center—Shreveport, Shreveport, LA, USA
e-mail: RScott1@lsuhsc.edu

W. Doerfler, J. Casadesús (eds.), *Epigenetics of Infectious Diseases*, Epigenetics and Human Health, DOI 10.1007/978-3-319-55021-3_4

4.1 Epigenetic Regulation of the EBV Life Cycle

4.1.1 EBV Life Cycle

The Epstein–Barr virus (EBV) double-stranded DNA genome is linear within the virion and lacks major epigenetic modifications (Szyf et al. 1985; Fernandez et al. 2009). EBV is primarily shed and transmitted in the saliva of infected persons (Niederman et al. 1976; Hadinoto et al. 2009) using epithelial cells and naïve B cells to carry out the viral life cycle. Upon entering a host cell, the viral genome circularizes via recombination of the terminal repeats (TR) located at the ends of the viral genome and is maintained as an episome in infected cells (Lindahl et al. 1976). The EBV genome is chromatinized and progressively DNA methylated similar to what is found on the host genome (Kalla et al. 2010). The virus uses the host epigenetic machinery to establish various epigenetic states that regulate viral gene expression and that are integral to completion of the viral life cycle.

Epithelial cells support the replicative phase of the viral life cycle with production of progeny virions (Sixbey et al. 1984), whereas EBV infection of B cells supports the latent phase of the viral life cycle characterized by a set of increasingly restricted viral gene expression patterns. Epithelial cells may harbor latent EBV infection, but the nature of EBV epithelial latency is less understood than in B cells (Reusch et al. 2015). EBV latent genes assist in navigating naïve B cells through their differentiation program into long-lived memory B cells. EBV uses a set of latency programs characterized by silencing of viral gene expression as an immune evasion strategy required for viral persistence in the B-cell compartment (Miyashita et al. 1997). During latency III (growth program) in the naïve B cell, six Epstein–Barr virus nuclear antigens (EBNA) 1, 2, 3A, 3B, 3C, and LP and three latent membrane proteins (LMP) 1, 2A, and 2B are expressed. The EBNA proteins are involved in transcriptional regulation while the LMPs mimic the signaling necessary for B-cell maturation. In latency II (default program), four latency proteins are expressed: EBNA1, LMP1, LMP2A, and LMP2B. Latency I (latency program) in resting memory B cells is characterized by expression of EBNA1, required for the maintenance of the viral episome during cell division (Yates et al. 1984), and sometimes LMP2A (Decker et al. 1996; Babcock et al. 2000). Latency 0 is the absence of viral proteins and has been reported in nondividing memory B cells (Thorley-Lawson and Gross 2004). In addition, EBV has noncoding RNAs that are expressed during latency that include the EBV encoded RNAs (EBERs), *BamH1* fragment A rightward transcripts (BARTs), and viral microRNAs (miRNA) (Qu and Rowe 1992; Tierney et al. 1994). Rescue from latency and induction of the productive phase of the viral life cycle occurs upon terminal differentiation of memory B cells into plasma cells, typically induced by antigen stimulation of the cognate B-cell receptor (BCR) (Crawford and Ando 1986; Laichalk and Thorley-Lawson 2005). Virus released from the B cell can then infect epithelial cells where the virus can be amplified and spread cell to cell or infect a new host.

4.1.2 Epigenetic Control of Viral Promoter Usage

In B cells, the latency transcription patterns of EBV have been well described with differential promoter activities that are epigenetically regulated through DNA methylation and chromatin modifications. DNA methylation occurs on cytosine residues in the context of a CpG dinucleotide, and catalyzed by the DNA methyltransferases 1, 3A, and 3B. DNMT1 is the maintenance DNMT, while DNMTs 3A and 3B act as de novo DNMTs. EBV has an underrepresentation of CpG residues compared to other human herpesviruses (Honess et al. 1989; Karlin et al. 1994). This reduction of CpGs is suggestive of EBV adaptations commandeering the host DNA methylation response to regulate its life cycle. Methylated CpG residues in EBV genomes are thought to have mutated and lost over time and suggesting that the remaining CpG residues have important regulatory roles in the viral lifecycle. Indeed, DNA methylation of the EBV genome is observed in latency and required for completion of the productive phase of the viral lifecycle. Dense CpG methylation at promoter regions can inhibit the binding of activating transcription factors as well as recruit chromatin remodelers to compact the DNA and hinder transcription (reviewed in Schubeler 2015). Transcription of EBNA2 and EBNA-LP following initial B-cell infection is driven by the W promoter (Wp) (Woisetschlaeger et al. 1990). Wp-derived EBNA2 with the cellular factor recombination signal binding protein for immunoglobulin Kappa J Region (RBP-Jκ) activates the upstream C promoter (Cp), required for transcription of the six EBNA gene products (Ling et al. 1993). Over time, the Wp becomes hypermethylated and transcription occurs mostly through Cp (Tierney et al. 2000). In latency I and II, the Q promoter (Qp) is used for expression of EBNA1, which is required for the maintenance and replication of the EBV episome in latency (Rawlins et al. 1985). Transcriptional silencing of the other five EBNAs results from Cp DNA methylation (Paulson and Speck 1999). Both Wp and Cp are methylated, at least in part, by the cellular DNA methyltransferases (DNMTs) 3A and 3B (Tao et al. 2002; Leonard et al. 2011). Qp is protected from silencing by DNA methylation and repressive chromatin through the host boundary protein CTCF (CCCTC-binding factor), which demarcates the active Qp on one side and the chromatinized and DNA methylated W repeats on the other (Tempera et al. 2010). DNA methylation of Wp and Cp is critical for the promoter switch to Qp. Additionally, the lytic promoters tend to be highly methylated in latency (Paulson and Speck 1999).

In addition to DNA methylation control of viral promoter usage, chromatin modifications add another layer of regulation. The EBV genome accumulates a nucleosomal density similar to the host genome with active latent promoters being relatively nucleosome free and lytic promoters being more densely packed (Dyson and Farrell 1985). Repressive chromatin generally lacks histone acetylation and promotes the compaction of the DNA to inhibit transcription. Repressive chromatin is associated with various histone modifications that include histone 3 lysine 9 trimethylation (H3K9me3), histone 3 lysine 27 trimethylation (H3K27me3), and histone 4 lysine 20 trimethylation (H4K20me3). Actively transcribed regions can be associated with histone acetylation and histone 3 lysine 4 tri- and dimethylation

(H3K4me3/2) (reviewed in Zhou et al. 2011). H3K9me3 is abundant in the latency I genome and tends to overlap with regions of DNA methylation (Chau and Lieberman 2004). Furthermore, lytic promoters are also associated with the repressive mark H3K27me3 during latency (Arvey et al. 2013). Inversely, active promoters during latency III, like the LMP promoters, are associated with the active histone mark H3K4me2 and acetylation (Gerle et al. 2007). Several CTCF binding sites are located on the EBV genome (Arvey et al. 2013). CTCF is involved in organizing chromatin into higher ordered structures that separate condensed and decondensed chromatin into distinct transcriptional domains. CTCF binding upstream of EBV Cp and Qp forms distinct chromatin loops with the distal enhancer element of the origin of latent replication, OriP (origin of plasmid replication), in the respective latency program that reflects Qp or Cp activity (Chau et al. 2006). Moreover, the LMP promoters can also form a chromatin loop with OriP mediated by CTCF that enhances their transcription in early latency (Chen et al. 2014).

4.1.3 Epigenetic Regulation of Lytic Reactivation

Viral reactivation of latent EBV genomes generally occurs upon B-cell differentiation into plasma cells or in response to epithelial cell differentiation (Reusch et al. 2015). Engagement of the BCR by antigen or antibody cross-linking induces the signaling pathways phosphatidylinositol-3-kinase (PI3K), Ras-family GTPases, and phospholipase C gamma-2 (PLC) that lead to derepression and activation of viral immediate early genes: BZLF1 (Zta, Z, or Zebra) or BRLF1 (Rta or R) (Crawford and Ando 1986; Laichalk and Thorley-Lawson 2005). Other stimuli such as phorbol esters, histone deacetylase inhibitors, calcium ionophores, and chemotherapeutic agents can stimulate viral reactivation in vitro (reviewed in Kenney and Mertz 2014).

The bicistronic R promoter (Rp) can transcribe both BZLF1 and BRLF1 genes, while the Z promoter (Zp) transcribes the BZLF1 gene. Expression of either BZLF1 or BRLF1 is sufficient to initiate viral reactivation in some cases. However, both immediate early proteins cooperate in viral replication and full activation of lytic gene expression (Wille et al. 2013). Epigenetic modifications likely regulate Rp and Zp activity. DNA hypermethylation is observed at Rp and other lytic promoters in latently infected tumor samples (Li et al. 2012). Zp has fewer CpG residues and is more likely controlled by repressive chromatin such as H3K9me2/3, H4K20me3, and H3K27me3. During lytic reactivation, Zp is associated with the active chromatin marks histone acetylation and H3K4me3 (Fernandez et al. 2009; Murata et al. 2012).

BZLF1 is a bZIP transcription factor that binds as a homodimer to activator protein 1 (AP-1)-like consensus sites known as BZLF1 response elements (ZRE) (Reinke et al. 2010). There are approximately 469 BZLF1 binding sites on the EBV genome encompassing over 90% of EBV promoters (Flower et al. 2011). BZLF1 has multiple epigenetic modulatory functions that aid in the disruption of latency.

(1) BZLF1 is a chromatin reader that can preferentially bind and activate methylated promoters (Bhende et al. 2004; Bergbauer et al. 2010). CpG methylation of the viral genome is required for full activation of the lytic genes compared to virion DNA devoid of methylation (Kalla et al. 2012). (2) BZLF1 can function as a chromatin eraser as binding of BZLF to repressed lytic promoters leads to nucleosome eviction (Woellmer et al. 2012). (3) BZLF1 can act as a chromatin writer through interaction with the histone acetyltransferases (HATs), CREB-binding protein (CBP) and p300, to enhance the BZLF1 transcriptional activation of early viral promoters (Adamson and Kenney 1999; Zerby et al. 1999). In contrast, sumoylated BZLF1 can act as a repressor and recruit histone deacetylases (HDACs). Sumoylation may regulate BZLF1 after initial infection to promote the establishment of latency (Murata et al. 2010).

The second lytic transcription factor BRLF1 can activate many of the same lytic promoters as BZLF1. DNA methylation does not inhibit its binding but DNA methylation reduces BRLF1 acetylation of H3K9. BRLF1 also interacts with CBP and p300, implicating these interactions in the preferential histone acetylation of unmethylated DNA (Wille et al. 2013). The outcome of BZLF1 and BRLF1 viral reactivation is reversion of an epigenetically repressive state imposed during latency. Viral lytic replication results in newly synthesized viral genomes that lack epigenetic modifications (Fernandez et al. 2009). It is not clear why EBV, like other herpesviruses, is packaged devoid of epigenetic marks. Since DNA methylation has been shown to enhance BZLF1 transactivation activity, loss of DNA methylation may mark newly synthesized templates for packaging. Loss of epigenetic marks also suggests that encapsidated viral genomes are free of epigenetic memory and are not committed to an epigenetic state that could impact the outcome of B-cell or epithelial cell infections.

4.2 Viral Protein Interactions with the Host Epigenetic Machinery

4.2.1 EBNA1

EBNA1 is a multifunctional protein expressed in all latency states and during viral reactivation. EBNA1 functions include sequence-specific and non-specific DNA binding and protein–protein interactions that occur in the context of chromatin. EBNA1 binds at OriP and is required for the replication and maintenance of the viral episome during latency (Rawlins et al. 1985). EBNA1 can bind directly to the host chromatin through its High Mobility Group Box (HMGB)-like domain or through interactions with cellular proteins (Shire et al. 1999; Lin et al. 2008; Jourdan et al. 2012; Coppotelli et al. 2013).

EBNA1 also binds viral and cellular promoters with positive and negative transcriptional effects (Canaan et al. 2009; Dresang et al. 2009; Lu et al. 2010).

EBNA1 bound at OriP activates the viral Cp and LMP promoters, while EBNA1 can directly bind and repress Qp as an autoregulatory feedback loop to control EBNA1 expression (Ambinder et al. 1990). Although EBNA1 has greater affinity for viral DNA than cellular DNA (Horner et al. 1995), EBNA1 binding at multiple locations to the host genome can influence transcription (Canaan et al. 2009). In addition, expression of EBNA1 in EBV-negative cell lines leads to global transcription changes and loss of heterochromatin (Wood et al. 2007; Lu et al. 2010; Coppotelli et al. 2013). Together, these observations provide evidence that EBNA1 is capable of modifying viral and cellular transcription.

EBNA1's ability to act as a chromatin modifier involves various mechanisms described from studies investigating EBNA1's replication or transactivation functions. (1) EBNA1 is capable of directly evicting nucleosomes in an ATP-independent mechanism by destabilization of the octamer in a manner similar to the host HMGB proteins (Avolio-Hunter et al. 2001). (2) EBNA1 can interact with the host nucleosome chaperones nucleosome assembly protein 1 (NAP1), TAF-1β (also known as SET), and nucleophosmin to reposition nucleosomes and recruit the HAT cofactor p300 family members (Malik-Soni and Frappier 2012). (3) EBNA1 can interact with host chromatin modifiers like the arginine methyltransferase Protein Arginine Methyltransferase 5 (PRMT5) (Wang and Frappier 2009) and the ubiquitin ligase of histone 2B, ubiquitin-specific peptidase 7 (USP7) (Holowaty et al. 2003). The EBNA1/TAF-1β interaction has also been shown to negatively regulate replication from OriP, yet is required for transcriptional activation of EBNA1 binding the OriP. TAF-1β can recruit HATs and HDACs, suggesting a role for EBNA1 in recruiting chromatin writers to EBNA1 target sites (Wang and Frappier 2009). (4) EBNA1 has a role in the formation of higher order chromatin structures. EBNA1 can loop DNA between EBNA1 complexes and link OriP on different episomes, as well as form DNA loops with EBNA1 bound at distal sites (Frappier and O'Donnell 1991; Mackey et al. 1995). (6) EBNA1 bound to DNA protects DNA from becoming DNA methylated (Hsieh 1999).

4.2.2 EBNA2

EBNA2 is a transcriptional activator expressed during latency III that partially mimics constitutively active cellular Notch signaling (Sakai et al. 1998). EBNA2 does not bind to DNA itself, but rather interacts with cellular DNA binding proteins, like RBP-Jκ (Ling et al. 1993), and recruits coactivators to specific sites with EBNA2 response elements on the viral genome (Abbot et al. 1990) as well as the cellular genome (Wang et al. 1987; Knutson 1990). EBNA2 interacts with HAT co-activators p300, CBP, and P/CAF, leading to histone acetylation and active transcription (Jayachandra et al. 1999; Wang et al. 2000). EBNA2 also interacts with a subunit of the chromatin remodeling SWI/SNF complex, SMARCB1 (also known as hSNF5/Ini1), to presumably evict nucleosomes and activate transcription (Wu et al. 1996). Chromatin immunoprecipitation-sequencing (ChIP-seq) experiments have shown EBNA2 bound at over 5000 cellular sites, some at locations far from gene promoters (Zhao et al. 2011; McClellan et al. 2013). A subset of sites were classified as super-enhancers where EBV and cellular transcription factors converged to regulate gene expression of cellular oncogenes such as c-MYC and the apoptosis regulator, BCL2 (Zhou et al. 2015).

4.2.2.1 EBNA3 Family

The EBNA3 family (A, B, and C) are expressed in latency III and are nonredundant transcription factors that can activate or repress viral and cellular genes. The EBNA3s do not directly bind to DNA but like EBNA2 interact with cellular DNA binding and chromatin remodeling factors to regulate transcription (Jiang et al. 2014; Schmidt et al. 2015). All EBNA3s contain a RBP-Jκ binding site (Robertson et al. 1996) and can compete with EBNA2 for this cellular factor, usually leading to transcriptional repression of EBNA2 regulated genes. Many EBNA2 and EBNA3 binding sites overlap, but they are not often bound at the same time, suggesting that EBNA3s act as antagonists of EBNA2 (White et al. 2010). EBNA3A and C can interact with a number of cellular repressive factors, including the polycomb repressive complex 2 (PRC2) which is responsible for deposition of H3K27me3. EBNA3A and C have been associated with silencing of the tumor suppressor gene (TSG) $p16^{INK4a}$ (CDKN2A) and the proapoptotic gene BCL2-Like 11 (BC2L11/BIM) through recruitment of PRC2 subunits, the chromatin remodifier C-terminal binding protein (CtBP), and HDAC cofactor Sin3a to the promoters (Skalska et al. 2010; Paschos et al. 2012). EBNA3C was also shown to interact with prothymosin alpha and the HAT p300 in a complex that led to a decrease in p300 HAT activity (Cotter and Robertson 2000). In addition, EBNA3A and C repress transcription by modulating enhancer-promoter loop formation (McClellan et al. 2013). Removal of EBNA3A or C can reverse repression, albeit slowly (Harth-Hertle et al. 2013), suggesting that their continued expression is required for the maintenance of their repressive effects.

4.2.2.2 EBNA-LP

EBNA-LP is one of the first proteins expressed upon initial infection and acts as a transcriptional co-activator for EBNA2 (Portal et al. 2013). EBNA-LP co-activation may act through the removal of repressive chromatin factors from EBNA2 responsive promoters. HDAC4 and 5 are able to repress EBNA2-mediated activation of Cp and LMP promoters. Expression of EBNA-LP resulted in a loss of repression and re-localization of HDAC4 to the cytoplasm (Portal et al. 2006). Furthermore, EBNA-LP displaces the Promyelocytic Leukemia (PML) nuclear body member Sp100 and its interaction partner heterochromatin protein 1 alpha (HP1α), a repressive chromatin binding protein, from PML nuclear bodies during early infection (Ling et al. 2005).

4.2.3 The Latent Membrane Proteins

The LMPs consist of three protein members: LMP1, LMP2A, and LMP2B. The LMPs mimic cellular signaling to help drive naïve B cells into long-lived memory B cells and promote survival (Fruehling and Longnecker 1997; Uchida et al. 1999). LMP1 mimics CD40 T cell help signaling, while LMP2A mimics B cell receptor (BCR) signaling. LMP2B acts as an antagonist of LMP2A (Rovedo and Longnecker 2007). LMP1 and LMP2A signaling has been shown to upregulate and activate the cellular DNMTs (Tsai et al. 2002; Hino et al. 2009). Upregulation of the DNMTs by LMP1 involved activation of the c-Jun N-terminal kinase (JNK) signaling pathway (Tsai et al. 2006) and for LMP2A upregulation involved activation of signal transducer and activator of transcription 3 (STAT3) signaling (Hino et al. 2009). The increased DNMT levels and activity resulted in hypermethylation of CpG-rich regions of viral and cellular genes and their subsequent repression. LMP1 activation of the DNMTs correlated with silencing of the cellular genes E-cadherin (CDH1) (Tsai et al. 2002) and Retinoic acid receptor-beta 2 (RARB2) (Seo et al. 2008). LMP2A expression led to upregulation of the DNMTs with subsequent silencing of phosphatase and tensin homolog (PTEN) (Hino et al. 2009).

4.2.4 EBV Noncoding RNAs

Noncoding RNAs have been shown to be epigenetic regulators that can target DNA methylation and chromatin modifiers to regulate transcription at specific loci, best exemplified by the XIST noncoding RNA in the regulation of X chromosome inactivation (reviewed in Gendrel and Heard 2014). EBV encodes various types of noncoding RNAs with similar epigenetic regulatory potential. The EBERs (EBER1 and 2) are abundantly expressed nuclear viral transcripts used clinically for the detection of EBV in cancer (Rymo 1979; Howe and Steitz 1986; Khan et al. 1992). Recently, EBER2 was shown to bind nascent TR transcripts and recruit the B-cell transcriptional regulator PAX5 to the TR. EBER2 and PAX5 were required for repression of the LMPs (Lee et al. 2015). Since EBERs have been shown to be secreted in exosomes (Iwakiri et al. 2009), these small RNAs have the potential to epigenetically regulate uninfected, bystander cells.

EBV encodes approximately 44 miRNAs from two clusters, four from the BHRF1 region and 40 from the BART region (Cai et al. 2006; Grundhoff et al. 2006; Zhu et al. 2009). Briefly, miRNAs are short noncoding RNAs about 22 nucleotides long that target mRNA and can lead to degradation of the mRNA or translational repression depending on the degree of complementation (reviewed in Ha and Kim 2014). The miRNAs from the BHRF1 region (miR-BHRF1 1–4) are more highly expressed in B cells, particularly during latency III. The BART miRNAs have been detected in all latency types and cells, and are highly expressed in epithelial cells (Cai et al. 2006; Yang et al. 2013). Several viral transcripts, such as the LMPs, the viral polymerase BALF5, and the viral BCL2 homolog BHRF1, are known to be downregulated by viral miRNAs (Lo et al. 2007; Barth et al. 2008; Lung et al. 2009; Riley et al. 2012). Viral miRNAs also target a number of cellular genes with potential regulatory effects on the viral life cycle as well as detrimental effects to the host, which include apoptosis, cell signaling, cell cycle modulation, and the immune response (Kang et al. 2015). Viral miRNAs can be released in exosomes from infected cells and be taken up by neighboring uninfected cells with coordinated effects on transcription, bringing in the possibility of long-range effects in vivo (Pegtel et al. 2010).

4.3 Epigenetics of EBV-Associated Malignancies

Greater than 90% of the adult population worldwide harbor EBV. While most primary infections are asymptomatic, infection later in life is associated with the development of infectious mononucleosis. In some rare cases, latent EBV infection is also associated with B-cell and epithelial cell malignancies. EBV immortalization of infected B cells in vitro underscores the virus' tumorigenic potential. As an opportunistic virus, EBV-driven transformation often requires other contributing factors, such as malarial infection, immunesuppression, or genetic alterations (reviewed in Rickinson 2014). EBV-associated B-cell malignancies include Burkitt Lymphoma (BL), Hodgkin Lymphoma (HL), and post-transplant lymphoproliferative disease (PTLD), while nasopharyngeal (NPC) and gastric carcinoma (GC) are epithelial cancers associated with EBV. As described, EBV is armed with a number of virally encoded epigenetic modulators. EBV-induced epigenetic changes can likely substitute for genetic mutations required for tumor evolution. Indeed, EBV-associated cancers tend to be epigenetically distinct from EBV-negative matched tumors with DNA hypermethylation of TSGs frequently occurring in EBV-associated cancers.

4.3.1 Lymphoid Cells

Endemic BL, found in sub-Saharan Africa, is nearly 100% associated with EBV infection, while sporadic BL is only ~15% associated with EBV (Cohen et al. 2011). EBV-positive BLs are characterized by the restricted type I latency program (Rowe et al. 1987). All BLs have a characteristic c-MYC translocation, putting c-MYC under the control of immunoglobulin (Ig) heavy-chain or light-chain enhancer region (Zech et al. 1976). EBV-positive BLs show resistance to apoptosis compared to EBV-negative BLs (Kelly et al. 2006; Piccaluga et al. 2015). Furthermore, BL cell lines that lose EBV are more sensitive to apoptosis (Shimizu et al. 1994), providing evidence that the presence of EBV confers a specific growth advantage. Yet, EBV-positive BLs have fewer mutations than EBV-negative BLs (Giulino-Roth et al. 2012), suggesting that virally induced epigenetic silencing of TSGs might replace the requirement for mutation. Indeed, DNA hypermethylation of PRDM1/Blimp-1, involved in B-cell terminal differentiation with putative tumor suppressor activity, is more frequent in EBV-positive BL than EBV-negative BL (Zhang et al. 2014).

Overall about 50% of HLs are associated with EBV (Glaser et al. 1997). The malignant cells of HL, the Reed-Sternberg (RS) cells, are thought to originate from germinal center B cells and have nonfunctional immunoglobulin rearrangements (Brock et al. 2007). RS cells account for <2% of the tumor, the rest of the cells being nonmalignant lymphocytes and various other cell types (Harris et al. 1994). The EBV genome is found in the RS cells and is monoclonal, as determined by EBV TR number (Brousset et al. 1994). RS cells display a latency II viral gene expression pattern (Pallesen et al. 1991). Crippling mutations in the immunoglobulin gene in RS cells are almost always found in EBV-positive cases (Brauninger et al. 2006), suggesting that EBV plays a role in survival. Although TGS promoter hypermethylation was observed in HL, EBV-positive HL displayed a lower frequency of TSG hypermethylation than EBV-negative cases (Dhiab et al. 2015). In vitro EBV infection of germinal center B cells generates immortalized cell lines with a type III latency profile. Analysis of the DNA methylation state of EBV-immortalized germinal center B cells showed widespread DNA hyper- and hypomethylation at cellular promoters. EBV infection increased DNMT3A expression but reduced expression of DNMT1 and DNMT3B (Leonard et al. 2011). LMP1, a known activator of DNMTs, was sufficient ot reprogram germinal center B cells with characteristics of HL Reed-Sternberg cells (Vockerodt et al. 2008).

Post-transplant lymphoproliferative disease (PTLD) associated with EBV exhibits type III latency, and can be modeled by the in vitro immortalization of B cells by EBV (LCLs) (Thorley-Lawson and Gross 2004). EBV immortalization of B cells was associated with large-scale demethylation covering two-thirds of the genome (Hansen et al. 2014). DNA hypomethylation was unique to EBV infection and was not observed following B-cell activation with CD40L and interleukin-4. DNA hypermethylation of CpG islands was also observed but to a lesser degree in EBV immortalized B cells (Hansen et al. 2014; Hernando et al. 2014). A recent study similarly observed that EBV infection resulted in DNA methylation of TSGs over a course of 15 days. Both DNMT3A and B were upregulated over time, while DNMT1 levels did not change. Additionally, all HDACs were also upregulated (Saha et al. 2015). Concomitant changes in the chromatin landscape have been reported following EBV immortalization of B cells. Analysis of histone modifications showed that EBV reduced H3K27me3, H3K9me3, and H4K20me3 levels and altered the nuclear distribution of H3K27me3 and H3K9me3 in EBV-immortalized B cells. Genes that lost repressive chromatin belonged to cell division, positive regulation of proliferation, apoptosis, and transcription factors by gene ontology analysis (Hernando et al. 2014).

In sum, EBV infection of B cells leads to hypo-methylation and loss of repressive chromatin over large portions of the host genome, with localized hypermethylation and repressive chromatin marks deposited on specific genes. Such epigenetic changes are also seen in EBV-positive B-cell lymphomas, as early events in the development and progression of cancer. With the paradigm that epigenetic changes are heritable, virally induced epigenetic events could have long-lasting implications and be maintained in latent state or following loss of the virus. Some EBV-associated BL and HL, previously positive for EBV, have been shown to be EBV negative upon relapse (Nerurkar et al. 2000; Xue et al. 2002) suggesting that EBV epigenetic alterations may act as a "hit-and-run" mechanism in lymphomagenesis.

4.3.2 *Epithelial Cells*

Latent EBV infection is associated with almost 100% of NPC and approximately 10% of GC worldwide (Cohen et al. 2011). In both carcinomas, EBV infection is associated with an increased DNA hypermethylation with a relatively low frequency of mutation compared to EBV-negative tumors of the same type (Lin et al. 2014; Wang et al. 2014). This suggests that the EBV-dependent increased DNA methylation is playing a major role in the pathogenesis of these malignancies.

NPC is characterized by a latency II pattern of viral gene expression with variable detection of LMP1 (Brooks et al. 1992). Expression of LMP1 is linked to a greater metastasis rate and faster disease progression (Ozyar et al. 2004). EBERs and BART noncoding RNAs are highly expressed, with the EBV BART miRNAs constituting up to 20% of the total cellular miRNA content (Hitt et al. 1989; Wu et al. 1991; Zhu et al. 2009). NPC has a low number of mutations; for example, almost all NPCs have functional p53 and it is often highly expressed (Sheu et al. 2004). However, NPC is also associated with a greater frequency of promoter CpG hypermethylation (Table 4.1). Promoters of the tumor suppressors, $p16^{INK4a}$ and $p27^{Kip1}$ (CDKN1B), are almost always DNA hypermethylated and repressed (Baba et al. 2001; Kwong et al. 2002).

Although EBV is only associated with 10% of GC, it is still the most prevalent EBV-associated malignancy. EBV-associated gastric carcinoma (EBVaGC) displays a latency I viral gene expression program (Imai et al. 1994). Analysis of molecular features that included gene copy number, mutations, DNA methylation, and transcriptional profiles of mRNA and miRNAs classified GC into four subtypes. EBVaGC had unique properties with extreme DNA hypermethylation and fewer mutations compared to other EBV-negative GC (Cancer Genome Atlas Research 2014). Various studies have confirmed a greater frequency of CpG island hypermethylation (CIMP) in EBVaGC compared to EBV-negative GC (Table 4.1).

Several in vitro studies have provided evidence for EBV infection inducing a CIMP phenotype. Stable infection of epithelial cell lines with EBV is challenging

Table 4.1 EBV-dependent hypermethylated genes

Genes		References
GC	APBA2, **APC**, ARID1A, **CDH1**, **CDKN2A**, CDKN2B, **DAPK1**, DLC1, EPHB6, **FHIT**, GSTP1, HOXA10, IHH, IL15RA, IRF5, MARK1, MGMT, NEK9, NKX3-1, PLXND1, PTEN, **RASSF1**, RBP1, REC8, RUNX3, SSTR1, TIMP2, TIMP3, **TP73**, WNT5A, WWOX, ZMYND10	Kang et al. (2002), Sudo et al. (2004), Chang et al. (2006), Kang et al. (2008), Matsusaka et al. (2011), Wang et al. (2011), Liu et al. (2013), Okada et al. (2013), Saito et al. (2013), Zhao et al. (2013a), Zhao et al. (2013b), Liang et al. (2014), Dong et al. (2015), and He et al. (2015)
NPC	**APC**, CADM1, CALCA, CCNA1, **CDH1**, CDH13, **CDKN2A**, CDKN1B, CHFR, DAB2, DACT2, **DAPK1**, DCC, DKK2, DKK3, DLC1, ESR1, FEZF2, **FHIT**, FLOT1, HOXA2, IER3, KIF1A, LOX, PCDH10, PRDM2, PTPRG, RARB, RARRES1, RASAL1, **RASSF1**, RASSF2, SFRP1, SFRP2, SFRP4, SFRP5, THY1, TNXB, **TP73**, UCHL1, WIF1	Kwong et al. (2002), Hui et al. (2003), Wong et al. (2003), Li et al. (2004), Lung et al. (2005), Ying et al. 2006), Jin et al. (2007), Sun et al. (2007), Cheung et al. (2008), Yanatatsaneejit et al. (2008), Tong et al. (2010), Hutajulu et al. (2011), Loyo et al. (2011), Li et al. (2013), Shu et al. (2013), Sung et al. (2014), Dai et al. (2015), Li et al. (2015), and Yang et al. (2015)

List of hypermethylated genes reported in the literature as EBV specific in gastric carcinoma and nasopharyngeal carcinoma involved in cell development, cell cycle regulation, cell survival, growth, and movement as determined by Ingenuity Pathway Analysis
GC gastric carcinoma; *NPC* nasopharyngeal carcinoma
Bold: genes hypermethylated in both GC and NPC

as the virus is rapidly lost; therefore, it requires the use of recombinant viruses with selectable markers (Sixbey et al. 1983). EBV infection of GC cell lines induced silencing of PTEN by promoter hypermethylation that could be recapitulated by ectopic LMP2A expression in GC cell lines. LMP2A induced DNMT1 through a signaling axis that involved STAT3 activation (Hino et al. 2009). LMP1 has also been implicated in CIMP, where LMP1 activation of DNMT1, DNMT3A, and DNMT3B expression resulted in E-cadherin promoter methylation and silencing (Tsai et al. 2002). Similarly, we have shown that EBV infection of a lung carcinoma cell line displaying a latency II viral gene expression program resulted in a cadherin switch that involved loss of E-cadherin via promoter methylation/repressive chromatin modifications and increased N-cadherin expression. Removal of selection pressure allowed cells to naturally lose EBV allowing us to study the epigenetic heritability of such changes. EBV-negative transiently infected cells retained the cadherin switch, E-cadherin promoter methylation, and the enhanced invasiveness observed in the EBV-positive counterparts (Queen et al. 2013). In a second study, we examined the cellular methylome following EBV infection of hTERT-immortalized oral keratinocytes including a set of clones that lost EBV (Birdwell et al. 2014). Common among the latently infected keratinocytes and those that lost EBV was CIMP of 25 genes, including RARRES1, which is also hypermethylated in NPC (Yanatatsaneejit et al. 2008). EBV-negative clones maintained gene expression changes and a delayed induction of keratinocyte differentiation for over 20 passages. These findings support the notion that EBV epigenetic reprogramming can be maintained in the absence of viral gene expression or loss of the viral genome with long-lasting consequences to the cell.

4.4 Concluding Remarks

EBV latency and reactivation are central to lifelong viral persistence and involve strict regulation of viral gene transcription. EBV achieves its latent gene expression patterns through the use of the host epigenetic machinery and can overcome the epigenetically silenced state during reactivation. Many viral genes are able to recruit or modulate host epigenetic modifiers to refine viral gene transcription and tweak the cellular environment to the virus' advantage. While the process proceeds with little appreciable side effects, EBV-induced epigenetic alterations to the host likely contribute to the oncogenic process. EBV-associated cancers are marked by unique epigenetic changes without the high number of accompanying gene mutations that are seen in matching EBV-negative tumors. The epigenetic modifications include targeting of cellular TSGs for silencing through DNA methylation of promoters and/or repressive chromatin. Such epigenetic changes likely provide a selective advantage in the cancer context. However, it is unclear if virally induced epigenetic changes arise from nontargeted stochastic events due to viral disturbance of the cellular epigenetic machinery or are specifically targeted to a gene loci directed by transcriptional circuits that maintain the epigenetic state. In colorectal cancer, such instructive epigenetic modulation has been shown to depend on oncogene signaling, recruitment of co-repressors complexes, and DNMTs to specific promoter regions. Continued oncogene expression was required for the maintenance of promoter DNA methylation, which may not be the case to maintain phenotypes during EBV latency (Serra et al. 2014). Similarly, the potent oncogenic signaling activities of EBV proteins, such as LMP1 and LMP2, or viral activation of cellular oncogenes may instruct the deposition of DNA methylation to specific genes. The selective growth advantage imposed by such epigenetic changes may be maintained despite latency or loss of the viral genome. If so, such virally induced epigenetic modifications provide a basis for EBV and agents with similar activities in "hit-and-run" carcinogenesis.

Acknowledgements We would like to acknowledge all the excellent work that we could not cite due to space constraints. This work was supported by funding from the NIDCR 1R01DE025565 to RSS and a predoctoral fellowship to CEB.

References

Abbot SD, Rowe M, Cadwallader K, Ricksten A, Gordon J, Wang F, Rymo L, Rickinson AB (1990) Epstein-Barr virus nuclear antigen 2 induces expression of the virus-encoded latent membrane protein. J Virol 64(5):2126–2134

Adamson AL, Kenney S (1999) The Epstein-Barr virus BZLF1 protein interacts physically and functionally with the histone acetylase CREB-binding protein. J Virol 73(8):6551–6558
Ambinder RF, Shah WA, Rawlins DR, Hayward GS, Hayward SD (1990) Definition of the sequence requirements for binding of the EBNA-1 protein to its palindromic target sites in Epstein-Barr virus DNA. J Virol 64(5):2369–2379
Arvey A, Tempera I, Lieberman PM (2013) Interpreting the Epstein-Barr virus (EBV) epigenome using high-throughput data. Viruses 5(4):1042–1054
Avolio-Hunter TM, Lewis PN, Frappier L (2001) Epstein-Barr nuclear antigen 1 binds and destabilizes nucleosomes at the viral origin of latent DNA replication. Nucleic Acids Res 29(17):3520–3528
Baba Y, Tsukuda M, Mochimatsu I, Furukawa S, Kagata H, Satake K, Koshika S, Nakatani Y, Hara M, Kato Y, Nagashima Y (2001) Reduced expression of p16 and p27 proteins in nasopharyngeal carcinoma. Cancer Detect Prev 25(5):414–419
Babcock GJ, Hochberg D, Thorley-Lawson AD (2000) The expression pattern of Epstein-Barr virus latent genes in vivo is dependent upon the differentiation stage of the infected B cell. Immunity 13(4):497–506
Barth S, Pfuhl T, Mamiani A, Ehses C, Roemer K, Kremmer E, Jaker C, Hock J, Meister G, Grasser FA (2008) Epstein-Barr virus-encoded microRNA miR-BART2 down-regulates the viral DNA polymerase BALF5. Nucleic Acids Res 36(2):666–675
Bergbauer M, Kalla M, Schmeinck A, Gobel C, Rothbauer U, Eck S, Benet-Pages A, Strom TM, Hammerschmidt W (2010) CpG-methylation regulates a class of Epstein-Barr virus promoters. PLoS Pathog 6(9):e1001114
Bhende PM, Seaman WT, Delecluse HJ, Kenney SC (2004) The EBV lytic switch protein, Z, preferentially binds to and activates the methylated viral genome. Nat Genet 36(10): 1099–1104
Birdwell CE, Queen KJ, Kilgore P, Rollyson P, Trutschl M, Cvek U, Scott RS (2014) Genome-wide DNA methylation as an epigenetic consequence of Epstein-Barr virus infection of immortalized keratinocytes. J Virol 88(19):11442–11458
Brauninger A, Schmitz R, Bechtel D, Renne C, Hansmann ML, Kuppers R (2006) Molecular biology of Hodgkin's and Reed/Sternberg cells in Hodgkin's lymphoma. Int J Cancer 118(8): 1853–1861
Brock MV, Herman JG, Baylin SB (2007) Cancer as a manifestation of aberrant chromatin structure. Cancer J 13(1):3–8
Brooks L, Yao QY, Rickinson AB, Young LS (1992) Epstein-Barr virus latent gene transcription in nasopharyngeal carcinoma cells: coexpression of EBNA1, LMP1, and LMP2 transcripts. J Virol 66(5):2689–2697
Brousset P, Schlaifer D, Meggetto F, Bachmann E, Rothenberger S, Pris J, Delsol G, Knecht H (1994) Persistence of the same viral strain in early and late relapses of Epstein-Barr virus-associated Hodgkin's disease. Blood 84(8):2447–2451
Cai X, Schafer A, Lu S, Bilello JP, Desrosiers RC, Edwards R, Raab-Traub N, Cullen BR (2006) Epstein-Barr virus microRNAs are evolutionarily conserved and differentially expressed. PLoS Pathog 2(3):e23
Canaan A, Haviv I, Urban AE, Schulz VP, Hartman S, Zhang Z, Palejev D, Deisseroth AB, Lacy J, Snyder M, Gerstein M, Weissman SM (2009) EBNA1 regulates cellular gene expression by binding cellular promoters. Proc Natl Acad Sci USA 106(52):22421–22426
Cancer Genome Atlas Research N (2014) Comprehensive molecular characterization of gastric adenocarcinoma. Nature 513(7517):202–209
Chang MS, Uozaki H, Chong JM, Ushiku T, Sakuma K, Ishikawa S, Hino R, Barua RR, Iwasaki Y, Arai K, Fujii H, Nagai H, Fukayama M (2006) CpG island methylation status in gastric carcinoma with and without infection of Epstein-Barr virus. Clin Cancer Res 12(10):2995–3002
Chau CM, Lieberman PM (2004) Dynamic chromatin boundaries delineate a latency control region of Epstein-Barr virus. J Virol 78(22):12308–12319

Chau CM, Zhang XY, McMahon SB, Lieberman PM (2006) Regulation of Epstein-Barr virus latency type by the chromatin boundary factor CTCF. J Virol 80(12):5723–5732

Chen HS, Martin KA, Lu F, Lupey LN, Mueller JM, Lieberman PM, Tempera I (2014) Epigenetic deregulation of the LMP1/LMP2 locus of Epstein-Barr virus by mutation of a single CTCF-cohesin binding site. J Virol 88(3):1703–1713

Cheung AK, Lung HL, Hung SC, Law EW, Cheng Y, Yau WL, Bangarusamy DK, Miller LD, Liu ET, Shao JY, Kou CW, Chua D, Zabarovsky ER, Tsao SW, Stanbridge EJ, Lung ML (2008) Functional analysis of a cell cycle-associated, tumor-suppressive gene, protein tyrosine phosphatase receptor type G, in nasopharyngeal carcinoma. Cancer Res 68(19):8137–8145

Chih-Lin Hsieh (1999) Evidence that protein binding specifies sites of DNA demethylation. Mol Cell Biol 19(1):46–56

Cohen JI, Fauci AS, Varmus H, Nabel GJ (2011) Epstein-Barr virus: an important vaccine target for cancer prevention. Sci Transl Med 3(107):107fs7

Coppotelli G, Mughal N, Callegari S, Sompallae R, Caja L, Luijsterburg MS, Dantuma NP, Moustakas A, Masucci MG (2013) The Epstein-Barr virus nuclear antigen-1 reprograms transcription by mimicry of high mobility group A proteins. Nucleic Acids Res 41(5):2950–2962

Cotter MA 2nd, Robertson ES (2000) Modulation of histone acetyltransferase activity through interaction of Epstein-Barr nuclear antigen 3C with prothymosin alpha. Mol Cell Biol 20(15): 5722–5735

Crawford DH, Ando I (1986) EB virus induction is associated with B-cell maturation. Immunology 59(3):405–409

Dai W, Cheung AK, Ko JM, Cheng Y, Zheng H, Ngan RK, Ng WT, Lee AW, Yau CC, Lee VH, Lung ML (2015) Comparative methylome analysis in solid tumors reveals aberrant methylation at chromosome 6p in nasopharyngeal carcinoma. Cancer Med 4(7):1079–1090

Decker LL, Klaman LD, Thorley-Lawson DA (1996) Detection of the latent form of Epstein-Barr virus DNA in the peripheral blood of healthy individuals. J Virol 70(5):3286–3289

Dhiab MB, Ziadi S, Mestiri S, Gacem RB, Ksiaa F, Trimeche M (2015) DNA methylation patterns in EBV-positive and EBV-negative Hodgkin lymphomas. Cell Oncol (Dordr) 38(6):453–462

Dong SM, Lee HG, Cho SG, Kwon SH, Yoon H, Kwon HJ, Lee JH, Kim H, Park PG, Kim H, Hayward SD, Park JH, Lee JM (2015) Hypermethylation of the interferon regulatory factor 5 promoter in Epstein-Barr virus-associated gastric carcinoma. J Microbiol 53(1):70–76

Dresang LR, Vereide DT, Sugden B (2009) Identifying sites bound by Epstein-Barr virus nuclear antigen 1 (EBNA1) in the human genome: defining a position-weighted matrix to predict sites bound by EBNA1 in viral genomes. J Virol 83(7):2930–2940

Dyson PJ, Farrell PJ (1985) Chromatin structure of Epstein-Barr virus. J Gen Virol 66(Pt 9): 1931–1940

Fernandez AF, Rosales C, Lopez-Nieva P, Grana O, Ballestar E, Ropero S, Espada J, Melo SA, Lujambio A, Fraga MF, Pino I, Javierre B, Carmona FJ, Acquadro F, Steenbergen RD, Snijders PJ, Meijer CJ, Pineau P, Dejean A, Lloveras B, Capella G, Quer J, Buti M, Esteban JI, Allende H, Rodriguez-Frias F, Castellsague X, Minarovits J, Ponce J, Capello D, Gaidano G, Cigudosa JC, Gomez-Lopez G, Pisano DG, Valencia A, Piris MA, Bosch FX, Cahir-McFarland E, Kieff E, Esteller M (2009) The dynamic DNA methylomes of double-stranded DNA viruses associated with human cancer. Genome Res 19(3):438–451

Flower K, Thomas D, Heather J, Ramasubramanyan S, Jones S, Sinclair AJ (2011) Epigenetic control of viral life-cycle by a DNA-methylation dependent transcription factor. PLoS One 6(10):e25922

Frappier L, O'Donnell M (1991) Epstein-Barr nuclear antigen 1 mediates a DNA loop within the latent replication origin of Epstein-Barr virus. Proc Natl Acad Sci USA 88(23):10875–10879

Fruehling S, Longnecker R (1997) The immunoreceptor tyrosine-based activation motif of Epstein-Barr virus LMP2A is essential for blocking BCR-mediated signal transduction. Virology 235(2):241–251

Gendrel AV, Heard E (2014) Noncoding RNAs and epigenetic mechanisms during X-chromosome inactivation. Annu Rev Cell Dev Biol 30:561–580

Gerle B, Koroknai A, Fejer G, Bakos A, Banati F, Szenthe K, Wolf H, Niller HH, Minarovits J, Salamon D (2007) Acetylated histone H3 and H4 mark the upregulated LMP2A promoter of Epstein-Barr virus in lymphoid cells. J Virol 81(23):13242–13247

Giulino-Roth L, Wang K, MacDonald TY, Mathew S, Tam Y, Cronin MT, Palmer G, Lucena-Silva N, Pedrosa F, Pedrosa M, Teruya-Feldstein J, Bhagat G, Alobeid B, Leoncini L, Bellan C, Rogena E, Pinkney KA, Rubin MA, Ribeiro RC, Yelensky R, Tam W, Stephens PJ, Cesarman E (2012) Targeted genomic sequencing of pediatric Burkitt lymphoma identifies recurrent alterations in antiapoptotic and chromatin-remodeling genes. Blood 120(26):5181–5184

Glaser SL, Lin RJ, Stewart SL, Ambinder RF, Jarrett RF, Brousset P, Pallesen G, Gulley ML, Khan G, O'Grady J, Hummel M, Preciado MV, Knecht H, Chan JK, Claviez A (1997) Epstein-Barr virus-associated Hodgkin's disease: epidemiologic characteristics in international data. Int J Cancer 70(4):375–382

Grundhoff A, Sullivan CS, Ganem D (2006) A combined computational and microarray-based approach identifies novel microRNAs encoded by human gamma-herpesviruses. RNA 12(5): 733–750

Ha M, Kim VN (2014) Regulation of microRNA biogenesis. Nat Rev Mol Cell Biol 15(8): 509–524

Hadinoto V, Shapiro M, Sun CC, Thorley-Lawson DA (2009) The dynamics of EBV shedding implicate a central role for epithelial cells in amplifying viral output. PLoS Pathog 5(7): e1000496

Hansen KD, Sabunciyan S, Langmead B, Nagy N, Curley R, Klein G, Klein E, Salamon D, Feinberg AP (2014) Large-scale hypomethylated blocks associated with Epstein-Barr virus-induced B-cell immortalization. Genome Res 24(2):177–184

Harris NL, Jaffe ES, Stein H, Banks PM, Chan JK, Cleary ML, Delsol G, De Wolf-Peeters C, Falini B, Gatter KC et al (1994) A revised European-American classification of lymphoid neoplasms: a proposal from the International Lymphoma Study Group. Blood 84(5): 1361–1392

Harth-Hertle ML, Scholz BA, Erhard F, Glaser LV, Dolken L, Zimmer R, Kempkes B (2013) Inactivation of intergenic enhancers by EBNA3A initiates and maintains polycomb signatures across a chromatin domain encoding CXCL10 and CXCL9. PLoS Pathog 9(9):e1003638

He D, Zhang YW, Zhang NN, Zhou L, Chen JN, Jiang Y, Shao CK (2015) Aberrant gene promoter methylation of p16, FHIT, CRBP1, WWOX, and DLC-1 in Epstein-Barr virus-associated gastric carcinomas. Med Oncol 32(4):92

Hernando H, Islam AB, Rodriguez-Ubreva J, Forne I, Ciudad L, Imhof A, Shannon-Lowe C, Ballestar E (2014) Epstein-Barr virus-mediated transformation of B cells induces global chromatin changes independent to the acquisition of proliferation. Nucleic Acids Res 42(1): 249–263

Hino R, Uozaki H, Murakami N, Ushiku T, Shinozaki A, Ishikawa S, Morikawa T, Nakaya T, Sakatani T, Takada K, Fukayama M (2009) Activation of DNA methyltransferase 1 by EBV latent membrane protein 2A leads to promoter hypermethylation of PTEN gene in gastric carcinoma. Cancer Res 69(7):2766–2774

Hitt MM, Allday MJ, Hara T, Karran L, Jones MD, Busson P, Tursz T, Ernberg I, Griffin BE (1989) EBV gene expression in an NPC-related tumour. EMBO J 8(9):2639–2651

Holowaty MN, Zeghouf M, Wu H, Tellam J, Athanasopoulos V, Greenblatt J, Frappier L (2003) Protein profiling with Epstein-Barr nuclear antigen-1 reveals an interaction with the herpesvirus-associated ubiquitin-specific protease HAUSP/USP7. J Biol Chem 278(32): 29987–29994

Honess RW, Gompels UA, Barrell BG, Craxton M, Cameron KR, Staden R, Chang YN, Hayward GS (1989) Deviations from expected frequencies of CpG dinucleotides in herpesvirus DNAs may be diagnostic of differences in the states of their latent genomes. J Gen Virol 70(Pt 4): 837–855

Horner D, Lewis M, Farrell PJ (1995) Novel hypotheses for the roles of EBNA-1 and BHRF1 in EBV-related cancers. Intervirology 38(3–4):195–205

Howe JG, Steitz JA (1986) Localization of Epstein-Barr virus-encoded small RNAs by in situ hybridization. Proc Natl Acad Sci USA 83(23):9006–9010

Hui AB, Lo KW, Kwong J, Lam EC, Chan SY, Chow LS, Chan AS, Teo PM, Huang DP (2003) Epigenetic inactivation of TSLC1 gene in nasopharyngeal carcinoma. Mol Carcinog 38(4): 170–178

Hutajulu SH, Indrasari SR, Indrawati LP, Harijadi A, Duin S, Haryana SM, Steenbergen RD, Greijer AE, Middeldorp JM (2011) Epigenetic markers for early detection of nasopharyngeal carcinoma in a high risk population. Mol Cancer 10:48

Imai S, Koizumi S, Sugiura M, Tokunaga M, Uemura Y, Yamamoto N, Tanaka S, Sato E, Osato T (1994) Gastric carcinoma: monoclonal epithelial malignant cells expressing Epstein-Barr virus latent infection protein. Proc Natl Acad Sci USA 91(19):9131–9135

Iwakiri D, Zhou L, Samanta M, Matsumoto M, Ebihara T, Seya T, Imai S, Fujieda M, Kawa K, Takada K (2009) Epstein-Barr virus (EBV)-encoded small RNA is released from EBV-infected cells and activates signaling from Toll-like receptor 3. J Exp Med 206(10): 2091–2099

Jayachandra S, Low KG, Thlick AE, Yu J, Ling PD, Chang Y, Moore PS (1999) Three unrelated viral transforming proteins (vIRF, EBNA2, and E1A) induce the MYC oncogene through the interferon-responsive PRF element by using different transcription coadaptors. Proc Natl Acad Sci USA 96(20):11566–11571

Jiang S, Willox B, Zhou H, Holthaus AM, Wang A, Shi TT, Maruo S, Kharchenko PV, Johannsen EC, Kieff E, Zhao B (2014) Epstein-Barr virus nuclear antigen 3C binds to BATF/IRF4 or SPI1/IRF4 composite sites and recruits Sin3A to repress CDKN2A. Proc Natl Acad Sci USA 111:421–426

Jin H, Wang X, Ying J, Wong AH, Cui Y, Srivastava G, Shen ZY, Li EM, Zhang Q, Jin J, Kupzig S, Chan AT, Cullen PJ, Tao Q (2007) Epigenetic silencing of a Ca(2+)-regulated Ras GTPase-activating protein RASAL defines a new mechanism of Ras activation in human cancers. Proc Natl Acad Sci USA 104(30):12353–12358

Jourdan N, Jobart-Malfait A, Dos Reis G, Quignon F, Piolot T, Klein C, Tramier M, Coppey-Moisan M, Marechal V (2012) Live-cell imaging reveals multiple interactions between Epstein-Barr virus nuclear antigen 1 and cellular chromatin during interphase and mitosis. J Virol 86(9):5314–5329

Kalla M, Schmeinck A, Bergbauer M, Pich D, Hammerschmidt W (2010) AP-1 homolog BZLF1 of Epstein-Barr virus has two essential functions dependent on the epigenetic state of the viral genome. Proc Natl Acad Sci USA 107(2):850–855

Kalla M, Gobel C, Hammerschmidt W (2012) The lytic phase of Epstein-Barr virus requires a viral genome with 5-methylcytosine residues in CpG sites. J Virol 86(1):447–458

Kang GH, Lee S, Kim WH, Lee HW, Kim JC, Rhyu MG, Ro JY (2002) Epstein-Barr virus-positive gastric carcinoma demonstrates frequent aberrant methylation of multiple genes and constitutes CpG island methylator phenotype-positive gastric carcinoma. Am J Pathol 160(3): 787–794

Kang GH, Lee S, Cho NY, Gandamihardja T, Long TI, Weisenberger DJ, Campan M, Laird PW (2008) DNA methylation profiles of gastric carcinoma characterized by quantitative DNA methylation analysis. Lab Invest 88(2):161–170

Kang D, Skalsky RL, Cullen BR (2015) EBV BART microRNAs target multiple pro-apoptotic cellular genes to promote epithelial cell survival. PLoS Pathog 11(6):e1004979

Karlin S, Doerfler W, Cardon LR (1994) Why is CpG suppressed in the genomes of virtually all small eukaryotic viruses but not in those of large eukaryotic viruses? J Virol 68(5):2889–2897

Kelly GL, Milner AE, Baldwin GS, Bell AI, Rickinson AB (2006) Three restricted forms of Epstein-Barr virus latency counteracting apoptosis in c-myc-expressing Burkitt lymphoma cells. Proc Natl Acad Sci USA 103(40):14935–14940

Kenney SC, Mertz JE (2014) Regulation of the latent-lytic switch in Epstein-Barr virus. Semin Cancer Biol 26:60–68

Khan G, Coates PJ, Kangro HO, Slavin G (1992) Epstein Barr virus (EBV) encoded small RNAs: targets for detection by in situ hybridisation with oligonucleotide probes. J Clin Pathol 45(7): 616–620

Knutson JC (1990) The level of c-fgr RNA is increased by EBNA-2, an Epstein-Barr virus gene required for B-cell immortalization. J Virol 64(6):2530–2536

Kwong J, Lo KW, To KF, Teo PM, Johnson PJ, Huang DP (2002) Promoter hypermethylation of multiple genes in nasopharyngeal carcinoma. Clin Cancer Res 8(1):131–137

Laichalk LL, Thorley-Lawson DA (2005) Terminal differentiation into plasma cells initiates the replicative cycle of Epstein-Barr virus in vivo. J Virol 79(2):1296–1307

Lee N, Moss WN, Yario TA, Steitz JA (2015) EBV noncoding RNA binds nascent RNA to drive host PAX5 to viral DNA. Cell 160(4):607–618

Leonard S, Wei W, Anderton J, Vockerodt M, Rowe M, Murray PG, Woodman CB (2011) Epigenetic and transcriptional changes which follow Epstein-Barr virus infection of germinal center B cells and their relevance to the pathogenesis of Hodgkin's lymphoma. J Virol 85(18): 9568–9577

Li Z, Ren Y, Lin SX, Liang YJ, Liang HZ (2004) Association of E-cadherin and beta-catenin with metastasis in nasopharyngeal carcinoma. Chin Med J (Engl) 117(8):1232–1239

Li L, Su X, Choi GC, Cao Y, Ambinder RF, Tao Q (2012) Methylation profiling of Epstein-Barr virus immediate-early gene promoters, BZLF1 and BRLF1 in tumors of epithelial, NK- and B-cell origins. BMC Cancer 12:125

Li HP, Peng CC, Chung IC, Huang MY, Huang ST, Chen CC, Chang KP, Hsu CL, Chang YS (2013) Aberrantly hypermethylated Homeobox A2 derepresses metalloproteinase-9 through TBP and promotes invasion in Nasopharyngeal carcinoma. Oncotarget 4(11):2154–2165

Li L, Zhang Y, Fan Y, Sun K, Su X, Du Z, Tsao SW, Loh TK, Sun H, Chan AT, Zeng YX, Chan WY, Chan FK, Tao Q (2015) Characterization of the nasopharyngeal carcinoma methylome identifies aberrant disruption of key signaling pathways and methylated tumor suppressor genes. Epigenomics 7(2):155–173

Liang Q, Yao X, Tang S, Zhang J, Yau TO, Li X, Tang CM, Kang W, Lung RW, Li JW, Chan TF, Xing R, Lu Y, Lo KW, Wong N, To KF, Yu C, Chan FK, Sung JJ, Yu J (2014) Integrative identification of Epstein-Barr virus-associated mutations and epigenetic alterations in gastric cancer. Gastroenterology 147(6):1350–1362 e4

Lin A, Wang S, Nguyen T, Shire K, Frappier L (2008) The EBNA1 protein of Epstein-Barr virus functionally interacts with Brd4. J Virol 82(24):12009–12019

Lin DC, Meng X, Hazawa M, Nagata Y, Varela AM, Xu L, Sato Y, Liu LZ, Ding LW, Sharma A, Goh BC, Lee SC, Petersson BF, Yu FG, Macary P, Oo MZ, Ha CS, Yang H, Ogawa S, Loh KS, Koeffler HP (2014) The genomic landscape of nasopharyngeal carcinoma. Nat Genet 46(8): 866–871

Lindahl T, Adams A, Bjursell G, Bornkamm GW, Kaschka-Dierich C, Jehn U (1976) Covalently closed circular duplex DNA of Epstein-Barr virus in a human lymphoid cell line. J Mol Biol 102(3):511–530

Ling PD, Rawlins DR, Hayward SD (1993) The Epstein-Barr virus immortalizing protein EBNA-2 is targeted to DNA by a cellular enhancer-binding protein. Proc Natl Acad Sci USA 90(20): 9237–9241

Ling PD, Peng RS, Nakajima A, Yu JH, Tan J, Moses SM, Yang WH, Zhao B, Kieff E, Bloch KD, Bloch DB (2005) Mediation of Epstein-Barr virus EBNA-LP transcriptional coactivation by Sp100. EMBO J 24(20):3565–3575

Liu X, Wang Y, Wang X, Sun Z, Li L, Tao Q, Luo B (2013) Epigenetic silencing of WNT5A in Epstein-Barr virus-associated gastric carcinoma. Arch Virol 158(1):123–132

Lo AK, To KF, Lo KW, Lung RW, Hui JW, Liao G, Hayward SD (2007) Modulation of LMP1 protein expression by EBV-encoded microRNAs. Proc Natl Acad Sci USA 104(41): 16164–16169

Loyo M, Brait M, Kim MS, Ostrow KL, Jie CC, Chuang AY, Califano JA, Liegeois NJ, Begum S, Westra WH, Hoque MO, Tao Q, Sidransky D (2011) A survey of methylated candidate tumor suppressor genes in nasopharyngeal carcinoma. Int J Cancer 128(6):1393–1403

Lu F, Wikramasinghe P, Norseen J, Tsai K, Wang P, Showe L, Davuluri RV, Lieberman PM (2010) Genome-wide analysis of host-chromosome binding sites for Epstein-Barr Virus Nuclear Antigen 1 (EBNA1). Virol J 7:262

Lung HL, Bangarusamy DK, Xie D, Cheung AK, Cheng Y, Kumaran MK, Miller L, Liu ET, Guan XY, Sham JS, Fang Y, Li L, Wang N, Protopopov AI, Zabarovsky ER, Tsao SW, Stanbridge EJ, Lung ML (2005) THY1 is a candidate tumour suppressor gene with decreased expression in metastatic nasopharyngeal carcinoma. Oncogene 24(43):6525–6532

Lung RW, Tong JH, Sung YM, Leung PS, Ng DC, Chau SL, Chan AW, Ng EK, Lo KW, To KF (2009) Modulation of LMP2A expression by a newly identified Epstein-Barr virus-encoded microRNA miR-BART22. Neoplasia 11(11):1174–1184

Mackey D, Middleton T, Sugden B (1995) Multiple regions within EBNA1 can link DNAs. J Virol 69(10):6199–6208

Malik-Soni N, Frappier L (2012) Proteomic profiling of EBNA1-host protein interactions in latent and lytic Epstein-Barr virus infections. J Virol 86(12):6999–7002

Matsusaka K, Kaneda A, Nagae G, Ushiku T, Kikuchi Y, Hino R, Uozaki H, Seto Y, Takada K, Aburatani H, Fukayama M (2011) Classification of Epstein-Barr virus-positive gastric cancers by definition of DNA methylation epigenotypes. Cancer Res 71(23):7187–7197

McClellan MJ, Wood CD, Ojeniyi O, Cooper TJ, Kanhere A, Arvey A, Webb HM, Palermo RD, Harth-Hertle ML, Kempkes B, Jenner RG, West MJ (2013) Modulation of enhancer looping and differential gene targeting by Epstein-Barr virus transcription factors directs cellular reprogramming. PLoS Pathog 9(9):e1003636

Miyashita EM, Yang B, Babcock GJ, Thorley-Lawson DA (1997) Identification of the site of Epstein-Barr virus persistence in vivo as a resting B cell. J Virol 71(7):4882–4891

Murata T, Hotta N, Toyama S, Nakayama S, Chiba S, Isomura H, Ohshima T, Kanda T, Tsurumi T (2010) Transcriptional repression by sumoylation of Epstein-Barr virus BZLF1 protein correlates with association of histone deacetylase. J Biol Chem 285(31):23925–23935

Murata T, Kondo Y, Sugimoto A, Kawashima D, Saito S, Isomura H, Kanda T, Tsurumi T (2012) Epigenetic histone modification of Epstein-Barr virus BZLF1 promoter during latency and reactivation in Raji cells. J Virol 86(9):4752–4761

Nerurkar AY, Vijayan P, Srinivas V, Soman CS, Dinshaw KA, Advani SH, Magrath I, Bhatia K, Naresh KN (2000) Discrepancies in Epstein-Barr virus association at presentation and relapse of classical Hodgkin's disease: impact on pathogenesis. Ann Oncol 11(4):475–478

Niederman JC, Miller G, Pearson HA, Pagano JS, Dowaliby JM (1976) Infectious mononucleosis. Epstein-Barr-virus shedding in saliva and the oropharynx. N Engl J Med 294(25):1355–1359

Okada T, Nakamura M, Nishikawa J, Sakai K, Zhang Y, Saito M, Morishige A, Oga A, Sasaki K, Suehiro Y, Hinoda Y, Sakaida I (2013) Identification of genes specifically methylated in Epstein-Barr virus-associated gastric carcinomas. Cancer Sci 104(10):1309–1314

Ozyar E, Ayhan A, Korcum AF, Atahan IL (2004) Prognostic role of Ebstein-Barr virus latent membrane protein-1 and interleukin-10 expression in patients with nasopharyngeal carcinoma. Cancer Invest 22(4):483–491

Pallesen G, Hamilton-Dutoit SJ, Rowe M, Young LS (1991) Expression of Epstein-Barr virus latent gene products in tumour cells of Hodgkin's disease. Lancet 337(8737):320–322

Paschos K, Parker GA, Watanatanasup E, White RE, Allday MJ (2012) BIM promoter directly targeted by EBNA3C in polycomb-mediated repression by EBV. Nucleic Acids Res 40(15): 7233–7246

Paulson EJ, Speck SH (1999) Differential methylation of Epstein-Barr virus latency promoters facilitates viral persistence in healthy seropositive individuals. J Virol 73(12):9959–9968

Pegtel DM, Cosmopoulos K, Thorley-Lawson DA, van Eijndhoven MA, Hopmans ES, Lindenberg JL, de Gruijl TD, Wurdinger T, Middeldorp JM (2010) Functional delivery of viral miRNAs via exosomes. Proc Natl Acad Sci USA 107(14):6328–6333

Piccaluga PP, Navari M, Falco G, Ambrosio MR, Lazzi S, Fuligni F, Bellan C, Rossi M, Sapienza MR, Laginestra MA, Etebari M, Rogena EA, Tumwine L, Tripodo C, Gibellini D, Consiglio J, Croce C, Pileri SA, Leoncini L (2015) Virus-encoded microRNA contributes to the molecular profile of EBV-positive Burkitt lymphomas. Oncotarget 7(1):224–240

Portal D, Rosendorff A, Kieff E (2006) Epstein-Barr nuclear antigen leader protein coactivates transcription through interaction with histone deacetylase 4. Proc Natl Acad Sci USA 103(51): 19278–19283

Portal D, Zhou H, Zhao B, Kharchenko PV, Lowry E, Wong L, Quackenbush J, Holloway D, Jiang S, Lu Y, Kieff E (2013) Epstein-Barr virus nuclear antigen leader protein localizes to promoters and enhancers with cell transcription factors and EBNA2. Proc Natl Acad Sci USA 110(46):18537–18542

Qu L, Rowe DT (1992) Epstein-Barr virus latent gene expression in uncultured peripheral blood lymphocytes. J Virol 66(6):3715–3724

Queen KJ, Shi M, Zhang F, Cvek U, Scott RS (2013) Epstein-Barr virus-induced epigenetic alterations following transient infection. Int J Cancer 132(9):2076–2086

Rawlins DR, Milman G, Hayward SD, Hayward GS (1985) Sequence-specific DNA binding of the Epstein-Barr virus nuclear antigen (EBNA-1) to clustered sites in the plasmid maintenance region. Cell 42(3):859–868

Reinke AW, Grigoryan G, Keating AE (2010) Identification of bZIP interaction partners of viral proteins HBZ, MEQ, BZLF1, and K-bZIP using coiled-coil arrays. Biochemistry 49(9): 1985–1997

Reusch JA, Nawandar DM, Wright KL, Kenney SC, Mertz JE (2015) Cellular differentiation regulator BLIMP1 induces Epstein-Barr virus lytic reactivation in epithelial and B cells by activating transcription from both the R and Z promoters. J Virol 89(3):1731–1743

Rickinson AB (2014) Co-infections, inflammation and oncogenesis: future directions for EBV research. Semin Cancer Biol 26:99–115

Riley KJ, Rabinowitz GS, Yario TA, Luna JM, Darnell RB, Steitz JA (2012) EBV and human microRNAs co-target oncogenic and apoptotic viral and human genes during latency. EMBO J 31(9):2207–2221

Robertson ES, Lin J, Kieff E (1996) The amino-terminal domains of Epstein-Barr virus nuclear proteins 3A, 3B, and 3C interact with RBPJ(kappa). J Virol 70(5):3068–3074

Rovedo M, Longnecker R (2007) Epstein-Barr virus latent membrane protein 2B (LMP2B) modulates LMP2A activity. J Virol 81(1):84–94

Rowe M, Rowe DT, Gregory CD, Young LS, Farrell PJ, Rupani H, Rickinson AB (1987) Differences in B cell growth phenotype reflect novel patterns of Epstein-Barr virus latent gene expression in Burkitt's lymphoma cells. EMBO J 6(9):2743–2751

Rymo L (1979) Identification of transcribed regions of Epstein-Barr virus DNA in Burkitt lymphoma-derived cells. J Virol 32(1):8–18

Saha A, Jha HC, Upadhyay SK, Robertson ES (2015) Epigenetic silencing of tumor suppressor genes during in vitro Epstein-Barr virus infection. Proc Natl Acad Sci USA 112(37): E5199–E5207

Saito M, Nishikawa J, Okada T, Morishige A, Sakai K, Nakamura M, Kiyotoki S, Hamabe K, Okamoto T, Oga A, Sasaki K, Suehiro Y, Hinoda Y, Sakaida I (2013) Role of DNA methylation in the development of Epstein-Barr virus-associated gastric carcinoma. J Med Virol 85(1):121–127

Sakai T, Taniguchi Y, Tamura K, Minoguchi S, Fukuhara T, Strobl LJ, Zimber-Strobl U, Bornkamm GW, Honjo T (1998) Functional replacement of the intracellular region of the Notch1 receptor by Epstein-Barr virus nuclear antigen 2. J Virol 72(7):6034–6039

Schmidt SC, Jiang S, Zhou H, Willox B, Holthaus AM, Kharchenko PV, Johannsen EC, Kieff E, Zhao B (2015) Epstein-Barr virus nuclear antigen 3A partially coincides with EBNA3C genome-wide and is tethered to DNA through BATF complexes. Proc Natl Acad Sci USA 112(2):554–559

Schubeler D (2015) Function and information content of DNA methylation. Nature 517(7534): 321–326

Seo SY, Kim EO, Jang KL (2008) Epstein-Barr virus latent membrane protein 1 suppresses the growth-inhibitory effect of retinoic acid by inhibiting retinoic acid receptor-beta2 expression via DNA methylation. Cancer Lett 270(1):66–76

Serra RW, Fang M, Park SM, Hutchinson L, Green MR (2014) A KRAS-directed transcriptional silencing pathway that mediates the CpG island methylator phenotype. Elife 3:e02313

Sheu LF, Chen A, Lee HS, Hsu HY, Yu DS (2004) Cooperative interactions among p53, bcl-2 and Epstein-Barr virus latent membrane protein 1 in nasopharyngeal carcinoma cells. Pathol Int 54(7):475–485

Shimizu N, Tanabe-Tochikura A, Kuroiwa Y, Takada K (1994) Isolation of Epstein-Barr virus (EBV)-negative cell clones from the EBV-positive Burkitt's lymphoma (BL) line Akata: malignant phenotypes of BL cells are dependent on EBV. J Virol 68(9):6069–6073

Shire K, Ceccarelli DF, Avolio-Hunter TM, Frappier L (1999) EBP2, a human protein that interacts with sequences of the Epstein-Barr virus nuclear antigen 1 important for plasmid maintenance. J Virol 73(4):2587–2595

Shu XS, Li L, Ji M, Cheng Y, Ying J, Fan Y, Zhong L, Liu X, Tsao SW, Chan AT, Tao Q (2013) FEZF2, a novel 3p14 tumor suppressor gene, represses oncogene EZH2 and MDM2 expression and is frequently methylated in nasopharyngeal carcinoma. Carcinogenesis 34(9):1984–1993

Sixbey JW, Vesterinen EH, Nedrud JG, Raab-Traub N, Walton LA, Pagano JS (1983) Replication of Epstein-Barr virus in human epithelial cells infected in vitro. Nature 306(5942):480–483

Sixbey JW, Nedrud JG, Raab-Traub N, Hanes RA, Pagano JS (1984) Epstein-Barr virus replication in oropharyngeal epithelial cells. N Engl J Med 310(19):1225–1230

Skalska L, White RE, Franz M, Ruhmann M, Allday MJ (2010) Epigenetic repression of p16 (INK4A) by latent Epstein-Barr virus requires the interaction of EBNA3A and EBNA3C with CtBP. PLoS Pathog 6(6):e1000951

Sudo M, Chong JM, Sakuma K, Ushiku T, Uozaki H, Nagai H, Funata N, Matsumoto Y, Fukayama M (2004) Promoter hypermethylation of E-cadherin and its abnormal expression in Epstein-Barr virus-associated gastric carcinoma. Int J Cancer 109(2):194–199

Sun D, Zhang Z, Van do N, Huang G, Ernberg I, Hu L (2007) Aberrant methylation of CDH13 gene in nasopharyngeal carcinoma could serve as a potential diagnostic biomarker. Oral Oncol 43(1):82–87

Sung FL, Cui Y, Hui EP, Li L, Loh TK, Tao Q, Chan AT (2014) Silencing of hypoxia-inducible tumor suppressor lysyl oxidase gene by promoter methylation activates carbonic anhydrase IX in nasopharyngeal carcinoma. Am J Cancer Res 4(6):789–800

Szyf M, Eliasson L, Mann V, Klein G, Razin A (1985) Cellular and viral DNA hypomethylation associated with induction of Epstein-Barr virus lytic cycle. Proc Natl Acad Sci USA 82(23): 8090–8094

Tao Q, Huang H, Geiman TM, Lim CY, Fu L, Qiu GH, Robertson KD (2002) Defective de novo methylation of viral and cellular DNA sequences in ICF syndrome cells. Hum Mol Genet 11(18):2091–2102

Tempera I, Wiedmer A, Dheekollu J, Lieberman PM (2010) CTCF prevents the epigenetic drift of EBV latency promoter Qp. PLoS Pathog 6(8):e1001048

Thorley-Lawson DA, Gross A (2004) Persistence of the Epstein-Barr virus and the origins of associated lymphomas. N Engl J Med 350(13):1328–1337

Tierney RJ, Steven N, Young LS, Rickinson AB (1994) Epstein-Barr virus latency in blood mononuclear cells: analysis of viral gene transcription during primary infection and in the carrier state. J Virol 68(11):7374–7385

Tierney RJ, Kirby HE, Nagra JK, Desmond J, Bell AI, Rickinson AB (2000) Methylation of transcription factor binding sites in the Epstein-Barr virus latent cycle promoter Wp coincides with promoter down-regulation during virus-induced B-cell transformation. J Virol 74(22): 10468–10479

Tong JH, Ng DC, Chau SL, So KK, Leung PP, Lee TL, Lung RW, Chan MW, Chan AW, Lo KW, To KF (2010) Putative tumour-suppressor gene DAB2 is frequently down regulated by promoter hypermethylation in nasopharyngeal carcinoma. BMC Cancer 10:253

Tsai CN, Tsai CL, Tse KP, Chang HY, Chang YS (2002) The Epstein-Barr virus oncogene product, latent membrane protein 1, induces the downregulation of E-cadherin gene expression via activation of DNA methyltransferases. Proc Natl Acad Sci USA 99(15):10084–10089

Tsai CL, Li HP, Lu YJ, Hsueh C, Liang Y, Chen CL, Tsao SW, Tse KP, Yu JS, Chang YS (2006) Activation of DNA methyltransferase 1 by EBV LMP1 involves c-Jun NH(2)-terminal kinase signaling. Cancer Res 66(24):11668–11676

Uchida J, Yasui T, Takaoka-Shichijo Y, Muraoka M, Kulwichit W, Raab-Traub N, Kikutani H (1999) Mimicry of CD40 signals by Epstein-Barr virus LMP1 in B lymphocyte responses. Science 286(5438):300–303

Vockerodt M, Morgan SL, Kuo M, Wei W, Chukwuma MB, Arrand JR, Kube D, Gordon J, Young LS, Woodman CB, Murray PG (2008) The Epstein-Barr virus oncoprotein, latent membrane protein-1, reprograms germinal centre B cells towards a Hodgkin's Reed-Sternberg-like phenotype. J Pathol 216(1):83–92

Wang S, Frappier L (2009) Nucleosome assembly proteins bind to Epstein-Barr virus nuclear antigen 1 and affect its functions in DNA replication and transcriptional activation. J Virol 83(22):11704–11714

Wang F, Gregory CD, Rowe M, Rickinson AB, Wang D, Birkenbach M, Kikutani H, Kishimoto T, Kieff E (1987) Epstein-Barr virus nuclear antigen 2 specifically induces expression of the B-cell activation antigen CD23. Proc Natl Acad Sci USA 84(10):3452–3456

Wang L, Grossman SR, Kieff E (2000) Epstein-Barr virus nuclear protein 2 interacts with p300, CBP, and PCAF histone acetyltransferases in activation of the LMP1 promoter. Proc Natl Acad Sci USA 97(1):430–435

Wang K, Kan J, Yuen ST, Shi ST, Chu KM, Law S, Chan TL, Kan Z, Chan AS, Tsui WY, Lee SP, Ho SL, Chan AK, Cheng GH, Roberts PC, Rejto PA, Gibson NW, Pocalyko DJ, Mao M, Xu J, Leung SY (2011) Exome sequencing identifies frequent mutation of ARID1A in molecular subtypes of gastric cancer. Nat Genet 43(12):1219–1223

Wang K, Yuen ST, Xu J, Lee SP, Yan HH, Shi ST, Siu HC, Deng S, Chu KM, Law S, Chan KH, Chan AS, Tsui WY, Ho SL, Chan AK, Man JL, Foglizzo V, Ng MK, Chan AS, Ching YP, Cheng GH, Xie T, Fernandez J, Li VS, Clevers H, Rejto PA, Mao M, Leung SY (2014) Whole-genome sequencing and comprehensive molecular profiling identify new driver mutations in gastric cancer. Nat Genet 46(6):573–582

White RE, Groves IJ, Turro E, Yee J, Kremmer E, Allday MJ (2010) Extensive co-operation between the Epstein-Barr virus EBNA3 proteins in the manipulation of host gene expression and epigenetic chromatin modification. PLoS One 5(11):e13979

Wille CK, Nawandar DM, Panfil AR, Ko MM, Hagemeier SR, Kenney SC (2013) Viral genome methylation differentially affects the ability of BZLF1 versus BRLF1 to activate Epstein-Barr virus lytic gene expression and viral replication. J Virol 87(2):935–950

Woellmer A, Arteaga-Salas JM, Hammerschmidt W (2012) BZLF1 governs CpG-methylated chromatin of Epstein-Barr virus reversing epigenetic repression. PLoS Pathog 8(9):e1002902

Woisetschlaeger M, Yandava CN, Furmanski LA, Strominger JL, Speck SH (1990) Promoter switching in Epstein-Barr virus during the initial stages of infection of B lymphocytes. Proc Natl Acad Sci USA 87(5):1725–1729

Wong TS, Tang KC, Kwong DL, Sham JS, Wei WI, Kwong YL, Yuen AP (2003) Differential gene methylation in undifferentiated nasopharyngeal carcinoma. Int J Oncol 22(4):869–874

Wood VH, O'Neil JD, Wei W, Stewart SE, Dawson CW, Young LS (2007) Epstein-Barr virus-encoded EBNA1 regulates cellular gene transcription and modulates the STAT1 and TGFbeta signaling pathways. Oncogene 26(28):4135–4147

Wu TC, Mann RB, Epstein JI, MacMahon E, Lee WA, Charache P, Hayward SD, Kurman RJ, Hayward GS, Ambinder RF (1991) Abundant expression of EBER1 small nuclear RNA in nasopharyngeal carcinoma. A morphologically distinctive target for detection of Epstein-Barr virus in formalin-fixed paraffin-embedded carcinoma specimens. Am J Pathol 138(6): 1461–1469

Wu DY, Kalpana GV, Goff SP, Schubach WH (1996) Epstein-Barr virus nuclear protein 2 (EBNA2) binds to a component of the human SNF-SWI complex, hSNF5/Ini1. J Virol 70(9):6020–6028

Xue SA, Labrecque LG, Lu QL, Ong SK, Lampert IA, Kazembe P, Molyneux E, Broadhead RL, Borgstein E, Griffin BE (2002) Promiscuous expression of Epstein-Barr virus genes in Burkitt's lymphoma from the central African country Malawi. Int J Cancer 99(5):635–643

Yanatatsaneejit P, Chalermchai T, Kerekhanjanarong V, Shotelersuk K, Supiyaphun P, Mutirangura A, Sriuranpong V (2008) Promoter hypermethylation of CCNA1, RARRES1, and HRASLS3 in nasopharyngeal carcinoma. Oral Oncol 44(4):400–406

Yang HJ, Huang TJ, Yang CF, Peng LX, Liu RY, Yang GD, Chu QQ, Huang JL, Liu N, Huang HB, Zhu ZY, Qian CN, Huang BJ (2013) Comprehensive profiling of Epstein-Barr virus-encoded miRNA species associated with specific latency types in tumor cells. Virol J 10:314

Yang X, Dai W, Kwong DL, Szeto CY, Wong EH, Ng WT, Lee AW, Ngan RK, Yau CC, Tung SY, Lung ML (2015) Epigenetic markers for noninvasive early detection of nasopharyngeal carcinoma by methylation-sensitive high resolution melting. Int J Cancer 136(4):E127–E135

Yates J, Warren N, Reisman D, Sugden B (1984) A cis-acting element from the Epstein-Barr viral genome that permits stable replication of recombinant plasmids in latently infected cells. Proc Natl Acad Sci USA 81(12):3806–3810

Ying J, Li H, Seng TJ, Langford C, Srivastava G, Tsao SW, Putti T, Murray P, Chan AT, Tao Q (2006) Functional epigenetics identifies a protocadherin PCDH10 as a candidate tumor suppressor for nasopharyngeal, esophageal and multiple other carcinomas with frequent methylation. Oncogene 25(7):1070–1080

Zech L, Haglund U, Nilsson K, Klein G (1976) Characteristic chromosomal abnormalities in biopsies and lymphoid-cell lines from patients with Burkitt and non-Burkitt lymphomas. Int J Cancer 17(1):47–56

Zerby D, Chen CJ, Poon E, Lee D, Shiekhattar R, Lieberman PM (1999) The amino-terminal C/H1 domain of CREB binding protein mediates zta transcriptional activation of latent Epstein-Barr virus. Mol Cell Biol 19(3):1617–1626

Zhang T, Ma J, Nie K, Yan J, Liu Y, Bacchi CE, Queiroga EM, Gualco G, Sample JT, Orazi A, Knowles DM, Tam W (2014) Hypermethylation of the tumor suppressor gene PRDM1/Blimp-1 supports a pathogenetic role in EBV-positive Burkitt lymphoma. Blood Cancer J 4:e261

Zhao B, Zou J, Wang H, Johannsen E, Peng CW, Quackenbush J, Mar JC, Morton CC, Freedman ML, Blacklow SC, Aster JC, Bernstein BE, Kieff E (2011) Epstein-Barr virus exploits intrinsic B-lymphocyte transcription programs to achieve immortal cell growth. Proc Natl Acad Sci USA 108(36):14902–14907

Zhao J, Liang Q, Cheung KF, Kang W, Dong Y, Lung RW, Tong JH, To KF, Sung JJ, Yu J (2013a) Somatostatin receptor 1, a novel EBV-associated CpG hypermethylated gene, contributes to the pathogenesis of EBV-associated gastric cancer. Br J Cancer 108(12):2557–2564

Zhao J, Liang Q, Cheung KF, Kang W, Lung RW, Tong JH, To KF, Sung JJ, Yu J (2013b) Genome-wide identification of Epstein-Barr virus-driven promoter methylation profiles of human genes in gastric cancer cells. Cancer 119(2):304–312

Zhou VW, Goren A, Bernstein BE (2011) Charting histone modifications and the functional organization of mammalian genomes. Nat Rev Genet 12(1):7–18

Zhou H, Schmidt SC, Jiang S, Willox B, Bernhardt K, Liang J, Johannsen EC, Kharchenko P, Gewurz BE, Kieff E, Zhao B (2015) Epstein-Barr virus oncoprotein super-enhancers control B cell growth. Cell Host Microbe 17(2):205–216

Zhu JY, Pfuhl T, Motsch N, Barth S, Nicholls J, Grasser F, Meister G (2009) Identification of novel Epstein-Barr virus microRNA genes from nasopharyngeal carcinomas. J Virol 83(7): 3333–3341

Chapter 5
Epigenetic Programming by Microbial Pathogens and Impacts on Acute and Chronic Disease

Michael J. Mahan, Douglas M. Heithoff, Lucien Barnes V, and Robert L. Sinsheimer

Abstract Epigenetic programming of the pathogen and the host can have a marked influence on the development and progression of acute and chronic disease. Bacterial pathogenesis may be viewed as a developmental program similar to that of cell differentiation and development in eukaryotes. Bacterial epigenetic programming is imparted by DNA methylation, whereby the virulence traits expressed by a pathogen may depend on the cumulative interactions between the microbe and its environment. Such bacterial "memory" provides a means for adaptation to the varied subsequent microenvironments encountered during the infective process. DNA methylation can affect DNA–protein interactions and resultant gene expression by altering DNA thermodynamic stability and curvature and by methyl-group-mediated steric hindrance. Some of these epigenetic interactions can form heritable DNA methylation patterns in the microbial genome that control gene expression in their progeny cells. Microbes can also stimulate heritable changes in the host epigenome via infection-associated alterations to host epigenetic determinants including DNA methylation, histone modifications, chromatin-associated complexes, and noncoding RNA-mediated silencing. The resultant changes in host chromatin remodeling and gene expression may be localized and/or systemic due to direct microbe-to-host cell communication or via dissemination of microbial-host signaling. Thus, the role of epigenetics in host–microbe interactions may be the nexus of many pathological syndromes even though there may be no apparent direct link between infection and disease, providing the basis for the development of novel therapeutics and diagnostic tests for diseases with epigenomic determinants.

Keywords Epigenetics • DNA methylation • Epigenetic disease • Infectious disease • Epigenetic programming • Epigenome • Bacterial memory • Epigenetic host-microbe interactions • Epigenetic memory

---

M.J. Mahan (✉) • D.M. Heithoff • L. Barnes V • R.L. Sinsheimer
Department of Molecular, Cellular and Developmental Biology, University of California, Santa Barbara, CA 93106, USA
e-mail: michael.mahan@lifesci.ucsb.edu

W. Doerfler, J. Casadesús (eds.), *Epigenetics of Infectious Diseases*, Epigenetics and Human Health, DOI 10.1007/978-3-319-55021-3_5

5.1 Introduction

Deciphering the mechanisms that govern epigenetic programming in the pathogen and the host is crucial to the development of new therapeutic approaches to control acute and chronic disease. For instance, pathogenic *Escherichia coli* utilize heritable DNA methylation patterns to control pili production via a phase variation mechanism, whereby individual cells either express pili (phase-ON) or not (phase-OFF), resulting in periods of attachment and detachment that are critical for progression of an ascending urinary tract infection (Low and Casadesús 2008; Marinus and Casadesús 2009). Microbial infection can also stimulate epigenetic changes in the host epigenome, potentially leading to a variety of human diseases including cancer and autoimmune disorders (Bierne et al. 2012; Dawson and Kouzarides 2012; Elinav et al. 2013; Feinberg and Tycko 2004; Stein 2011). Despite this knowledge, the role of epigenetic modifications on pathogen virulence is poorly understood, and the role of host epigenetic modifications that contribute to, or result from, infectious diseases are only just beginning to be elucidated.

5.2 DNA Methylation

DNA methylation is a fundamental epigenetic process that provides a means to impart additional information to the genomic sequence. In bacteria, DNA methylation occurs at the N^6 position of adenine (6mA) and the C^5 or N^4 positions of cytosine (5mC; 4mC), and these modifications are catalyzed by DNA methyltransferases (Noyer-Weidner and Trautner 1992; Palmer and Marinus 1994; Sánchez-Romero et al. 2015; Wion and Casadesús 2006). Such epigenetic information can alter the timing and targeting of cellular events including transcription, transposition, chromosomal replication, and DNA repair. The most common DNA modification in eukaryotes is 5mC, which is involved in a variety of processes including gene regulation, genomic imprinting, X-chromosome inactivation, and epigenetic memory maintenance (Jones 2012; Jones and Takai 2001; Smith and Meissner 2013). Additionally, 6mA has been recently reported as a possible epigenetic mark in eukaryotes that plays a potential role in transcription and epigenetic inheritance (Luo et al. 2015).

5.2.1 *Bacterial Restriction Modification Systems*

DNA methylation is a standard means by which restriction-modification (R-M) systems serve to protect bacterial cells from foreign DNA (viruses, transposons, plasmids) (Kobayashi et al. 1999; Meselson et al. 1972; Roberts and Macelis 2001). In most R-M systems, base methylation by a methyltransferase (on adenine or cytosine) prevents DNA cleavage of host DNA by cognate restriction enzymes. Recent evidence suggests that the role of R-M-associated methyltransferases is not restricted to protecting host genomes as the lack of certain R-M systems alters the gene expression pattern of the cell, suggesting a role in epigenetic control of gene expression (Fang et al. 2012; Furuta et al. 2014; Sánchez-Romero et al. 2015; Vasu and Nagaraja 2013). *E. coli* O104:H4 is a hemolytic uremic syndrome (HUS)-linked outbreak strain that harbors multiple active adenine methyltransferases—some of which are associated with R-M systems (Fang et al. 2012). *E. coli* O104:H4 also contains a lysogenic lambdoid phage, fStx104, which encodes Shiga toxin (the cause of HUS), and an R-M system that engenders both the production of Shiga toxin and alteration of the bacterium's transcriptome. These findings indicate that DNA methyltransferases associated with R-M systems can make considerable contributions to bacterial virulence (discussed further below).

5.2.2 *Solitary DNA Methyltransferases*

Some bacterial DNA methyltransferases lack cognate restriction enzymes and thus are not part of R-M systems. These solitary methyltransferases play roles in cellular regulatory events including those that control bacterial gene regulation, cell-cycle events, and virulence (Low et al. 2001; Marinus and Casadesús 2009; Reisenauer et al. 1999).

5.2.2.1 Dam Methylase

DNA adenine methylase (Dam) is a solitary methyltransferase of *Gammaproteobacteria* (*e.g.*, *E. coli* and *Salmonella*) that methylates the N^6 position of adenine in the sequence "GATC" of the bacterial genome and plays a role in the timing and targeting of many cellular events by influencing the interactions of regulatory proteins with DNA (Casadesús and Low 2006; Løbner-Olesen et al. 2005; Low and Casadesús 2008; Low et al. 2001; Marinus and Casadesús 2009). DNA adenine methylation can affect DNA–protein interactions at GATC sequences by altering DNA thermodynamic stability and curvature and by methyl-group-mediated steric hindrance (Wion and Casadesús 2006). There are about 130 molecules of Dam per cell in *E. coli*, a level that allows sufficient time for some DNA–protein binding between DNA synthesis and the methylation of GATC sequences within newly synthesized DNA (Boye et al. 1992). Competition between Dam and DNA-binding proteins resulted in the formation of ~35 nonmethylated GATC sequences in the *E. coli* genome (Hale et al. 1994; Ringquist and Smith 1992; Tavazoie and Church 1998; Wang and Church 1992). The actual number of nonmethylated sites at any one time is dependent on bacterial growth rate and growth phase, supporting the hypothesis that the DNA-binding proteins are in competition with Dam at these sites to control the timing and targeting of cellular regulatory events.

5.2.2.2 Cytosine Methylases

Although the role of cytosine methylases has been generally associated with R-M systems, recent evidence suggests that this view may need to be broadened (Marinus and Casadesús 2009; Sánchez-Romero et al. 2015). DNA cytosine methylase (Dcm) is a solitary methyltransferase of *Gammaproteobacteria* that methylates the internal cytosine in the CCA/TGG motif at the C5 position (5mC) (Bigger et al. 1973; Kahramanoglou et al. 2012). *E. coli dcm* mutants display increased expression of the stress response sigma factor, RpoS, suggesting cytosine methylation may be involved in gene expression and the stress response (Kahramanoglou et al. 2012). Further, the absence of a solitary cytosine methyltransferase (5mC), HpyA-VIBM, in *Helicobacter pylori* alters the expression of genes involved in motility, adhesion, and virulence (Kumar et al. 2012).

5.2.2.3 CcrM Methylase

The role of DNA adenine methylation in cell-cycle-related events has been extensively studied in *Caulobacter crescentus*, serving as a model organism for bacterial cell-cycle regulation and development (Gonzalez et al. 2014; Marczynski and Shapiro 2002; McAdams and Shapiro 2003; Reisenauer et al. 1999). *C. crescentus* is a member of the *Alphaproteobacteria*, which includes *Brucella abortus* (brucellosis), *Agrobacterium tumefaciens* (crown gall disease), and *Sinorhizobium meliloti* (nitrogen-fixation). It has a dimorphic life cycle, spending part of its life cycle as a non-replicating motile swarmer cell and the other as a replicating sessile stalked cell. Many of the cellular events leading to differentiation into these morphological stages are modulated by the solitary cell-cycle regulated methyltransferase, CcrM, which methylates the N^6 position of adenine in the sequence GANTC (Marczynski and Shapiro 2002; McAdams and Shapiro 2003; Reisenauer et al. 1999). The *C. crescentus* chromosomal methylation state (unmethylated, hemimethylated, fully methylated) controls a regulatory cascade that couples DNA replication and the expression of cell-cycle master regulators, which facilitate progression of the *Caulobacter* cell cycle (Collier et al. 2006).

5.3 Dam Methylation Modulates the Timing and Targeting of Cellular Processes

Dam plays a role in the timing and targeting of many cellular processes including DNA repair, DNA replication, transposition, conjugation, as well as those specifically involved in bacterial virulence (Løbner-Olesen et al. 2005; Low and Casadesús 2008; Low et al. 2001; Marinus and Casadesús 2009; Sánchez-Romero et al. 2015).

5.3.1 Dam Controls DNA Repair and Replication

Errors that occur during replication are corrected by methyl-directed mismatch repair that can distinguish base mismatches on the newly synthesized strand. Such DNA strand discrimination is accomplished using hemimethylated DNA that arises after passage of the replication fork, whereby the parental strand is methylated at Dam-target sequences (GATC sites) and the newly synthesized strand is non-methylated (Pukkila et al. 1983). DNA base mismatches on newly synthesized DNA are recognized and removed by the MutHLS DNA mismatch repair proteins, and the errors are corrected using the parental strand as a template (Iyer et al. 2006). Subsequently during the cell cycle, the newly synthesized strand is methylated by Dam at GATC sites resulting in fully methylated DNA. Dam levels are controlled primarily at the transcriptional level (Løbner-Olesen et al. 2003) and the absence, or overproduction, of Dam leads to an increase in spontaneous mutation frequency due to the lack of hemimethylated DNA needed for strand discrimination during DNA mismatch repair (Heithoff et al. 2007; Herman and Modrich 1982; Marinus and Morris 1974).

The timing of DNA replication is controlled by a competition between Dam and DNA-binding proteins that recognize hemimethylated DNA. SeqA binds specifically to several hemimethylated GATC sites at and near the origin of replication (*oriC*), delaying their methylation by Dam (Kang et al. 1999; Lu et al. 1994). The sequestration of these hemimethylated sites by SeqA delays further replication fork initiation since it represses transcription of the replication initiator (*dnaA)* and inhibits DnaA binding at *oriC* as both processes operate optimally at fully methylated GATC sites (Marinus and Casadesús 2009). Additionally, SeqA acts at hemimethylated sites to play a role in nucleoid structure, organization, and partitioning into daughter cells (Bach et al. 2003; Helgesen et al. 2015; Joshi et al. 2013; Skarstad and Katayama 2013). Thus, competition between Dam and DNA-binding proteins controls the timing and targeting of many cellular events that are critical to the cell cycle.

5.3.2 Dam Controls Bacterial Gene Expression

Dam methylation of GATC sites can control gene expression via altering the affinity of DNA-binding proteins to regulatory sequences, as described in the following examples (Casadesús and Low 2006; Løbner-Olesen et al. 2005; Low and Casadesús 2008; Low et al. 2001; Marinus and Casadesús 2009; Sánchez-Romero et al. 2015).

Initiation of DNA Replication Sequestration of hemimethylated GATC sites by SeqA at the *oriC* region delays replication fork initiation via *dnaA* transcriptional repression and DnaA-binding inhibition at *oriC* (discussed above). *Implications*: Maintenance of hemimethylated DNA near the *oriC* region limits the number of replication forks that can initiate before cell division.

Transposition Tn*10* transposition occurs upon the generation of hemimethylated GATC sites in the transposase promoter (Roberts et al. 1985). The transposase promoter is only active when the transposase-coding strand is methylated and the noncoding strand is not methylated. *Implications*: Transposition is repressed through most of the cell cycle, preventing high-level transposition that would otherwise cause detrimental effects to the genome. Transposition is limited to one copy while the other copy remains in the original location.

Conjugal Plasmid Transfer Stimulation of the *tra* operon for conjugal transfer of the *Salmonella* virulence plasmid occurs upon generation of hemimethylated GATC sites within the upstream regulatory sequences for *traJ* expression, a transcriptional activator of the *tra* operon. Methylation of the noncoding strand (but not the coding strand) stimulates binding of the leucine-responsive regulatory protein (Lrp), with resultant *traJ* transcription, and conjugal transfer of the methylated noncoding single-stranded DNA into the recipient bacterium (Camacho and Casadesús 2002; Camacho and Casadesús 2005). *Implications*: Conjugal transfer is repressed through most of the cell cycle via a *traJ* epigenetic switch, thereby modulating the considerable metabolic and energetic cost of mating functions to the cell. Recipient cells are competent for conjugation since the noncoding, methylated strand serves as a template for DNA replication, reproducing the DNA methylation pattern that permits Lrp binding.

Cell Invasion *Salmonella* invasion of human epithelial cells is impaired in the absence of Dam methylation (Garcia-Del Portillo et al. 1999). Binding of the HdfR regulatory protein to unmethylated GATC sites in regulatory sequences for the *std* fimbrial operon stimulates StdEF-mediated repression of invasion determinants encoded on *Salmonella* Pathogenicity Island I (SPI-1) (Jakomin et al. 2008; López-Garrido and Casadesús 2012). *Implications*: Methylation state of invasion-associated regulatory sequences ensures bacterial invasion of only appropriate cells/cellular compartments that contribute to the onset and progression of infection.

Pili Phase Variation Dam methylation controls the production of *E. coli* pyelonephritis-associated pili (*pap*) via a phase-variation mechanism that results in cells that either express or do not express the pili. Dam is in competition with two transcriptional activators (Lrp, PapI) for GATC sites in upstream regulatory sequences for *pap* expression, forming DNA methylation patterns that can be inherited in progeny populations similar to that observed in eukaryotes (Low and Casadesús 2008; Marinus and Casadesús 2009). *Implications*: Phase variation (ON-OFF) control of pili adherence via Dam methylation results in periods of bacterial attachment and detachment, facilitating uropathogenic *E. coli* progression from the bladder to kidney, resulting in pyelonephritis.

5.4 DNA Methylation Plays an Essential Role in Bacterial Virulence

DNA methylation has been shown to play a role or has been implicated in the virulence of many bacterial pathogens (Casadesús and Low 2006; Heusipp et al. 2007; Low et al. 2001; Marinus and Casadesús 2009; Sánchez-Romero et al. 2015). Representative examples are discussed below, including pathogens that utilize solitary or R-M methyltransferases to modulate bacterial virulence.

5.4.1 DNA Methylation Controls Bacterial Pathogenesis

Salmonella spp. Nontyphoidal *Salmonella* (NTS) is the greatest foodborne-disease burden in the United States, with greater than one million illnesses annually (Gilliss et al. 2011; Scallan et al. 2011). *Salmonella enterica* infection can result in any of four disease syndromes: enterocolitis/diarrhea, bacteremia, typhoid fever, and chronic asymptomatic carriage (Coburn et al. 2007). Many serovars infect both humans and animals, with the particular syndrome a function of the serovar (serotypic variant), strain virulence, and host susceptibility (Coburn et al. 2007; Heithoff et al. 2012). Dam methylation plays an essential role in *Salmonella* virulence (Garcia-Del Portillo et al. 1999; Heithoff et al. 1999). The lack or overproduction of Dam confers significant virulence attenuation (10,000-fold) in murine models of typhoid fever. Dam methylation is involved in the invasion of nonphagocytic cells, M-cell cytotoxicity, bile resistance, envelope stability, cell motility, fimbrial, O-antigen and cytotoxin production, systemic dissemination, and the elicitation of host innate and adaptive immune responses (Badie et al. 2007; Garcia-Del Portillo et al. 1999; Heithoff et al. 1999, 2001, 2007, 2008; López-Garrido and Casadesús 2010; Pucciarelli et al. 2002; Sarnacki et al. 2009; Shtrichman et al. 2002; Simon et al. 2007). *Implications*: Dam methylation controls the production of many factors underlying microbial virulence (adhesins, invasins, toxins) and impacts host–pathogen interactions that compromise host immunity. *Salmonella dam* mutants are capable of eliciting cross-protection against a diversity of salmonellae and are well-tolerated when applied as modified live vaccines in mice (Heithoff et al. 2001, 2008, 2015), poultry (Dueger et al. 2001, 2003a), sheep (Mohler et al. 2011) and calves (Dueger et al. 2003b; Mohler et al. 2006, 2008). Induction of immunity is rapid, and the vaccine can be delivered in drinking water for low-cost and low-stress vaccination of livestock populations (Mohler et al. 2011, 2012).

Yersinia spp. *Yersinia pseudotuberculosis* and *enterocolitica* are zoonotic foodborne pathogens that can cause severe disease in humans including gastroenteritis, mesenteric lymphadenitis, and septicemia (Galindo et al. 2011; Tauxe 2015). Many pathogenic strains infect both humans and animals whereby the particular syndrome is a function of the serotype, strain virulence, and host susceptibility. The *dam* gene is essential in certain strains of *Yersinia* species, and the lack or overproduction of Dam in *Y. pseudotuberculosis* leads to severe virulence attenuation in murine models of bacteremia (Julio et al. 2001; Kubicek-Sutherland et al. 2014; Taylor et al. 2005) and confers protection to heterologous *Y. pseudotuberculosis* or *Y. pestis* challenge (Julio et al. 2001; Kubicek-Sutherland et al. 2014; Taylor et al. 2005). Dam overproducing *Y. pseudotuberculosis* ectopically secrete several *Yersinia* outer proteins (e.g., YopE cytotoxin) as well as LcrV, a low-calcium-responsive virulence factor normally involved in Yop synthesis, localization, and suppression of host inflammatory activities (Badie et al. 2004; Julio et al. 2001, 2002). Dam overproducing *Y. enterocolitica* confer altered invasion, motility, and composition of the lipopolysaccharide (LPS) O-antigen and also display ectopic Yop secretion via increased ClpP protease degradation of the LcrG regulatory protein that normally blocks Yop secretion (Fälker et al. 2005, 2006, 2007). *Implications*: Dam methylation controls the strict environmental regulation of *Yersinia* virulence function synthesis and localization, serving to modulate bacterial pathogenesis and host inflammatory activities.

Enterohemorrhagic E. coli EHEC are a subgroup of Shiga toxin-producing *E. coli* (STEC) that can cause severe intestinal disease [*i.e.*, hemorrhagic colitis [HC] and hemolytic uremic syndrome [HUS] (Hartland and Leong 2013; Mahan et al. 2013)]. EHEC intestinal adherence requires the delivery of the type III secretion system (TTSS) effector proteins Tir and $EspF_U$ into the host cell and expression of the bacterial outer membrane adhesin, intimin. Increased adherence exhibited by *dam* mutant EHEC was correlated with increased protein levels of Tir, $EspF_U$, and intimin (Campellone et al. 2007). Dam also controls the maintenance of lysogeny for a bacteriophage (933W) that encodes Shiga toxin (Stx-2), which inhibits protein synthesis (via ribosomal inactivation) and leads to renal toxicity in HUS patients (Murphy et al. 2008). *Implications*: Dam methylation modulates EHEC intestinal adherence and Shiga toxin production during infection.

Brucella abortus *B. abortus* is an intracellular pathogen and the causative agent of brucellosis, a zoonotic disease that causes abortions and stillbirths in livestock and acute febrile illness in humans, which may progress to chronically debilitating disease (World Health Organization 2006). It is also designated as a select agent with the potential for bioterrorism due to the chronic nature of disease in livestock and humans and its ability to undergo aerosolization (Centers for Disease Control and Prevention 2015). *B. abortus* is a member of the *Alphaproteobacteria* which have defined morphological stages that are modulated by the solitary cell cycle-regulated DNA methyltransferase, CcrM (Marczynski and Shapiro 2002). CcrM is essential for viability in *B. abortus*, and its overexpression attenuates replication within murine macrophages (Robertson et al. 2000). *Implications*: CcrM methylation may play a role in intracellular replication of the bacterium within phagocytes, a key virulence characteristic for both acute and chronic cases of brucellosis.

Mycobacterium tuberculosis *M. tuberculosis* infections cause nine million active cases and 1.5 million tuberculosis deaths annually, with one-third of the world's population having latent infections (World Health Organization 2014a). Although there are no predicted *dam* homologues, the *M. tuberculosis* solitary DNA methyltransferase, MamA, plays a role in gene expression and fitness during hypoxia, and different methyltransferases are observed in different lineages of *M. tuberculosis* (Shell et al. 2013). *Implications*: DNA methylation may play a role in *M. tuberculosis* lineage-specific differences in preferences for distinct host environments and different disease courses in humans.

Haemophilus influenzae Non-typeable *H. influenzae* (NTHi) is a major cause of middle ear (otitis media) infections in children (Haggard 2008). NTHi contains R-M systems comprised of a methyltransferase (*mod*) and a restriction endonuclease (*res*) (Srikhanta et al. 2010). Phase variable (ON-OFF) switching of *mod* alleles (due to the presence of tandem repeats in the corresponding *mod* genes) regulates the expression of multiple proteins that are involved in antibiotic resistance, biofilm formation, and immune evasion. Recent studies indicate that *mod* switching to the ON orientation was highly selected in a chinchilla model of otitis media, and ON phase-variants formed more robust biofilms in vitro (Atack et al. 2015). These findings suggest that *mod* is involved in bacterial virulence, immune evasion, and niche adaptation. Several other human pathogens contain phase-variable R-M systems, including *H. pylori* (atrophic gastritis), *Neisseria meningitidis* (meningitis), *N. gonorrhoeae* (gonorrhea), and *Moraxella catarrhalis* (otitis media) (Atack et al. 2015; Srikhanta et al. 2010). *Implications*: Phase-variable R-M systems that modulate microbial virulence traits may be shared across the microbial realm.

5.5 Perspectives: Epigenetic Programming of the Pathogen and Disease Susceptibility

Bacterial pathogenesis can be regarded as a developmental program (Casadesús and D'Ari 2002; Mahan et al. 2010) similar to eukaryotic cell differentiation and development (Bird 2002, 2007; Jaenisch and Bird 2003). Bacterial epigenetic programming is imparted by DNA methylation, whereby the virulence traits expressed are dependent on the aggregate of interactions between the microbe and its environment. Thus, the bacterial epigenome provides a means for bacterial "memory," engendering the capacity for adaption to the disparate microenvironments encountered as the infection proceeds due to dissemination to new host sites, tissue breakdown, inflammation, and immune clearance mechanisms. Thus, a microbial population may comprise a spectrum of genotypically identical cells with significant phenotypic differences in virulence traits since pathogenicity may be a reflection of cumulative exposure to selective pressures within host(s) and environments experienced during the microbial life cycle. Epigenetic programming may provide insights into the virulence disparities of closely related strains that exhibit marked differences with regard to pathogenicity, host range, and preferences for distinct host environments and different disease courses in humans.

5.6 Microbial Infection, Epigenetic Reprogramming, and Human Disease

Microbe-associated changes in the host epigenome can play a significant role in human disease via chromatin remodeling and resultant transcriptional reprogramming driven by host DNA methylation, histone modifications, chromatin-associated complexes, and noncoding RNA-mediated silencing (Bannister and Kouzarides 2011; Bierne et al. 2012; Dawson and Kouzarides 2012; Herceg et al. 2013; Paschos and Allday 2010). DNA methylation occurs at the 5′ position of cytosines within CpG dinucleotides, and can recruit protein complexes that can alter chromatin structure or affect the binding of transcription factors with resultant gene silencing (Bird 2002, 2007; Dawson and Kouzarides 2012; Jones and Takai 2001). Histone modifications (e.g., methylation, acetylation, phosphorylation, ubiquitination) can alter chromatin structure and affect gene expression (Bannister and Kouzarides 2011; Dawson and Kouzarides 2012). Noncoding RNA-mediated silencing involves microRNAs (noncoding, 18–25 nucleotides) that target mRNAs and negatively control gene expression (He and Hannon 2004; Sato et al. 2011).

Such transcriptional reprogramming can alter host defense genes involved in TLR (Toll-like receptor), MAPK (mitogen-activated protein kinase), interferon (IFN), and NF-kB signaling pathways (Gómez-Díaz et al. 2012; Paschos and Allday 2010; Stein 2011). For instance, *M. tuberculosis* inhibits IFN-γ-induced chromatin remodeling by TLR2 and MAPK signaling (via inhibition of histone acetylation), leading to reduced expression of several immune genes and resultant persistence of chronic infections (Pennini et al. 2006). Influenza virus suppresses the antiviral response via the production of a histone mimic that serves as a sink for a host transcription factor (hPAF1) involved in antiviral gene expression (Marazzi et al. 2012). Conversely, some microbe-associated epigenome changes are protective. Following acute viral infection, chromosome remodeling is implicated in the formation of memory CD8+ T cells that provide the host with long-term protective immunity against the pathogen (Youngblood et al. 2010).

5.6.1 Microbial Infection and Cancer

Microbe-associated cancers account for a significant proportion (>20%) of all human cancers (Moore and Chang 2010; zur Hausen 2009). The molecular basis involves microbe-stimulated changes in the host epigenome, with resultant changes in host chromatin remodeling, gene expression, and metabolism (Dawson and Kouzarides 2012; Esteller 2008; Feinberg and Tycko 2004; Herceg et al. 2013; Stein 2011).

Hepatitis B Virus (HBV) Liver cancer is the second leading cause of cancer death worldwide (World Health Organization 2014b, 2015). The risk for liver cancer is increased 100-fold in individuals with chronic HBV infection (Fernandez et al. 2009), and recent estimates indicate that ~250 million individuals in the human population are chronically infected with HBV (Schweitzer et al. 2015). HBV persists in host cells by the nuclear accumulation of covalently closed circular DNA (cccDNAs) that serve as a template for transcription of all viral mRNAs and are organized into minichromosomes by histones and nonhistone viral and cellular proteins (Grimm et al. 2011; Protzer 2015). High viral loads in patients with chronic hepatitis correlate with hyperacetylation of histone H3 and H4 bound to cccDNA in liver biopsy samples (Pollicino et al. 2006), allowing access of the HBV cccDNA chromatin-like structure to liver-specific transcription factors and subsequent replication (Quasdorff et al. 2008). The HBx regulatory protein (Kew 2011) relieves chromatin-mediated transcriptional repression of HBV cccDNA that involves the histone methyltransferase, SETDB1 (Rivière et al. 2015). HBx also upregulates several DNA methyltransferases (DNMTs), resulting in increased promoter methylation and repression of tumor-suppressor genes encoding p16, a cyclin-dependent kinase inhibitor that functions in cell-cycle arrest and cellular senescence, and E-cadherin, a cell–cell adhesion molecule that affects tumor invasiveness (Fernandez and Esteller 2010; Jung et al. 2007; Tian et al. 2013). Patient samples from various stages of HBV infection show increased methylation of the HBV genome as an acute infection transitions to a chronic infection and during the subsequent progression to premalignant lesions and cancer (Fernandez and Esteller 2010; Fernandez et al. 2009; Stein 2011). Additionally, microRNA (miR-152), whose normal function is to downregulate DNMT1, is downregulated in patients with HBV-associated liver cancer, thus causing DNA hypermethylation (Huang et al. 2010; Saito et al. 2014). These findings suggest a tumor-suppressive role of miR-152, and therapeutic use of this microRNA may reduce aberrant DNA methylation.

Human Papillomavirus (HPV) Genital human papillomavirus (HPV) is the most commonly diagnosed sexually transmitted infection in the United States and is associated with 95% of cervical and anal cancers and 60% of oropharyngeal cancers. (Centers for Disease Control and Prevention 2012; Gilmer 2015). Through preexisting lesions, HPV infects the basal (lower) layer of the stratified cervical epithelium, and viral genomes are maintained as episomal DNA in the nuclei of infected cells (Kajitani et al. 2012). HPV oncoproteins E6 and E7 inactivate p53 and retinoblastoma (pRb) tumor-suppressing proteins, respectively, resulting in aberrant proliferation and delayed differentiation of infected host cells (Münger et al. 2004). The productive phase of the lifecycle (genome amplification, virion assembly/release) occurs in upper layers of the cervical epithelium that are terminally differentiated. In infections with HPV "high-risk" invasive serotypes (16 and 18), progression of the disease is associated with increased methylation of the HPV genome and considerable suppression of E-cadherin (Anayannis et al. 2015; Fernandez and Esteller 2010; Fernandez et al. 2009; Sun et al. 2011; Wilson et al. 2013). E-cadherin is utilized by Langerhans cells (antigen processing/presentation) to move through stratified epithelium, and its reduction may impact HPV clearance and the length of persistent infections. In a keratinocyte cell line, the HPV E7 oncoprotein is necessary for E-cadherin downregulation via augmentation of host DNMT1 levels and resultant E-cadherin repression (Laurson et al. 2010). DNMT inhibition (via 5-aza-deoxycytidine administration) restored E-cadherin levels, suggesting that epigenetic intervention may have utility in combating persistent infections via restoring influx of Langerhans cells to infected tissue. Further, epigenetic alterations to the viral genome via methylation of viral promoter regions have been implicated in HPV E6 and E7 expression during a transforming infection (Steenbergen et al. 2014). The overall consequence of deregulated expression of E6 and E7 in proliferating cells is chromosomal instability, leading to accumulation of lesions in host cell cancer genes and subsequent progression toward cancer (Korzeniewski et al. 2011).

Epstein–Barr Virus (EBV) EBV is carried in the vast majority (>90%) of the human population as an asymptomatic lifelong infection, yet it is also correlated with several nonmalignant and malignant diseases (Odumade et al. 2011; Rickinson et al. 2014; Thompson and Kurzrock 2004; Thorley-Lawson 2015). EBV causes mononucleosis and many human tumors of B cell, T cell, and epithelial origin such as Burkett's lymphoma, Hodgkin's disease, gastric carcinoma, nasopharyngeal carcinoma, and lymphoproliferative tumors in immunocompromised individuals. The EBV lifecycle involves infection of oropharyngeal cells; host colonization through growth-transforming latent infection of B cells within oropharyngeal lymphoid tissues; long-term persistence within recirculating memory B cells as a silent latent infection; and reactivation to the viral lytic phase and subsequent infection of naïve host cells (Rickinson et al. 2014). Many of these events are driven by epigenetic reprogramming of the pathogen and host, whereby B cell growth transformation is facilitated by several latent proteins, including EBV nuclear antigens (EBNAs) and latent membrane proteins (LMPs), followed by regulated shutdown of latent protein expression that ultimately results in latency in recirculating B cells (Hammerschmidt 2015; Paschos and Allday 2010). Increased methylation of the EBV genome occurs as an acute infection transitions to chronic infection and during subsequent development and progression of cancer (Fernandez and Esteller 2010; Fernandez et al. 2009), and infection of B lymphocytes or nasopharyngeal carcinoma cell lines results in the expression of several DNMTs (Schmeinck 2011; Tsai et al. 2006). During the latent phase, EBV lytic genes are transcriptionally silenced by histone methyltransferase EZH2, a component of the Polycomb Repressive Complex 2, PRC2; and these "histone marks" are erased upon lytic phase induction (Hammerschmidt 2015; Woellmer et al. 2012). These findings indicate that epigenetic modifications of viral DNA determine viral latency.

Helicobacter pylori Stomach cancer is the third leading cause of cancer death worldwide (World Health Organization 2014b, 2015). *H. pylori* is a gastric pathogen that colonizes approximately 50% of the world's population (Wroblewski et al. 2010), associated with 65% of gastric cancers, and classified as a class I carcinogen (Polk and Peek 2010; World Health Organization 2014b). In patients infected with *H. pylori*, aberrant methylation and repression of tumor-suppressor genes (E-cadherin, p16) was linked with increased gastric cancer risk (Kaise et al. 2008; Maekita et al. 2006; Nakajima et al. 2006; Yoshida et al. 2013). In a gerbil model of gastric cancer, *H. pylori* infection was shown to be causally involved in the induction of aberrant methylation in the host epigenome, which was associated with the upregulation of several inflammation-related genes (*CXCL2*, *IL-1β*, *NOS2*, *TNF-α*) (Niwa et al. 2010). Methylation decreased upon bacterial clearance but remained significantly higher than that observed in uninfected control animals. Suppressing inflammation with the immunosuppressive drug, cyclosporin A, prevented aberrant methylation without affecting colonization, indicating that epigenetic modifications occurred as a consequence of inflammation rather than the infection itself. These studies revealed an "epigenetic field defect" whereby increased DNA methylation that arises as a result of infection marks a region with higher risk for transformation (Niwa et al. 2010; Stein 2011). Thus, DNA methylation has potential clinical utility as a biomarker for the risk of malignant transformation for a number of cancers, offering new therapeutic opportunities that target and monitor epigenetic changes (discussed below).

5.7 Concluding Remarks: Microbial Infection and Its Impact on the Host Epigenome and Disease

The origin of some diseases may have a microbial component even though there may be no apparent direct link between infection and disease. How does this occur and what are the possible implications? Microbial infection can trigger heritable changes in the host epigenome that lead to profound differences in disease susceptibility, host cell metabolism, inflammation, and immune responses, and some of these responses may be maintained long after microbial clearance (Bierne et al. 2012; Davis et al. 2011; Stein 2011). The primary challenge toward establishing a causal link between infection-associated changes in the host epigenome and disease origin is the considerable interplay between epigenetic, genetic (mutational), and nonmicrobial (*e.g.*, carcinogen) risk factors that cloud the assignment of primary versus secondary events leading to disease development and progression (Fig. 5.1). Microbes cause cancer directly via harboring oncogenes that contribute to cell transformation or indirectly through chronic inflammation whereby ultimately carcinogenic mutations are generated in host cells (Moore and Chang 2010; Parsonnet 1999; zur Hausen 2001). Additionally, microbial and nonmicrobial associated alterations in host epigenetic determinants influence many biological

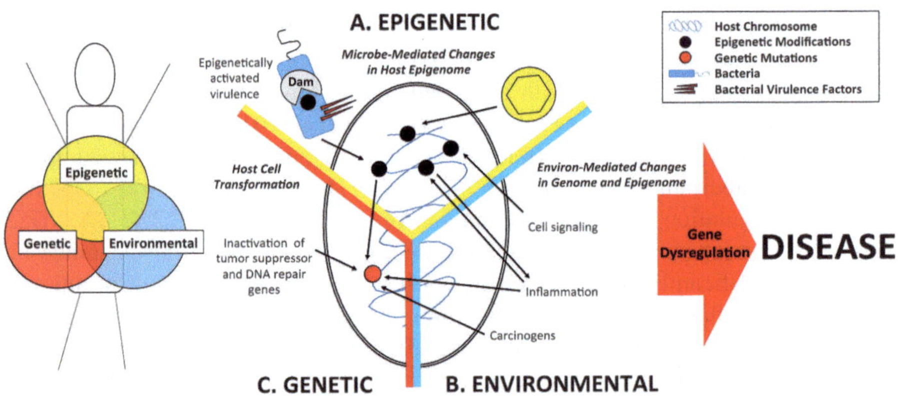

Fig. 5.1 Epigenetic programming of the pathogen and the host can stimulate the development and progression of acute and chronic disease. (**a**) The epigenome of pathogenic microbes can be modified to stimulate the production of virulence determinants (via Dam; host DNMTs). Pathogenic bacteria can modify the host epigenome (*dark circles*) via DNA methylation, histone modifications, chromatin-associated complexes, and noncoding RNA mediated silencing. (**b**) Environmental inputs can alter disease susceptibility by stimulating genetic (mutational; *red circles*) or epigenetic changes (nonmutational; *dark circles*) in the host genome via carcinogen exposure, cell signaling, and inflammation. (**c**) Chronic disease (*e.g.*, cancer) can be stimulated directly by genetic changes in the host genome caused by exposure to carcinogens, microbial oncogenes, and chronic inflammation or indirectly via epigenetic changes in the host genome by inactivation of tumor-suppressor genes and/or DNA-repair genes, which predispose the genome to mutation. Such complex interactions between genetic, epigenetic, and environmental inputs result in host gene dysregulation and human disease

processes that are fundamental to the development of cancer including the inactivation of tumor-suppressor genes and/or DNA-repair genes, which predispose the genome to mutation (Baylin and Herman 2000; Dawson and Kouzarides 2012; Herceg et al. 2013; Moore and Chang 2010; Paschos and Allday 2010; Romani et al. 2015; Stein 2011).

Despite these challenges, significant advances have been made toward establishing a direct link between microbe-associated changes in the host epigenome and cancer, and it remains a possibility that certain disorders are a consequence of chronic inflammation with microbial origin (Bierne et al. 2012; Costenbader et al. 2012; Elinav et al. 2013; Feinberg and Tycko 2004; Grivennikov et al. 2010; Herceg et al. 2013; Liu et al. 2008; Portela and Esteller 2010; Schett et al. 2013; Ushijima and Hattori 2012; Van Vliet et al. 2007; Wilson 2008). For example, the gut microbiome (the largest reservoir of microbes in the body) stimulates host epigenome changes that are linked to inflammatory bowel disease (Khor et al. 2011; Knights et al. 2013; Kostic et al. 2014; Ventham et al. 2013). Since there are ~100 trillion microbial cells in the gastrointestinal tract—roughly ten times more than the cells in the human body—the gut microbiome has the capacity to produce a variety of compounds that can impact host genomic/ epigenomic processes and metabolism (Bianconi et al. 2013; Garagnani et al. 2013; Shenderov 2012; Stilling et al. 2014). Examples include microbial structural components and metabolites (*e.g.*, peptides, polysaccharides, endotoxins, short-chain fatty acids, co-factors) that are potential epigenomic modifiers, which can affect gene expression and metabolism in the host via transcriptional reprogramming of host signaling pathways (Gómez-Díaz et al. 2012; Knights et al. 2013; Paschos and Allday 2010; Stein 2011). Notably, host–gut microbe interactions can lead to considerable systemic signaling, involving many organs and organ systems, including the central nervous system (Stilling et al. 2014). Thus, the role of epigenetics in host–microbe interactions leading to pathological syndromes—with the potential of the disruption of homeostasis due to pathogen exposure—provides the foundation for the development of new medicines and diagnostic tests for diseases with epigenomic determinants.

The significant challenge of epigenetic therapies lies in the lack of specificity—and the global hypomethylation achieved by DNMT inhibitors—which may be detrimental to developing an effective treatment. Notwithstanding, cancer treatment applications include administration of small molecules that inhibit epigenetic factors (Dawson and Kouzarides 2012; Romani et al. 2015), risk assessments that link the degree of aberrant DNA methylation to the likelihood of cell transformation (Niwa et al. 2010; Stein 2011), and gene therapy targeting epigenetic factors (Yao et al. 2015). The use of "epigenetic modifier drugs" may extend beyond cancer to other epigenetically based diseases as evidenced by their current testing in noncancer clinical trials (*e.g.*, irritable bowel syndrome, Alzheimer disease, cardiovascular disease, thalassemia, psoriasis) (Romani et al. 2015). Additionally, combinational therapies—comprising epigenetic modifier drugs and antimicrobials—may prove useful in combating infectious diseases and associated disease manifestations such as blood clotting and inflammation that can cause severe tissue damage and organ failure leading to death (Grewal et al. 2013; Herceg et al. 2013; Moore and Chang 2010; Schleithoff et al. 2012; Yang et al. 2015).

Acknowledgments The scope of this chapter and its space limitations have unfortunately resulted in the inability to separately cite many of the original publications that have contributed substantially to the field. We sincerely apologize to the authors of these publications. This work was supported by G. Harold & Leila Y. Mathers Foundation and Santa Barbara Cottage Hospital Research Program (to M.J.M).

References

Anayannis N, Schlecht N, Belbin T (2015) Epigenetic mechanisms of human papillomavirus-associated head and neck cancer. Arch Pathol Lab Med. doi:10.5858/arpa.2014-0554-RA

Atack J, Srikhanta Y, Fox K, Jurcisek J, Brockman K, Clark T, Boitano M, Power P, Jen F-C, McEwan A (2015) A biphasic epigenetic switch controls immunoevasion, virulence and niche adaptation in non-typeable *Haemophilus influenzae*. Nat Commun 6:7828. doi:10.1038./ncomms8828

Bach T, Krekling M, Skarstad K (2003) Excess SeqA prolongs sequestration of oriC and delays nucleoid segregation and cell division. EMBO J 22:315–323

Badie G, Heithoff D, Mahan M (2004) LcrV synthesis is altered by DNA adenine methylase overproduction in Yersinia pseudotuberculosis and is required to confer immunity in vaccinated hosts. Infect Immun 72:6707–6710

Badie G, Heithoff D, Sinsheimer R, Mahan M (2007) Altered levels of Salmonella DNA adenine methylase are associated with defects in gene expression, motility, flagellar synthesis, and bile resistance in the pathogenic strain 14028 but not in the laboratory strain LT2. J Bacteriol 189:1556–1564

Bannister A, Kouzarides T (2011) Regulation of chromatin by histone modifications. Cell Res 21:381–395

Baylin S, Herman J (2000) DNA hypermethylation in tumorigenesis: epigenetics joins genetics. Trends Genet 16:168–174

Bianconi E, Piovesan A, Facchin F, Beraudi A, Casadei R, Frabetti F, Vitale L, Pelleri M, Tassani S, Piva F (2013) An estimation of the number of cells in the human body. Ann Hum Biol 40:463–471

Bierne H, Hamon M, Cossart P (2012) Epigenetics and bacterial infections. Cold Spring Harb Perspect Med 2:a010272

Bigger C, Murray K, Murray N (1973) Recognition sequence of a restriction enzyme. Nature 244:7–10

Bird A (2002) DNA methylation patterns and epigenetic memory. Genes Dev 16:6–21

Bird A (2007) Perceptions of epigenetics. Nature 447:396–398

Boye E, Marinus M, Løbner-Olesen A (1992) Quantitation of Dam methyltransferase in Escherichia coli. J Bacteriol 174:1682–1685

Camacho E, Casadesús J (2002) Conjugal transfer of the virulence plasmid of Salmonella enterica is regulated by the leucine-responsive regulatory protein and DNA adenine methylation. Mol Microbiol 44:1589–1598

Camacho E, Casadesús J (2005) Regulation of traJ transcription in the Salmonella virulence plasmid by strand-specific DNA adenine hemimethylation. Mol Microbiol 57:1700–1718

Campellone K, Roe A, Løbner-Olesen A, Murphy K, Magoun L, Brady M, Donohue-Rolfe A, Tzipori S, Gally D, Leong J (2007) Increased adherence and actin pedestal formation by dam-deficient enterohaemorrhagic Escherichia coli O157: H7. Mol Microbiol 63:1468–1481

Casadesús J, D'Ari R (2002) Memory in bacteria and phage. Bioessays 24:512–518

Casadesús J, Low D (2006) Epigenetic gene regulation in the bacterial world. Microbiol Mol Biol Rev 70:830–856

Centers for Disease Control and Prevention (2012) Human papillomavirus-associated cancers-United States, 2004–2008. Morb Mortal Wkly Rep 61:258

Centers for Disease Control and Prevention (2015) Humans and Brucella species. http://www.cdc.gov/brucellosis/clinicians/brucella-species.html

Coburn B, Grassl G, Finlay B (2007) Salmonella, the host and disease: a brief review. Immunol Cell Biol 85:112–118

Collier J, Murray SR, Shapiro L (2006) DnaA couples DNA replication and the expression of two cell cycle master regulators. EMBO J 25:346–356

Costenbader K, Gay S, Alarcón-Riquelme M, Iaccarino L, Doria A (2012) Genes, epigenetic regulation and environmental factors: which is the most relevant in developing autoimmune diseases? Autoimmun Rev 11:604–609

Davis B, Wen H, Ting J (2011) The inflammasome NLRs in immunity, inflammation, and associated diseases. Annu Rev Immunol 29:707

Dawson M, Kouzarides T (2012) Cancer epigenetics: from mechanism to therapy. Cell 150:12–27

Dueger EL, House JK, Heithoff DM, Mahan MJ (2001) *Salmonella* DNA adenine methylase mutants elicit protective immune responses to homologous and heterologous serovars in chickens. Infect Immun 69:7950–7954

Dueger EL, House JK, Heithoff DM, Mahan MJ (2003a) Salmonella DNA adenine methylase mutants prevent colonization of newly hatched chickens by homologous and heterologous serovars. Int J Food Microbiol 80:153–159

Dueger EL, House JK, Heithoff DM, Mahan MJ (2003b) Salmonella DNA adenine methylase mutants elicit early and late onset protective immune responses in calves. Vaccine 21:3249–3258

Elinav E, Nowarski R, Thaiss C, Hu B, Jin C, Flavell R (2013) Inflammation-induced cancer: crosstalk between tumours, immune cells and microorganisms. Nat Rev Cancer 13:759–771

Esteller M (2008) Epigenetics in cancer. N Engl J Med 358:1148–1159

Fälker S, Schilling J, Schmidt M, Heusipp G (2007) Overproduction of DNA adenine methyltransferase alters motility, invasion, and the lipopolysaccharide O-antigen composition of Yersinia enterocolitica. Infect Immun 75:4990–4997

Fälker S, Schmidt M, Heusipp G (2005) DNA methylation in Yersinia enterocolitica: role of the DNA adenine methyltransferase in mismatch repair and regulation of virulence factors. Microbiology 151:2291–2299

Fälker S, Schmidt M, Heusipp G (2006) Altered Ca2+ regulation of Yop secretion in Yersinia enterocolitica after DNA adenine methyltransferase overproduction is mediated by Clp-dependent degradation of LcrG. J Bacteriol 188:7072

Fang G, Munera D, Friedman D, Mandlik A, Chao M, Banerjee O, Feng Z, Losic B, Mahajan M, Jabado O (2012) Genome-wide mapping of methylated adenine residues in pathogenic Escherichia coli using single-molecule real-time sequencing. Nat Biotechnol 30:1232–1239

Feinberg A, Tycko B (2004) The history of cancer epigenetics. Nat Rev Cancer 4:143–153

Fernandez A, Esteller M (2010) Viral epigenomes in human tumorigenesis. Oncogene 29:1405–1420

Fernandez A, Rosales C, Lopez-Nieva P, Graña O, Ballestar E, Ropero S, Espada J, Melo S, Lujambio A, Fraga M (2009) The dynamic DNA methylomes of double-stranded DNA viruses associated with human cancer. Genome Res 19:438–451

Furuta Y, Namba-Fukuyo H, Shibata T, Nishiyama T, Shigenobu S, Suzuki Y, Sugano S, Hasebe M, Kobayashi I (2014) Methylome diversification through changes in DNA methyltransferase sequence specificity. PLoS Genet 10:e1004272

Galindo C, Rosenzweig J, Kirtley M, Chopra A (2011) Pathogenesis of Y. enterocolitica and Y. pseudotuberculosis in human yersiniosis. J Pathog 2011:1–16

Garagnani P, Pirazzini C, Franceschi C (2013) Colorectal cancer microenvironment: among nutrition, gut microbiota, inflammation and epigenetics. Curr Pharm Des 19:765–778

Garcia-Del Portillo F, Pucciarelli MG, Casadesus J (1999) DNA adenine methylase mutants of *Salmonella typhimurium* show defects in protein secretion, cell invasion, and M cell cytotoxicity. Proc Natl Acad Sci U S A 96:11578–11583

Gilliss D, Cronquist A, Cartter M, Tobin-D'Angelo M, Blythe D, Smith K, Lathrop S, Birkhead G, Cieslak P (2011) Vital signs: incidence and trends of infection with pathogens transmitted commonly through food—foodborne diseases active surveillance Network, 10 U.S. Sites, 1996–2010. Morb Mortal Wkly Rep 60:749–755

Gilmer L (2015) Human papillomavirus vaccine update. Prim Care Clin Off Pract 42:17–32

Gómez-Díaz E, Jordà M, Peinado M, Rivero A (2012) Epigenetics of host–pathogen interactions: the road ahead and the road behind. PLoS Pathog 8:e1003007

Gonzalez D, Kozdon J, McAdams H, Shapiro L, Collier J (2014) The functions of DNA methylation by CcrM in Caulobacter crescentus: a global approach. Nucleic Acids Res 42:3720–3735

Grewal P, Aziz P, Uchiyama S, Rubio G, Lardone R, Le D, Varki N, Nizet V, Marth J (2013) Inducing host protection in pneumococcal sepsis by preactivation of the Ashwell-Morell receptor. Proc Natl Acad Sci U S A 110:20218–20223

Grimm D, Thimme R, Blum H (2011) HBV life cycle and novel drug targets. Hepatol Int 5:644–653

Grivennikov S, Greten F, Karin M (2010) Immunity, inflammation, and cancer. Cell 140:883–899

Haggard M (2008) Otitis media: prospects for prevention. Vaccine 26:G20–G24

Hale W, Van der Woude M, Low D (1994) Analysis of nonmethylated GATC sites in the Escherichia coli chromosome and identification of sites that are differentially methylated in response to environmental stimuli. J Bacteriol 176:3438–3441

Hammerschmidt W (2015) The epigenetic life cycle of Epstein–Barr virus. Curr Top Microbiol Immunol 390(Pt 1):103–117

Hartland E, Leong J (2013) Enteropathogenic and enterohemorrhagic E. coli: ecology, pathogenesis, and evolution. Front Cell Infect Microbiol 3:15–20

He L, Hannon G (2004) MicroRNAs: small RNAs with a big role in gene regulation. Nat Rev Genet 5:522–531

Heithoff D, Badie G, Julio S, Enioutina E, Daynes R, Sinsheimer R, Mahan M (2007) In vivo-selected mutations in methyl-directed mismatch repair suppress the virulence attenuation of Salmonella dam mutant strains following intraperitoneal, but not oral, infection of naïve mice. J Bacteriol 189:4708–4717

Heithoff DM, Enioutina EY, Bareyan D, Daynes RA, Mahan MJ (2008) Conditions that diminish myeloid-derived suppressor cell activities stimulate cross-protective immunity. Infect Immun 76:5191–5199

Heithoff DM, Enioutina EY, Daynes RA, Sinsheimer RL, Low DA, Mahan MJ (2001) *Salmonella* DNA adenine methylase mutants confer cross-protective immunity. Infect Immun 69:6725–6730

Heithoff D, House J, Thomson P, Mahan M (2015) Development of a Salmonella cross-protective vaccine for food animal production systems. Vaccine 33:100–107

Heithoff DM, Shimp WR, House JK, Xie Y, Weimer BC, Sinsheimer RL, Mahan MJ (2012) Intraspecies variation in the emergence of hyperinfectious bacterial strains in nature. PLoS Pathog 8:e1002647

Heithoff DM, Sinsheimer RL, Low DA, Mahan MJ (1999) An essential role for DNA adenine methylation in bacterial virulence. Science 284:967–970

Helgesen E, Fossum-Raunehaug S, Sætre F, Schink K, Skarstad K (2015) Dynamic Escherichia coli SeqA complexes organize the newly replicated DNA at a considerable distance from the replisome. Nucleic Acids Res 43:2730–2743

Herceg Z, Lambert M-P, van Veldhoven K, Demetriou C, Vineis P, Smith M, Straif K, Wild C (2013) Towards incorporating epigenetic mechanisms into carcinogen identification and evaluation. Carcinogenesis 34:1955–1967

Herman G, Modrich P (1982) Escherichia coli dam methylase. Physical and catalytic properties of the homogeneous enzyme. J Biol Chem 257:2605–2612

Heusipp G, Fälker S, Schmidt M (2007) DNA adenine methylation and bacterial pathogenesis. Int J Med Microbiol 297:1–7
Huang J, Wang Y, Guo Y, Sun S (2010) Down-regulated microRNA-152 induces aberrant DNA methylation in hepatitis B virus–related hepatocellular carcinoma by targeting DNA methyltransferase 1. Hepatology 52:60–70
Iyer R, Pluciennik A, Burdett V, Modrich P (2006) DNA mismatch repair: functions and mechanisms. Chem Rev 106:302–323
Jaenisch R, Bird A (2003) Epigenetic regulation of gene expression: how the genome integrates intrinsic and environmental signals. Nat Genet 33:245–254
Jakomin M, Chessa D, Bäumler A, Casadesús J (2008) Regulation of the Salmonella enterica std fimbrial operon by DNA adenine methylation, SeqA, and HdfR. J Bacteriol 190:7406–7413
Jones P (2012) Functions of DNA methylation: islands, start sites, gene bodies and beyond. Nat Rev Genet 13:484–492
Jones P, Takai D (2001) The role of DNA methylation in mammalian epigenetics. Science 293:1068–1070
Joshi M, Magnan D, Montminy T, Lies M, Stepankiw N, Bates D (2013) Regulation of sister chromosome cohesion by the replication fork tracking protein SeqA. PLoS Genet 9:e1003673
Julio S, Heithoff D, Provenzano D, Klose K, Sinsheimer R, Low D, Mahan M (2001) DNA Adenine methylase is essential for viability and plays a role in the pathogenesis of Yersinia pseudotuberculosis and Vibrio cholerae. Infect Immun 69:7610–7615
Julio S, Heithoff D, Sinsheimer R, Low D, Mahan M (2002) DNA adenine methylase overproduction in Yersinia pseudotuberculosis alters YopE expression and secretion and host immune responses to infection. Infect Immun 70:1006–1009
Jung J, Arora P, Pagano J, Jang K (2007) Expression of DNA methyltransferase 1 is activated by hepatitis B virus X protein via a regulatory circuit involving the p16INK4a-cyclin D1-CDK 4/6-pRb-E2F1 pathway. Cancer Res 67:5771–5778
Kahramanoglou C, Prieto A, Khedkar S, Haase B, Gupta A, Benes V, Fraser G, Luscombe N, Seshasayee A (2012) Genomics of DNA cytosine methylation in Escherichia coli reveals its role in stationary phase transcription. Nat Comm 3:886
Kaise M, Yamasaki T, Yonezawa J, Miwa J, Ohta Y, Tajiri H (2008) CpG island hypermethylation of tumor-tuppressor genes in H. pylori-infected non-Neoplastic gastric mucosa is linked with gastric cancer risk. Helicobacter 13:35–41
Kajitani N, Satsuka A, Kawate A, Sakai H (2012) Productive lifecycle of human papillomaviruses that depends upon squamous epithelial differentiation. Front Microbiol 3:1–12
Kang S, Lee H, Han J, Hwang D (1999) Interaction of SeqA and Dam methylase on the hemimethylated origin of Escherichia coli chromosomal DNA replication. J Biol Chem 274:11463–11468
Kew MC (2011) Hepatitis B virus x protein in the pathogenesis of hepatitis B virus-induced hepatocellular carcinoma. J Gastroenterol Hepatol 26:144–152
Khor B, Gardet A, Xavier R (2011) Genetics and pathogenesis of inflammatory bowel disease. Nature 474:307–317
Knights D, Lassen K, Xavier R (2013) Advances in inflammatory bowel disease pathogenesis: linking host genetics and the microbiome. Gut 62:1505–1510
Kobayashi I, Nobusato A, Kobayashi-Takahashi N, Uchiyama I (1999) Shaping the genome–restriction–modification systems as mobile genetic elements. Curr Opin Genet Dev 9:649–656
Korzeniewski N, Spardy N, Duensing A, Duensing S (2011) Genomic instability and cancer: lessons learned from human papillomaviruses. Cancer Lett 305:113–122
Kostic A, Xavier R, Gevers D (2014) The microbiome in inflammatory bowel disease: current status and the future ahead. Gastroenterology 146:1489–1499
Kubicek-Sutherland J, Heithoff D, Ersoy S, Shimp W, Mahan M (2014) Immunization with a DNA adenine methylase over-producing Yersinia pseudotuberculosis vaccine confers robust cross-protection against heterologous pathogenic serotypes. Vaccine 32:1451–1459

Kumar R, Mukhopadhyay A, Ghosh P, Rao D (2012) Comparative transcriptomics of H. pylori strains AM5, SS1 and their hpyAVIBM deletion mutants: possible roles of cytosine methylation. PLoS One 7(8):e42303

Laurson J, Khan S, Chung R, Cross K, Raj K (2010) Epigenetic repression of E-cadherin by human papillomavirus 16 E7 protein. Carcinogenesis 31:918–926

Liu L, Li Y, Tollefsbol T (2008) Gene-environment interactions and epigenetic basis of human diseases. Curr Issues Mol Biol 10:25

Løbner-Olesen A, Marinus M, Hansen F (2003) Role of SeqA and Dam in Escherichia coli gene expression: a global/microarray analysis. Proc Natl Acad Sci U S A 100:4672–4677

Løbner-Olesen A, Skovgaard O, Marinus M (2005) Dam methylation: coordinating cellular processes. Curr Opin Microbiol 8:154–160

López-Garrido J, Casadesús J (2010) Regulation of Salmonella enterica pathogenicity island 1 by DNA adenine methylation. Genetics 184:637–649

López-Garrido J, Casadesús J (2012) Crosstalk between virulence loci: regulation of Salmonella enterica pathogenicity island 1 (SPI-1) by products of the std fimbrial operon. PLoS One 7: e30499

Low D, Casadesús J (2008) Clocks and switches: bacterial gene regulation by DNA adenine methylation. Curr Opin Microbiol 11:106–112

Low D, Weyand N, Mahan M (2001) Roles of DNA adenine methylation in regulating bacterial gene expression and virulence. Infect Immun 69:7197–7204

Lu M, Campbell J, Boye E, Kleckner N (1994) SeqA: a negative modulator of replication initiation in E. coli. Cell 77:413–426

Luo G-Z, Blanco M, Greer E, He C, Shi Y (2015) DNA N6-methyladenine: a new epigenetic mark in eukaryotes? Nat Rev Mol Cell Biol 16:705–710

Maekita T, Nakazawa K, Mihara M, Nakajima T, Yanaoka K, Iguchi M, Arii K, Kaneda A, Tsukamoto T, Tatematsu M (2006) High levels of aberrant DNA methylation in Helicobacter pylori–infected gastric mucosae and its possible association with gastric cancer risk. Clin Can Res 12:989–995

Mahan MJ, Kubicek-Sutherland JZ, Heithoff DM (2013) Rise of the microbes. Virulence 4:213–222

Mahan M, Sinsheimer R, Shimp W, Heithoff D (2010) Covert operations: the adaptable plan of attack deployed by pathogenic bacteria. In: Maloy S, Hughes K, Casadesús J (eds) The lure of bacterial genetics: a tribute to John Roth. American Society for Microbiology Press, Washington, DC, pp 185–200

Marazzi I, Ho J, Kim J, Manicassamy B, Dewell S, Albrecht R, Seibert C, Schaefer U, Jeffrey K, Prinjha R (2012) Suppression of the antiviral response by an influenza histone mimic. Nature 483:428–433

Marczynski G, Shapiro L (2002) Control of chromosome replication in Caulobacter crescentus. Annu Rev Microbiol 56:625–656

Marinus M, Casadesús J (2009) Roles of DNA adenine methylation in host–pathogen interactions: mismatch repair, transcriptional regulation, and more. FEMS Microbiol Rev 33:488–503

Marinus M, Morris N (1974) Biological function for 6-methyladenine residues in the DNA of Escherichia coli K12. J Mol Biol 85:309–322

McAdams H, Shapiro L (2003) A bacterial cell-cycle regulatory network operating in time and space. Science 301:1874–1877

Meselson M, Yuan R, Heywood J (1972) Restriction and modification of DNA. Annu Rev Biochem 41:447–466

Mohler V, Heithoff D, Mahan M, Hornitzky M, Thomson P, House J (2012) Development of a novel in-water vaccination protocol for DNA adenine methylase deficient Salmonella enterica serovar Typhimurium vaccine in adult sheep. Vaccine 30:1481–1491

Mohler VL, Heithoff DM, Mahan MJ, Walker KH, Hornitzky MA, Gabor L, Thomson PC, Thompson A, House JK (2011) Protective immunity conferred by a DNA adenine methylase deficient Salmonella enterica serovar Typhimurium vaccine when delivered in-water to sheep challenged with Salmonella enterica serovar Typhimurium. Vaccine 29:3571–3582

Mohler V, Heithoff D, Mahan M, Walker K, Hornitzky M, McConnell C, Shum L, House J (2006) Cross-protective immunity in calves conferred by a DNA adenine methylase deficient Salmonella enterica serovar Typhimurium vaccine. Vaccine 24:1339

Mohler V, Heithoff D, Mahan M, Walker K, Hornitzky M, Shum L, Makin K, House J (2008) Cross-protective immunity conferred by a DNA adenine methylase deficient Salmonella enterica serovar Typhimurium vaccine in calves challenged with *Salmonella* serovar Newport. Vaccine 26:1751–1758

Moore P, Chang Y (2010) Why do viruses cause cancer? Highlights of the first century of human tumour virology. Nat Rev Cancer 10:878–889

Münger K, Baldwin A, Edwards K, Hayakawa H, Nguyen C, Owens M, Grace M, Huh K (2004) Mechanisms of human papillomavirus-induced oncogenesis. J Virol 78:11451–11460

Murphy K, Ritchie J, Waldor M, Løbner-Olesen A, Marinus M (2008) Dam methyltransferase is required for stable lysogeny of the Shiga toxin (Stx2)-encoding bacteriophage 933W of enterohemorrhagic Escherichia coli O157: H7. J Bacteriol 190:438–441

Nakajima T, Maekita T, Oda I, Gotoda T, Yamamoto S, Umemura S, Ichinose M, Sugimura T, Ushijima T, Saito D (2006) Higher methylation levels in gastric mucosae significantly correlate with higher risk of gastric cancers. Cancer Epidemiol Biomarkers Prev 15:2317–2321

Niwa T, Tsukamoto T, Toyoda T, Mori A, Tanaka H, Maekita T, Ichinose M, Tatematsu M, Ushijima T (2010) Inflammatory processes triggered by Helicobacter pylori infection cause aberrant DNA methylation in gastric epithelial cells. Cancer Res 70:1430–1440

Noyer-Weidner M, Trautner T (1992) Methylation of DNA in prokaryotes. In: Jost JP, Saluz H (eds) DNA methylation: molecular biology and biological significance. Birkhauser Basel, Switzerland, pp 39–108

Odumade O, Hogquist K, Balfour H (2011) Progress and problems in understanding and managing primary Epstein-Barr virus infections. Clin Microbiol Rev 24:193–209

Palmer B, Marinus M (1994) The dam and dcm strains of Escherichia coli—a review. Gene 143:1–12

Parsonnet J (1999) Microbes and malignancy: infection as a cause of human cancers. Oxford University Press, USA

Paschos K, Allday M (2010) Epigenetic reprogramming of host genes in viral and microbial pathogenesis. Trends Microbiol 18:439–447

Pennini M, Pai R, Schultz D, Boom W, Harding C (2006) Mycobacterium tuberculosis 19-kDa lipoprotein inhibits IFN-γ-induced chromatin remodeling of MHC2TA by TLR2 and MAPK signaling. J Immunol 176:4323–4330

Polk D, Peek R (2010) Helicobacter pylori: gastric cancer and beyond. Nat Rev Cancer 10:403–414

Pollicino T, Belloni L, Raffa G, Pediconi N, Squadrito G, Raimondo G, Levrero M (2006) Hepatitis B virus replication is regulated by the acetylation status of hepatitis B virus cccDNA-bound H3 and H4 histones. Gastroenterology 130:823–837

Portela A, Esteller M (2010) Epigenetic modifications and human disease. Nat Biotechnol 28:1057–1068

Protzer U (2015) Hepatitis: epigenetic control of HBV by HBx protein-releasing the break? Nat Rev Gastroenterol Hepatol 12. doi:10.1016/j.jhep.2015.1006.1023

Pucciarelli M, Prieto A, Casadesús J, García-del Portillo F (2002) Envelope instability in DNA adenine methylase mutants of Salmonella enterica. Microbiology 148:1171–1182

Pukkila P, Peterson J, Herman G, Modrich P, Meselson M (1983) Effects of high levels of DNA adenine methylation on methyl-directed mismatch repair in Escherichia coli. Genetics 104:571–582

Quasdorff M, Hösel M, Odenthal M, Zedler U, Bohne F, Gripon P, Dienes HP, Drebber U, Stippel D, Goeser T (2008) A concerted action of HNF4α and HNF1α links hepatitis B virus replication to hepatocyte differentiation. Cell Microbiol 10:1478–1490

Reisenauer A, Kahng L, McCollum S, Shapiro L (1999) Bacterial DNA methylation: a cell cycle regulator? J Bacteriol 181:5135–5139

Rickinson A, Long H, Palendira U, Münz C, Hislop A (2014) Cellular immune controls over Epstein–Barr virus infection: new lessons from the clinic and the laboratory. Trends Immunol 35:159–169

Ringquist S, Smith C (1992) The Escherichia coli chromosome contains specific, unmethylated dam and dcm sites. Proc Natl Acad Sci U S A 89:4539–4543

Rivière L, Gerossier L, Ducroux A, Dion S, Deng Q, Michel M-L, Buendia M-A, Hantz O, Neuveut C (2015) HBx relieves chromatin-mediated transcriptional repression of hepatitis B viral cccDNA involving SETDB1 histone methyltransferase. J Hepatol 63:1093–1102

Roberts D, Hoopes B, McClure W, Kleckner N (1985) IS10 transposition is regulated by DNA adenine methylation. Cell 43:117–130

Roberts R, Macelis D (2001) REBASE—restriction enzymes and methylases. Nucleic Acids Res 29:268–269

Robertson G, Reisenauer A, Wright R, Jensen R, Jensen A, Shapiro L, Roop R (2000) The Brucella abortus CcrM DNA methyltransferase is essential for viability, and its overexpression attenuates intracellular replication in murine macrophages. J Bacteriol 182:3482–3489

Romani M, Pistillo M, Banelli B (2015) Environmental epigenetics: crossroad between public health, lifestyle, and cancer prevention. BioMed Res Int 2015:587983

Saito Y, Hibino S, Saito H (2014) Alterations of epigenetics and microRNA in hepatocellular carcinoma. Hepatol Res 44:31–42

Sánchez-Romero M, Cota I, Casadesús J (2015) DNA methylation in bacteria: from the methyl group to the methylome. Curr Opin Microbiol 25:9–16

Sarnacki S, Marolda C, Llana M, Giacomodonato M, Valvano M, Cerquetti M (2009) Dam methylation controls O-antigen chain length in Salmonella enterica serovar enteritidis by regulating the expression of Wzz protein. J Bacteriol 191:6694–6700

Sato F, Tsuchiya S, Meltzer S, Shimizu K (2011) MicroRNAs and epigenetics. Febs J 278:1598–1609

Scallan E, Hoekstra RM, Angulo FJ, Tauxe RV, Widdowson MA, Roy SL, Jones JL, Griffin PM (2011) Foodborne illness acquired in the United States—major pathogens. Emerg Infect Dis 17:7–15

Schett G, Elewaut D, McInnes I, Dayer J-M, Neurath M (2013) How cytokine networks fuel inflammation: toward a cytokine-based disease taxonomy. Nat Med 19:822–824

Schleithoff C, Völter-Mahlknecht S, Dahmke I, Mahlknecht U (2012) On the epigenetics of vascular regulation and disease. Clin Epigenetics 4. http://www.clinicalepigeneticsjournal.com/content/4/1/7

Schmeinck A (2011) Acquisition and loss of chromatin modifications during an Epstein-Barr Virus infection. Dissertaion, Ludwig Maximilians-University, Munich

Schweitzer A, Horn J, Mikolajczyk R, Krause G, Ott JJ (2015) Estimations of worldwide prevalence of chronic hepatitis B virus infection: a systematic review of data published between 1965 and 2013. Lancet 246:1546–1555

Shell S, Prestwich E, Baek S-H, Shah R, Sassetti C, Dedon P, Fortune S (2013) DNA methylation impacts gene expression and ensures hypoxic survival of Mycobacterium tuberculosis. PLoS Pathog 9:e1003419

Shenderov B (2012) Gut indigenous microbiota and epigenetics. Microb Ecol Health Dis 23. doi:10.3402/mehd.v23i0.17195

Shtrichman R, Heithoff D, Mahan M, Samuel C (2002) Tissue selectivity of interferon-stimulated gene expression in mice infected with Dam+ versus Dam− Salmonella enterica serovar Typhimurium strains. Infect Immun 70:5579–5588

Simon R, Heithoff D, Mahan M, Samuel C (2007) Comparison of tissue-selective proinflammatory gene induction in mice infected with wild-type, DNA adenine methylase-deficient, and flagellin-deficient Salmonella enterica. Infect immun 75:5627–5639

Skarstad K, Katayama T (2013) Regulating DNA replication in bacteria. Cold Spring Harb Perspect Biol 5:a012922

Smith Z, Meissner A (2013) DNA methylation: roles in mammalian development. Nat Rev Genet 14:204–220

Srikhanta Y, Fox K, Jennings M (2010) The phasevarion: phase variation of type III DNA methyltransferases controls coordinated switching in multiple genes. Nat Rev Microbiol 8:196–206

Steenbergen R, Snijders P, Heideman D, Meijer C (2014) Clinical implications of (epi) genetic changes in HPV-induced cervical precancerous lesions. Nat Rev Cancer 14:395–405

Stein R (2011) Epigenetics—the link between infectious diseases and cancer. JAMA 305:1484–1485

Stilling R, Dinan T, Cryan J (2014) Microbial genes, brain & behaviour–epigenetic regulation of the gut–brain axis. Genes Brain Behav 13:69–86

Sun C, Reimers L, Burk R (2011) Methylation of HPV16 genome CpG sites is associated with cervix precancer and cancer. Gynecol Oncol 121:59–63

Tauxe R (2015) Epidemiology of yersiniosis. In: Calderwood SB, section editor, Bloom A, deputy editor (eds) http://www.uptodate.com/contents/epidemiology-of-yersiniosis. Accessed 14 Aug 2015

Tavazoie S, Church G (1998) Quantitative whole-genome analysis of DNA-protein interactions by in vivo methylase protection in E. coli. Nat Biotechnol 16:566–571

Taylor V, Titball R, Oyston P (2005) Oral immunization with a dam mutant of Yersinia pseudotuberculosis protects against plague. Microbiology 151:1919–1926

Thompson P, Kurzrock R (2004) Epstein-Barr virus and cancer. Clin Can Res 10:803–821

Thorley-Lawson D (2015) EBV persistence—Introducing the virus. In: Münz C (ed) Epstein Barr Virus Volume 1. Springer, Switzerland, pp 151–209

Tian Y, Yang W, Song J, Wu Y, Ni B (2013) Hepatitis B virus X protein-induced aberrant epigenetic modifications contributing to human hepatocellular carcinoma pathogenesis. Mol Cell Biol 33:2810–2816

Tsai C-L, Li H-P, Lu Y-J, Hsueh C, Liang Y, Chen C-L, Tsao S, Tse K-P, Yu J-S, Chang Y-S (2006) Activation of DNA methyltransferase 1 by EBV LMP1 involves c-Jun NH2-terminal kinase signaling. Cancer Res 66:11668–11676

Ushijima T, Hattori N (2012) Molecular pathways: involvement of helicobacter pylori–triggered inflammation in the formation of an epigenetic field defect, and its usefulness as cancer risk and exposure markers. Clin Can Res 18:923–929

Van Vliet J, Oates N, Whitelaw E (2007) Epigenetic mechanisms in the context of complex diseases. Cell Mol Life Sci 64:1531–1538

Vasu K, Nagaraja V (2013) Diverse functions of restriction-modification systems in addition to cellular defense. Microbiol Mol Biol Rev 77:53–72

Ventham N, Kennedy N, Nimmo E, Satsangi J (2013) Beyond gene discovery in inflammatory bowel disease: the emerging role of epigenetics. Gastroenterology 145:293–308

Wang M, Church G (1992) A whole genome approach to in vivo DNA-protein interactions in E. coli. Nature 360:606–610

Wilson A (2008) Epigenetic regulation of gene expression in the inflammatory response and relevance to common diseases. J Periodontol 79:1514–1519

Wilson G, Lechner M, Köferle A, Caren H, Butcher L, Feber A, Fenton T, Jay A, Boshoff C, Beck S (2013) Integrated virus-host methylome analysis in head and neck squamous cell carcinoma. Epigenetics 8:953–961

Wion D, Casadesús J (2006) N6-methyl-adenine: an epigenetic signal for DNA–protein interactions. Nat Rev Microbiol 4:183–192

Woellmer A, Arteaga-Salas J, Hammerschmidt W (2012) BZLF1 governs CpG-methylated chromatin of Epstein-Barr virus reversing epigenetic repression. PLoS Pathog 8:e1002902

World Health Organization (2006) http://www.who.int/csr/resources/publications/Brucellosis.pdf. In: Corbel MJ, Elberg SS, Cosivi O (eds) Geneva, pp 1–102

World Health Organization (2014a) Global tuberculosis report 2014. http://apps.who.int/iris/bitstream/10665/137094/1/9789241564809_eng.pdf?ua=1. WHO Press, Geneva, pp 1–171

World Health Organization (2014b) World cancer report, 2014. In: Stewart B, Wild C (eds) WHO Report. WHO Press, Geneva
World Health Organization (2015) Cancer. Fact sheet No. 297. WHO Press, Geneva
Wroblewski L, Peek R, Wilson K (2010) Helicobacter pylori and gastric cancer: factors that modulate disease risk. Clin Microbiol Rev 23:713–739
Yang W, Aziz P, Heithoff D, Mahan M, Smith J, Marth J (2015) An intrinsic mechanism of secreted protein aging and turnover. Proc Natl Acad Sci U S A 112:13657–13662
Yao S, He Z, Chen C (2015) CRISPR/Cas9-mediated genome editing of epigenetic factors for cancer therapy. Hum Gene Ther 26:463–471
Yoshida T, Kato J, Maekita T, Yamashita S, Enomoto S, Ando T, Niwa T, Deguchi H, Ueda K, Inoue I (2013) Altered mucosal DNA methylation in parallel with highly active Helicobacter pylori-related gastritis. Gastric Cancer 16:488–497
Youngblood B, Davis C, Ahmed R (2010) Making memories that last a lifetime: heritable functions of self-renewing memory CD8 T cells. Int Immunol. doi:10.1093/intimm/dxq437
zur Hausen H (2001) Oncogenic DNA viruses. Oncogene 20:7820–7823
zur Hausen H (2009) The search for infectious causes of human cancers: where and why. Virology 392:1–10

Chapter 6
Cross Talk Between Bacteria and the Host Epigenetic Machinery

Hélène Bierne

Abstract Multidisciplinary approaches combining microbiology, cell biology, and genetics have improved our understanding of bacterial diseases by elucidating mechanisms employed by bacteria to manipulate eukaryotic cellular processes. In parallel, research on epigenetics has increased our knowledge about eukaryotic gene expression by providing a mechanistic basis for the amazing plasticity of the genome in response to developmental and environmental cues. These two fields of research have now converged, providing information about the ways in which bacteria shape the epigenome and the mechanisms by which the epigenetic machinery allows the host to respond to colonization by pathogenic or commensal bacteria. The study of this cross talk has revealed remarkable diversity in the mechanisms of action of bacteria on chromatin and has identified epigenetic regulators involved in host responsiveness to bacteria. One powerful strategy used by intracellular pathogens (e.g., *Anaplasma, Chlamydia, Ehrlichia, Legionella, Listeria, Mycobacteria, Mycoplasma, Shigella*) is the secretion of nucleomodulins that manipulate chromatin structure in the host nucleus. The effects of this dialog are often limited in time, causing transient gene expression changes. However, increasing evidence suggests that certain epigenetic changes triggered by bacterial molecules are long-lasting, leading to the priming of transcriptional responses and the reprogramming of genes involved in inflammation or tolerance, with consequences for reinfection and polymicrobial infections. In addition, the effects of bacteria on the host epigenome may ultimately modify the identity of the cell by breaking epigenetic barriers, leading to cell differentiation, dedifferentiation, or trans-differentiation, thereby potentially contributing to tissue remodeling and emergence of complex diseases.

Keywords Infectious diseases • Chromatin • Microbiota • Bacterial pathogenesis • Cellular reprogramming • Host–Pathogen interaction • Nucleomodulin • Patho-epigenetics

H. Bierne (✉)
Micalis Institute, INRA, AgroParisTech, Université Paris-Saclay, Equipe Epigénétique et Microbiologie Cellulaire, Jouy-en-Josas 78350, France
e-mail: helene.bierne@inra.fr

W. Doerfler, J. Casadesús (eds.), *Epigenetics of Infectious Diseases*, Epigenetics and Human Health, DOI 10.1007/978-3-319-55021-3_6

6.1 Introduction

Pathogenic bacteria have evolved a wide range of mechanisms for manipulating eukaryotic cell functions to their advantage (Diacovich and Gorvel 2010; Jimenez et al. 2016). In particular, by modifying the host cell transcriptional program, they disturb diverse cellular processes and take control of host defense systems. The commensal bacteria of the microbiota also affect host transcriptional gene networks by producing metabolites that influence the differentiation, proliferation, migration, and metabolic functions of mucosal cells (Brestoff and Artis 2013). Conversely, in conditions of microbial attack or colonization, host cells trigger various responses enabling them to tolerate or eliminate the invaders by mobilizing genes involved in key processes (e.g., immunity, cell death/survival, adhesion/motility, metabolism) (Jenner and Young 2005). Studies of the molecular basis of this cross talk are crucial for an understanding of infectious diseases and mucosal homeostasis. Research has long focused on the manipulation of transcription factors (e.g., NF-κB, FOS/JUN, IRFs, STATs, HIFs, SMADs) (Jenner and Young 2005; Bhavsar et al. 2007), through the bacteria-mediated deregulation of signaling pathways, or through posttranslational modifications (PTMs), activating, shutting down, or delocalizing these transcription factors. Another powerful means by which bacteria alter the expression of host genes has recently emerged from studies in different bacterial models: specific modifications of chromatin in the cell nucleus. This chapter will update a previous contribution dealing with the relationship between bacteria and chromatin regulation (Bierne et al. 2012) and will present new evidence for the epigenetic inheritance of bacterial imprints in cells and tissues. We will first recall the general principles of epigenetic regulation, and several examples will then be used to illustrate the diversity of mechanisms employed by bacteria and animal or human cells to mobilize the epigenetic machinery and modify the expression of susceptibility or resistance genes in the short or long term. The epigenetic control of adaptive immunity (Alvarez-Errico et al. 2015) and the effects of plant-associated bacteria on chromatin (Ma et al. 2011; Canonne and Rivas 2012; Holeski et al. 2012) will not be addressed here.

6.2 The Machinery of Chromatin Regulation

6.2.1 Chromatin Marks

In eukaryotic cells, DNA is wrapped around histone proteins to form nucleosomes, which are themselves packed with non-histone chromosomal proteins in the chromatin fiber. Chromatin condensation organizes and confines the genome into the tight space of the nucleus. More locally, the state of chromatin compaction plays a major role in nuclear processes by controlling the accessibility of DNA to the transcription, replication, and repair machineries. The regulation of chromatin structure is a dynamic process that involves DNA methylation (mostly on cytosines) (Klose and Bird 2006; Chen and Riggs 2011), histone PTMs (e.g., phosphorylation, methylation, acetylation, ubiquitination, SUMOylation, citrullination, ADP-ribosylation) (Kouzarides 2007; Sadakierska-Chudy and Filip 2015), and the sliding of nucleosomes along the DNA. Mechanisms for modifying and remodeling chromatin function together, controlling the formation of higher order chromatin structures that are either loosely packed and transcriptionally active (i.e., "euchromatin") or highly condensed and transcriptionally silent (i.e., "heterochromatin"). Different combinations of histone PTMs and DNA methylation patterns form a code that controls transcription by affecting either chromatin structure itself or the recruitment of DNA-binding transcription activators or repressors. Several chromatin marks are known to be associated essentially with transcriptional activation (e.g., H3K4me, H3S10p, and H3K14ac; all abbreviations are listed in Table 6.1), whereas others are associated with repression (e.g., H3K9me, H3K27me, and deacetylated histones). However, it is often difficult to interpret a specific chromatin signature for the prediction of gene expression outcomes. Some genes may carry both repressive and activating histone marks, and RNA polymerase II may constitutively bind their proximal promoters, preparing the gene for efficient future transcription while remaining silent (Mikkelsen et al. 2007). Such active chromatin states at sites of repressed transcription may "poise" transcripts for rapid activation in cells in which a rapid change in expression levels is required, during immune and metabolic responses, for example (Cuddapah et al. 2010; Rye et al. 2014).

Chromatin modifications also contribute to the alternative splicing of pre-mRNA, making it possible for a single eukaryotic gene to encode several proteins with different functions (Allemand et al. 2008; Hnilicova and Stanek 2011). An additional level of complexity is added by noncoding RNAs and RNA-binding proteins (Turner and Morris 2010; Kaikkonen et al. 2011; Sadakierska-Chudy and Filip 2015). Most of the genomic DNA of eukaryotes is transcribed, but only 1–2% of transcripts encode proteins. The vast majority of RNAs are thus noncoding RNAs (ncRNA) of various sizes, generated from exons, introns, enhancers, or intergenic regions, in sense or antisense orientation (Mattick and Makunin 2005; Tisseur et al. 2011). Antisense RNAs and small ncRNAs play diverse roles in chromatin regulation by recruiting chromatin-modifying enzymes (Faghihi and Wahlestedt 2009; Kaikkonen et al. 2011; Cao 2014) and/or acting as scaffolds localizing genes to specific subnuclear regions (Yang et al. 2011; Schmitz et al. 2010). Last but not least, RNA can itself be modified, particularly by adenosine methylation (Dominissini et al. 2012), which affects messenger RNA localization, stability, splicing, and translation (Meyer and Jaffrey 2014; Dominissini et al. 2016). The role of the RNA world in the relationship between microbes and epigenetic regulation is an emerging field of research worthy of consideration in its own right. This chapter focuses exclusively on the role of histone and DNA modifications.

Table 6.1 List of abbreviations

Name	Full name	Function
5mC	5-methylcytosine	Epigenetic mark
BAHD1	Bromo Adjacent Homology Domain containing 1	Scaffold and reader subunit of the BAHD1 complex
CHD	Chromodomain-helicase-DNA-binding protein	Nucleosome remodeler
CIITA	Class II, major histocompatibility complex, transactivator	Transcription factor
DNMT	DNA methyltransferase	DNA modifier (writer)
G9a/ EHMT2	Euchromatic histone-lysine N-methyltransferase 2	Histone modifier (writer)
GATAD 2A/2B	GATA Zinc Finger Domain Containing 2A/2B	Subunits of the NurD complex
H2A	Histone H2A	Core histone
H2AX	H2A histone family, member X	Histone variant
H3T3p	Histone H3 phosphorylated at threonine 3	Epigenetic mark
H3K4me	Histone H3 methylated at lysine 4	Epigenetic mark
H3K8ac	Histone H3 acetylated at lysine 8	Epigenetic mark
H3K9ac	Histone H3 acetylated at lysine 9	Epigenetic mark
H3K9me	Histone H3 methylated at lysine 9	Epigenetic mark
H3K9ac	Histone H3 acetylated at lysine 9	Epigenetic mark
H3S10p	Histone H3 phosphorylated at serine 10	Epigenetic mark
H3K14ac	Histone H3 acetylated at lysine 14	Epigenetic mark
H3K14me	Histone H3 methylated at lysine 14	Epigenetic mark
H3K18ac	Histone H3 acteylated at lysine 18	Epigenetic mark
H3K23ac	Histone H3 acetylated at lysine 23	Epigenetic mark
H3K27me	Histone H3 methylated at lysine 27	Epigenetic mark
HDM	Histone demethylase	Histone modifier (eraser)
HDAC	Histone deacetylase	Histone modifier (eraser)
HMT	Histone methyltransferase	Histone modifier (writer)
HP1	Heterochromatin protein 1	Chromatin reader
INO80	INOsitol requiring protein 80	Nucleosome remodeler
IKKα	IκB kinase alpha	Kinase
ISG	Interferon-stimulated gene	Immunity gene
ISWI	Imitation Switch (ISWI)	Nucleosome remodeler
JNK	c-Jun N-terminal kinase	Kinase
LPS	Lipopolysaccharide	Major component of the outer membrane of Gram-negative bacteria
LSD1	Lysine (K)-specific demethylase 1A	Histone modifier (eraser)
MAMPS	Microbe-associated molecular patterns	Bacterial molecules
MAPK	Mitogen-activated protein (MAP) kinase	Kinase
MBD1/2	Methyl-CpG-binding protein 1/2	5mC reader
MIER1/2/ 3	Mesoderm induction early response protein 1/2/3	Subunits of the BAHD1 complex

(continued)

Table 6.1 (continued)

Name	Full name	Function
MSK1/2	Mitogen- and Stress-activated Kinase 1/2	Kinase-Histone writer
MTA1/2/3	Metastasis-associated gene 1/2/3	Scaffold subunit of the NurD complex
MRN	Mre11-Rad50-Nbs1 (MRN) complex	DNA repair complex
NF-κB	nuclear factor kappa-light-chain-enhancer of activated B cells	Transcription factor
NOD1/2	Nucleotide-binding oligomerization domain-containing protein 1/2	Pattern-recognition receptor
NuRD	Nucleosome remodeling and histone deacetylase	Chromatin-remodeling complex
p300/CBP	E1A-binding protein p300/CREB-binding protein	Histone modifier (writer) and reader of histone modifications
PRC1/2	Polycomb repressive complex 1/2	Chromatin-repressive complex
RBBP4/7	Retinoblastoma Binding Protein 4/7	Subunit of the NurD complex
SATB1	Special AT-rich sequence-binding protein-1	Nuclear matrix attachment protein
SETDB1	SET domain, bifurcated 1	Histone modifier (writer)
SIN3A	SIN3 transcription regulator homolog A; Histone deacetylase complex subunit Sin3a	Scaffold subunit of the SIN3 complex
SIRT2	Sirtuin 2	Histone modifier (eraser)
SWI/SNF	SWItch/Sucrose Nonfermentable	Nucleosome remodeler
TET	Ten-eleven translocation	Enzyme converting 5-mC to 5hmC
TLR	Toll-Like-Receptor	Pattern-recognition receptor
TNF-R	Tumor necrosis factor-receptor	Cytokine receptor
VRK1	Vaccinia-Related kinase 1	Kinase

6.2.2 Chromatin Regulators

The molecular machinery controlling chromatin structure includes about 800 proteins with diverse functions (referenced in the EpiFactors database (Medvedeva et al. 2015); abbreviations for those mentioned here are listed in Table 6.1). ATP-dependent remodeling enzymes from the SWI/SNF, ISWI, CHD, and INO80/SWR families position the nucleosomes by catalyzing the movement of histone octamers relative to DNA, using energy from ATP hydrolysis to move, destabilize, evict, or reassemble nucleosomes (Langst and Manelyte 2015). Covalent modifications of chromatin are added or removed by a wide range of enzymes known as "writers", such as histone kinases, acetyltransferases (HATs), and methyltransferases (HMTs), and "erasers", such as histone phosphatases, deacetylases (HDACs), and demethylases (HDMs) (Fig. 6.1) (Zhou et al. 2011). In DNA methylation, the writers of 5-methylcytosine (5mC) are DNA methyltransferases (DNMTs), which either establish methylation (i.e., the "de novo" methyltransferases DNMT3a and DNMT3b) or copy methylation patterns

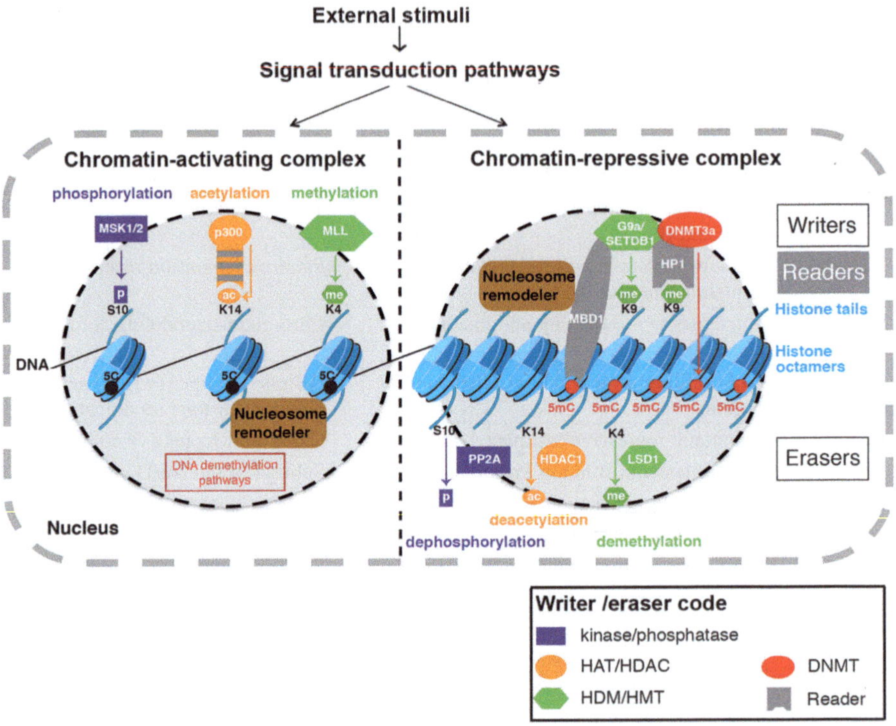

Fig. 6.1 The epigenetic machinery. Histone posttranslational modifications, histone sliding, and 5-cytosine methylation (5mC) control chromatin structure and gene expression. Chromatin is modified and remodeled by enzymes known as "writers," "erasers," and "remodelers." Examples of chromatin enzymes and marks are shown. Activating marks include serine phosphorylation on lysine 10 (S10p) and acetylation on lysine 14 (K14ac) of histone H3 and methylation on lysine 4 (K4me) of histone H4. Repressive marks include dephosphorylation, deacetylation, and demethylation of the same residues, as well as methylation of lysine 9 (K9me) of H3. Cytosine methylation is a repressive mark at promoter sequences and an activating mark at gene bodies. Its erasure involves a complex pathway with chemical modifications of 5mC, followed by passive demethylation or DNA repair. Epigenetic marks are recognized and interpreted by protein modules known as "readers." For instance, the bromodomain of the HAT p300 binds H3K14ac, the chromodomain of HP1 binds H3K9me, and the MBD domain of MBD1 binds 5mC. Writers, readers, erasers, and nucleosome remodelers act within large macromolecular complexes that open or close chromatin, leading to gene activation or repression. Cell signaling pathways triggered by external stimuli control interaction or stability of chromatin-activating or -repressive complex subunits and their combinatorial interaction with transcription factors (not shown). HAT, histone acetyltransferase; HMT, histone methyltransferase; HDAC, histone deacetylase; HDM, histone demethyltransferase; DNMT, DNA methyltransferases. All other abbreviations are listed in Table 6.1. Adapted from (Bierne et al. 2012)

onto the newly synthesized DNA strand during replication (i.e., the "maintenance" methyltransferase DNMT1). DNA methylation can be reversed passively, through a lack of 5mC copying during DNA replication, or actively through an active erasure process, involving intermediate chemical modifications of 5mC, followed by passive demethylation or DNA repair. Several groups of proteins, such as the ten-eleven translocation (TET) proteins and DNA glycosylases, have been shown to be involved in this complex process (Bhutani et al. 2011; Chen and Riggs 2011; Wu and Zhang 2011). 5mC and histone PTMs serve as signaling platforms for proteins known as "readers", which interact, stabilize, or modify other chromatin components (Fig. 6.1). Histone readers are docked onto specific PTMs via chromatin-binding modules, such as bromodomains (BRD), bromo-adjacent homology domains (BAH), chromodomains, 14-3-3, Tudor, or PHD domains (Taverna et al. 2007). Methyl-cytosine readers include methyl-DNA-binding domains (MBD), SET and RING-associated domains (SRA), and some specific zinc finger motifs (Sasai and Defossez 2009; Liu et al. 2013; Buck-Koehntop and Defossez 2013).

Writers, readers, and erasers are often modular proteins with several properties. The enzyme p300/CBP illustrates this well: it acts as a writer (via its HAT module), a reader (via its bromodomain), and an adaptor (via other modules). In addition, writers, readers, erasers, and remodelers often function as subunits of large macromolecular complexes assembled with scaffold proteins. NurD is a paradigm of such chromatin-remodeling complexes (Fig. 6.2). It contains MTA scaffolding proteins (MTA1, MTA2, MTA3) that bridge subunits involved in nucleosome remodeling (CHD3, CHD4), histone deacetylation (HDAC1, HDAC2), and demethylation (LSD1), binding to other subunits and histones (RBBP4, RBBP7, GATAD2A, GATAD2B), and the targeting of methylated DNA (MBD2) and transcription factors (MBD3) (Lai and Wade 2011). The combinatorial assembly of these subunits determines the function of NuRD in genomic targeting and in the mediation of cell type-specific transcriptional regulations, such as the repression of tumor suppressor genes.

6.2.3 Signaling to Chromatin

The modular, multifunctional, and combinatorial nature of this regulation ensures extremely precise temporal and spatial control over chromatin structure (Ram et al. 2011). The vast array and different combinations of histone PTMs coordinate the sequential recruitment of complexes in a regulatory process that reinforces or reverses existing histone PTMs (Latham and Dent 2007; Lee et al. 2010; Suganuma and Workman 2011). For instance, the Polycomb repressive complexes PRC2 and PRC1 are sequentially recruited, first to "write" H3K27me3 and then to "read" this mark to induce the mono-ubiquitylation of histone H2A, ultimately leading to chromatin compaction at target genes (Margueron and Reinberg 2011). Such cross talk also takes place between histone PTMs and DNA methylation (Cedar and Bergman 2009; Du et al. 2015), via the interaction of histone and DNA modifiers and readers (Li et al. 2006; Vire et al. 2006; Fujita et al. 2003; Ichimura et al. 2005). Cooperation between chromatin-remodeling complexes and DNMTs is particularly important in the maintenance of a specific chromatin state (Cai et al.

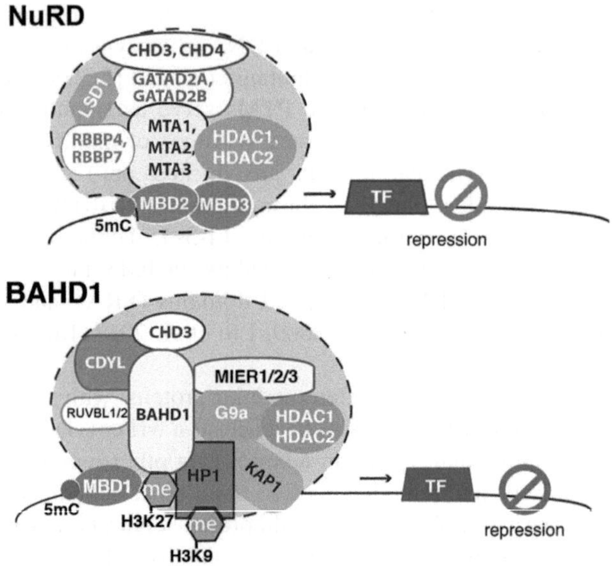

Fig. 6.2 The NuRD and BAHD1 chromatin-repressive complexes. NuRD and BAHD1 complexes are examples of chromatin-associated macromolecular complexes involved in gene repression. Both contain scaffold proteins (MTA1/2/3 and BAHD1-MIER1/2/3, respectively) that bridge subunits involved in histone deacetylation (HDAC1, HDAC2), nucleosome remodeling (CHD3, CHD4), and binding to methylcytosine (5mC) (MBD2 and MBD1, respectively). The BAHD1 complex also contains a HMT subunit (e.g., G9a) that "writes" the H3K9me mark to which the heterochromatin protein HP1 binds. In addition, the BAH domain of BAHD1 is a reader of the H3K27me mark. The function and targeting of these complexes to specific loci depend on the combinatorial assembly of the different subunits with transcription factors (TF) in response to external signals. Upon *Listeria* infection, the BAHD1 complex assembles at promoters of a set of interferon-stimulated genes (as shown in Fig. 6.3). BAHD1 also controls expression of metabolic genes. Adapted from (Lakisic et al. 2016)

2014). The spatial information required to guide chromatin regulators towards specific sites within the genome is provided by combinatorial interactions with DNA-bound transcriptional factors and/or ncRNAs.

Diverse PTMs induced by cell signaling pathways alter the interaction or stability of subunits of chromatin-associated complexes and transcription factors. Signal transduction information is thus translated into chromatin structure in response to various external signals (Mohammad and Baylin 2010; Arzate-Mejia et al. 2011). In particular, phosphorylation and sumoylation are important modifications in the function of chromatin-modifying complexes (Garcia-Dominguez and Reyes 2009; Baek 2011). For instance, phosphorylation of the transcription factor c-JUN by JNK kinase impairs the binding of c-JUN to the MBD3 subunit of the NuRD complex, thereby relieving repression of target genes (Aguilera et al. 2011). Several kinases of signal transduction pathways, such as JNK, MSK1/2, and IKKα, can directly phosphorylate histones (Baek 2011; Tiwari et al. 2011) or histone readers, such as HP1 (Hiragami-Hamada et al. 2011). The signaling molecules activated in cells in response to a wide range of stimuli thus have major effects on the language and syntax of chromatin, through their control of transcription factors and large chromatin-associated co-activator or co-repressor complexes (Fig. 6.1).

6.2.4 Epigenetic Inheritance

Some chromatin modifications remain stable in interphase cells and can be transmitted to daughter cells through mitosis, resulting in their persistence after the disappearance of the initiating signal. This transmission process, resulting in heritable changes in gene expression without altering the sequence of nucleotides in the DNA, defines epigenetic regulation (Riggs et al. 1996).

Epigenetic marks play a key role in cell differentiation, by enabling a cell to "remember" its transcriptional profile. Specific epigenetic signatures fix the identity of the cell while allowing it to respond to external signals. This plasticity explains how the DNA sequence of single cell, the zygote, can generate a huge number of different cell types (about 200 in the human body), most of which being highly differentiated and specialized, whereas others remaining undifferentiated and pluripotent for cell renewal. However, many of the epigenetic marks induced by cell signaling, DNA repair, or cell cycle transitions are short-lived and do not result in long-term memory. This has led to controversy concerning the use of the terms "epigenetic" and "epigenetic marks" to describe chromatin-associated processes and modifications that are not heritable. Adrian Bird has proposed a definition of epigenetic events accounting for both transient and stable modifications of epigenetic language: "a structural adaptation of chromosomal regions so as to register, signal or perpetuate altered activity states" (Bird 2007).

Intense efforts are currently focused on unraveling the mechanisms by which transient epigenetic changes are converted into epigenetic inheritance, particularly given the great importance of these processes in regenerative medicine and research on complex diseases, such as cancer and metabolic and autoimmune diseases. Novel technologies, such as genome-wide epigenomics, chromosome conformation capture (3C), and super-resolution microscopy, have highlighted the complex three-dimensional organization of the genome with large chromatin domains (Guelen et al. 2008; Padeken and Heun 2014; Mattout et al. 2015), the formation of chromosomal loops (Kohwi-Shigematsu et al. 2012; Noordermeer and Duboule 2013), and a nonrandom subnuclear localization of chromatin-associated complexes (Wani et al. 2016). Diverse elements, including enhancers and insulators, regulate topological domains, their boundary regions, and gene looping (Noordermeer and Duboule 2013). The formation of boundaries blocking the spread of heterochromatin is particularly critical for the maintenance of stable gene expression patterns. Recent studies compiling data from a hundred of human epigenomes have highlighted the importance of examining chromatin at the megabase scale and of defining epigenetic profiles for regulatory elements located at some distance from promoters (Roadmap Epigenomics et al. 2015; Romanoski et al. 2015). Such chromatin signatures constitute epigenetic barriers to transcription factor-mediated reprogramming processes. Here, it is worth recalling a groundbreaking discovery in 2006: exogenous expression of a cocktail of transcription factors (Oct4, Sox2, Klf4, and Myc) is sufficient to turn any cell of the body into a pluripotent stem cell (iPS) (Takahashi and Yamanaka 2006). However, this nuclear reprogramming is an inefficient process. Recently, it was reported that depletion of the MBD3 subunit of the NuRD complex greatly improves the efficiency of reprogramming (Rais et al. 2013). This highlights the need to reset the epigenetic landscape of differentiated cells, so they can go back to pluripotency.

In conclusion, the epigenome can change rapidly in response to developmental, physiological, or environmental stimuli, but its stability is also important for the maintenance of cell identity. The mechanisms underlying this plasticity are highly sophisticated. Many studies have shown that bacterial products affect these mechanisms, through the activation of signaling cascades or the direct targeting of chromatin and chromatin regulators in the nucleus, as reviewed below through several examples (Table 6.2). Assessing the magnitude of these effects is an emerging fundamental question, which will also be illustrated here.

6.3 Bacterial Effects on the Host Epigenome

6.3.1 Lessons from Listeria *and* Anaplasma

Listeria monocytogenes is a food contaminant causing listeriosis, a serious disease for immunocompromised individuals, fetuses, and newborns. This facultative intracellular bacterium is a powerful model to study various aspects of the molecular interactions between pathogen and mammalian cells (Hamon et al. 2006), especially as it invades many different cell types and reaches various organs, such as the liver, spleen, placenta, and brain. In addition, it triggers a wide range of innate immune responses and a potent protective T-cell response (Pamer 2004). *Anaplasma phagocytophilum,* the causative agent of human granulocytic anaplasmosis, is another interesting bacterial model to address fundamental questions in cellular microbiology, as it displays a remarkable tropism for neutrophils. This obligate intracellular pathogen survives in the hostile environment of the neutrophil by abrogating key antimicrobial functions. This property is partly attributed to *A. phagocytophilum's* ability to shape the transcriptional program of the host cell to its advantage (Borjesson et al. 2005; Lee et al. 2008; Sinclair et al. 2014). The studies on *Listeria* and *Anaplasma* have proven to be particularly suitable to identify mechanisms involved in chromatin modifications induced by microbial pathogens (Fig. 6.3).

Table 6.2 Examples of bacteria-mediated effects on chromatin regulation

Bacterial species	Bacterial factor	Effect
Anaplasma phagocytophilum	AnkA	Nucleomodulin. Binding to chromatin at AT-rich DNA sequences; silencing of the *CYBB* gene
	?	Increased expression of *HDAC1* and *DNMT3a* genes
	?	Genome-wide DNA methylation changes
Bacillus anthracis	LT	Toxin. Inhibition of MAPK and of H3S10p and downregulation of *IL-8* and *KC* genes
	BaSET	Nucleomodulin. Putative histone methyltransferase
Campylobacter rectus	?	Hypermethylation in the promoter region P0 of the *IGF2* gene in the murine placenta
Chlamydia trachomatis	NUE	Nucleomodulin. Histone methyltransferase
Chlamydia psittaci	SinC	Nucleomodulin. Binding to the nuclear inner membrane
Coxiella burnetii	Cbu1314	Nucleomodulin. Binding to chromatin
Ehrlichia chaffeensis	Ank 200 Trp120	Nucleomodulin. Binding to chromatin at Alu-Sx elements Nucleomodulin. Binding to G+C-rich motifs
Escherichia coli	NleC	Nucleomodulin. Protease that degrades the HAT p300/CBP (in EPEC and EHEC)
	?	DNA methylation and downregulation of *CDKN2A* (in UPEC)
	?	Change expression of the HMT EZH2 (PRC2 complex) (in UPEC)
Helicobacter pylori	?	Induction of H3 modifications
	?	Induction of DNA methylation in gastric mucosa
Legionella pneumophila	Flagellin	Histone acetylation in infected lung epithelial cells
	RomA/LegAS4	Nucleomodulin. Histone methyltransferase at H3K14 (in chromosomes) or H3K4 (in the nucleolus).
Listeria monocytogenes	MAMPS	PRR-induced signaling pathways leading to H4 acetylation and H3 phosphorylation/acetylation
	LLO	Toxin. Induction of a signaling pathway via K^+ efflux at the plasma membrane, leading to histone dephosphorylation and deacetylation
	InlB	Invasin. Activation of the Met-PI3K-Akt signaling pathway, leading to nuclear translocation of the sirtuin SIRT2 and histone H3K18 deacetylation
	?	Repression of ISGs by the BAHD1 chromatin-repressive complex (epithelial cells)
	LntA	Nucleomodulin. Inhibitor of BAHD1 and activator of ISGs (epithelial cells)
Moraxella catarrhalis	?	Induction of histone H3 and H4 modifications at the *IL8* promoter
	?	Reduction in expression and activity of HDAC1/2 in airway epithelial cells.

(continued)

Table 6.2 (continued)

Bacterial species	Bacterial factor	Effect
Mycobacterium tuberculosis	Rv1988	Nucleomodulin. Histone methyltransferase at H3R42
	Rv3423	Nucleomodulin. Histone acetyltransferase at H3K9/K14
	Rv2966c	Nucleomodulin. DNA methyltransferase that methylates cytosines in a non-CpG context
	?	Control of the chromatin-repressive complex SIN3a at ISGs
	?	Genome-wide DNA methylation changes
	?	Epigenetic reprogramming of monocytes
Mycoplasma hyorhinis	Mhy1	Nucleomodulin. DNA methyltransferase that methylates cytosines in a CG context
	Mhy2	Nucleomodulin. DNA methyltransferase that methylates cytosines in a CG context
	Mhy3	Nucleomodulin. DNA methyltransferase that methylates cytosines in a GATC context
Porphyromonas gingivalis	?	Reactivation of latent viruses via chromatin modification induced by butyrate
Shigella flexneri	OspF	Nucleomodulin. Downregulation of MAP-kinase in the nucleus by eliminylation. Inhibition of phosphorylation of H3S10 and HP1-γ. Downregulation of immune response genes.
	OspB	Nucleomodulin. Binding to Rb
	IpaH	Nucleomodulin. Ubiquitin ligase targeting a splicing factor
Bacterial product	Butyrate	Inhibition of HDAC activity
	Flagellin	Activation of NF-κB pathways leading to histone phosphorylation and acetylation
	LPS	Inducer of innate immunity via activation of TLR4-mediated responses and the production of pro-inflammatory cytokines; induces immunosuppression by chromatin modifications upon repeated challenge.
	2-aminoacetophenone	A bacterial quorum-sensing molecule that dampens host immune responses, by increasing HDAC1 expression and activity

Bacterial species are listed by alphabetical orders. References are in the text

6.3.1.1 Histone Modifications as a Host Response to Bacterial Molecular Patterns

A first effect of *L. monocytogenes* infection on chromatin has been described in human umbilical vein endothelial cells (HUVECs) cells, in which sensing of cytosolic bacterial molecules by pattern-recognition receptors, such as NOD1, activates MAP-kinases (MAPK). The downstream signaling pathway activates

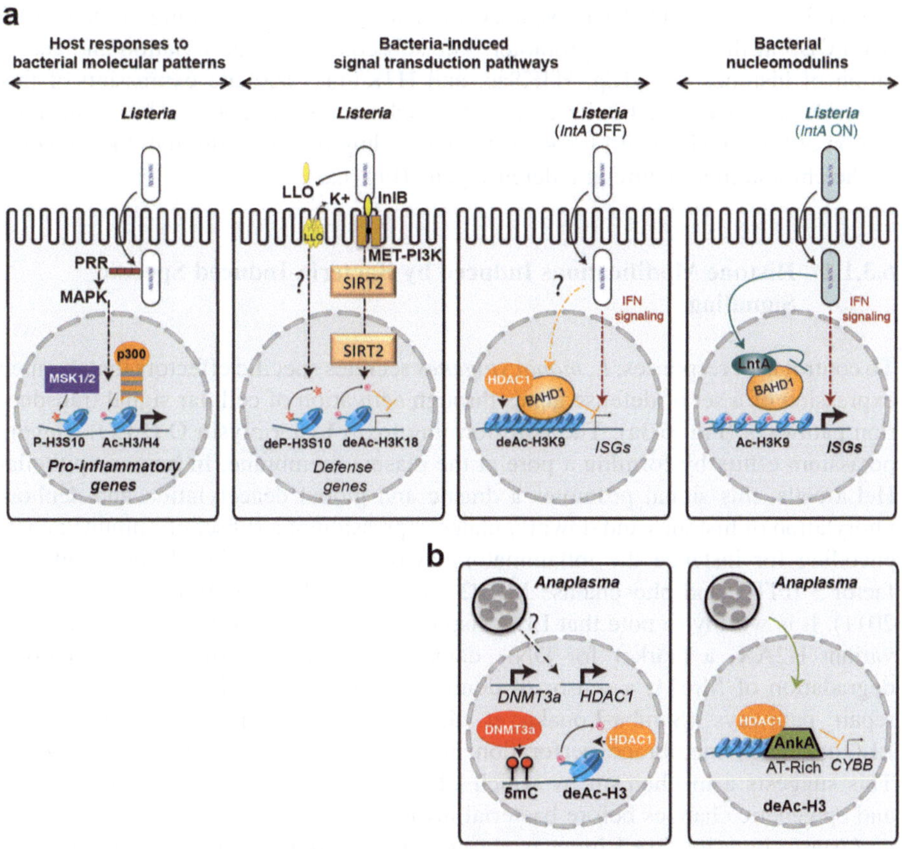

Fig. 6.3 Impacts of *Listeria monocytogenes* and *Anaplasma phagocytophilum* on chromatin regulation. (**a**) Detection of intracellular *L. monocytogenes* by pattern-recognition receptors (PRR) activates MAPK signaling pathways, leading to histone phosphorylation by MSK1/2 and histone acetylation by p300/CBP and transcriptional activation of pro-inflammatory genes. To control host genes, *L. monocytogenes* secretes effectors that activate signaling cascades or directly act on the chromatin-regulatory machinery. Toxin LLO induces dephosphorylation and deacetylation of histones via a signaling pathway involving potassium efflux at the plasma membrane. Invasin InlB activates the Met-PI3K pathway, leading to translocation of histone deacetylase SIRT2 into the nucleus and SIRT2-mediated H3K18 deacetylation and repression at a set of defense genes. In epithelial cells, *L. monocytogenes* infection induces interferon signaling pathways and recruitment of the BAHD1 repressive complex at interferon-stimulated genes by an unknown signal. When bacteria express the *lntA* gene, the nucleomodulin LntA enters the nucleus where it binds BAHD1, inhibits the BAHD1-HDAC1 silencing complex, restores H3K9 acetylation, and enhances the expression of ISGs. (**b**) *A. phagocytophilum* infection activates expression of *HDAC1* and *DNMT3a* genes and induces genome-wide DNA hypermethylation. *Anaplasma* secretes the nucleomodulin AnkA that binds host DNA at AT-rich motifs overlapping with nuclear matrix attachment regions. One AnkA targeted locus is the *CYBB* promoter, where AnkA recruits HDAC1, leading to *CYBB* silencing. Adapted from (Lebreton et al. 2012) and (Rennoll-Bankert and Dumler 2012)

MSK1/2-mediated H3S10 phosphorylation and increased binding of the HAT p300/CBP at the *IL-8* gene promoter. The subsequent phosphorylation and acetylation of histones (H3S10p, H4K8ac, and H3K14ac) activate expression of this pro-inflammatory gene (Opitz et al. 2006; Schmeck et al. 2008). This is an illustration of how the host cell responds to an invading pathogen through local change of the chromatin structure at a defense gene (Fig. 6.3).

6.3.1.2 Histone Modifications Induced by Bacteria-Induced Specific Signaling

To control host responses, *L. monocytogenes* secretes specific effectors that dampen expression of a set of defense genes through activation of cellular signal transduction pathways (Fig. 6.3a). The pore-forming toxin Listeriolysin O (LLO) triggers potassium efflux by forming a pore at the plasma membrane. In human epithelial HeLa cells, this signal promotes a drastic and global deacetylation and dephosphorylation of histones and downregulates expression of a subset of immune genes, encoding for instance the inflammatory cytokine CXCL2, interferon regulatory factor 3 IFIT3, and phosphatase MKP2 (Hamon et al. 2007; Hamon and Cossart 2011). It is worthy to note that LLO also increases phosphorylation of the histone variant H2AX, a marker for DNA damage. The mechanism at play involves degradation of Mre11, a sensor of double-strand DNA breaks involved in DNA repair pathways (Samba-Louaka et al. 2014). Thus, signaling responses to LLO-mediated membrane perforation impact both the genome and epigenome. This suggests a mechanism by which LLO may prime the host cell for genetic and epigenetic changes before bacterial invasion.

Listeria invasion itself impacts chromatin regulation through the action of the internalization protein InlB, a ligand of the tyrosine kinase receptor c-Met. InlB–cMet interaction activates the PI3K-Akt pathway, leading to relocalization of the histone deacetylase SIRT2 from the cytoplasm to the nucleus. SIRT2 represses expression of a set of genes during *Listeria* infection by catalyzing H3K18 deacetylation at their transcription start sites (Fig. 6.3a). A significant number of these genes are implicated in transcription regulation (i.e., *SMAD1, FOXM1, IRF2*) and cell signaling (*RASGRP1, MAPK14, PIK3R3, PTPNG, SOS1, VAV3, ABL1, CAMK26, MAP2K6, LEF1*). The inactivation of the *Sirt2* gene in a mouse model of listeriosis has demonstrated the importance of the SIRT2 regulation in *Listeria* infection (see Sect. 4.1). The link between Akt signaling and SIRT2 is intriguing and remains to be characterized.

6.3.1.3 Direct Control of the Chromatin-Regulatory Machinery by Bacterial Nucleomodulins: The LntA and AnkA Paradigms

Searching for *L. monocytogenes* effectors targeting intracellular organelles was an opportunity to discover a more active manner used by *L. monocytogenes* to subvert chromatin regulation processes. This approach led to the identification of *Listeria*-nuclear-targeted protein A (LntA), a small basic protein that translocates into the nucleus when expressed by intracellular *Listeria*. LntA interacts with a chromatin repressor, BAHD1 (Bierne et al. 2009; Lebreton et al. 2011), which is a core component of a chromatin-repressive complex that stimulates histone modifications, DNA methylation, and chromatin compaction (Libertini et al. 2015; Lakisic et al. 2016). The BAHD1-associated complex displays analogy with NuRD: it contains BAHD1 and MIER proteins that share structural features with the scaffold proteins MTAs of NuRD and bridge together chromatin writers, erasers, and readers mostly involved in gene repression (Fig. 6.2). As for NurD, the set of genes repressed by the BAHD1 complex depends on the cell type, as well as on the signal to which cells are submitted. Upon infection of epithelial cells with *L. monocytogenes*, BAHD1 represses Interferon-Stimulated Genes (ISGs) (Lebreton et al. 2011), which are important players in the innate immune response (Dussurget et al. 2014). When *L. monocytogenes* expresses *lntA*, the secreted factor LntA enters the nucleus and alleviates BAHD1 and HDAC1/2 binding to ISG promoters, leading to histone deacetylation and upregulation of ISG expression (Fig. 6.3a). LntA interacts directly with a central proline-rich region of BAHD1, via a surface patch containing a dilysine motif (K180/K181), located nearby a groove on the elbow region of LntA identified by crystallography (Lebreton et al. 2014). Mutation of this strategic dilysine abolishes LntA binding to BAHD1 and LntA-mediated stimulation of interferon responses upon infection. Inactivation or overexpression of *lntA* in bacteria, as well as knockdown of *Bahd1* in the mouse (see Sect. 4.1), alters the infectious process in vivo (Lebreton et al. 2011). However, the signaling pathways that govern BAHD1 and LntA synthesis and the loading of these factors onto chromatin nearby ISGs are unknown. Thus, several questions remain to be addressed to understand how the LntA-BAHD1 interplay modulates the interferon (IFN) response in time and space during bacterial colonization of the host.

The study of LntA enabled to define the family of nucleomodulins, which encompasses bacterial effectors acting on nuclear processes after translocation into the nucleus (Bierne and Cossart 2012). *A. phagocytophilum* produces several nucleomodulins (Sinclair et al. 2015a). The extensive characterization of one of them, Ankyrin A (AnkA), has provided other conceptual advances on mechanisms by which bacterial actors may act on chromatin. AnkA is a large bacterial effector characterized by a central region containing ankyrin (Ank) repeats (Park et al. 2004; Garcia-Garcia et al. 2009b). Interestingly, Ank repeats are commonly found in eukaryotic proteins, notably in several nuclear proteins that bind transcription factors. Following its secretion in the cytoplasm by a bacterial type IV secretion system (T4SS), AnkA enters the granulocyte nucleus, binds stretches of AT-rich DNA, and alters transcription of antimicrobial defense genes. In particular, AnkA represses *CYBB*, which encodes the subunit beta (NOX2) of the NADPH oxidase (Garcia-Garcia et al. 2009b; Rennoll-Bankert and Dumler 2012). The mechanism at play involves binding of AnkA to DNA in the *CYBB* promoter region, direct recruitment of HDAC1 by AnkA, and deacetylation of H3 (Rennoll-Bankert et al.

2015) (Fig. 6.3b). As a consequence, the pathogen obtains a significant fitness advantage as it prevents superoxide anion production by the NADPH oxidase and associated bactericidal effects.

AnkA not only binds the *CYBB* locus. It also targets several DNA regions rich in AT nucleotides on distinct chromosomes (Park et al. 2004; Garcia-Garcia et al. 2009b). Remarkably, AnkA-binding sites overlap within matrix attachment regions (MARs) that serve as attachment sites for nuclear matrix proteins and mediate structural organization of the chromatin within the nucleus (Rennoll-Bankert et al. 2015). AnkA is in this way a functional mimic of the host MAR-binding protein SATB1, which is known to bind to the *CYBB* promoter and represses transcription early during myeloid differentiation by recruiting HDACs (Wang et al. 2010). SATB1 has a wide action on chromatin, as it contributes to the formation of nuclear architectural platforms that anchor hundreds of gene loci and control large-scale transcriptional reprogramming (Kohwi-Shigematsu et al. 2012). This opens the fascinating hypothesis that bacterial effectors like AnkA could act as global genome organizers both acting in *cis* (locally) and *trans* (at a distance) to a target gene. By controlling the dynamics of chromosomal looping, they may change the three-dimensional structure of chromatin (Sinclair et al. 2014). Recent mapping of AnkA binding sites on the neutrophil genome by ChIP-seq further supports this concept of microbial factors acting as genome "re-organizers" (Dumler et al. 2016). Also in line with this idea, there is evidence that BAHD1-mediated heterochromatin formation plays a role in the spatial architecture of the genome (Libertini et al. 2015). Thus, LntA-mediated inhibition of BAHD1 might also change the structure of large domains involved in the co-regulations of ISGs upon *L. monocytogenes* infection.

6.3.1.4 Deregulation of Epigenetic Factor Genes and Genome-Wide Mediated Epigenetic Changes

Numerous neutrophil genes are differentially expressed during *A. phagocytophilum* infection (Borjesson et al. 2005; Lee et al. 2008; Sinclair et al. 2014). Several of them are downregulated, coinciding with HDAC1 binding and H3 deacetylation at their promoters (Garcia-Garcia et al. 2009a, b). However, most of them are upregulated (Borjesson et al. 2005), in agreement with the complex effects of infection on host gene expression. It was recently shown that DNA methylation levels in the neutrophil genome are profoundly altered after 24 h of infection with *A. phagocytophilum*. In particular, many regions within 3 kb from gene transcriptional start and termination sites and at intron–exon junctions become hypermethylated. In addition, expression of the *HDAC1* and *DNMT3A* genes is increased with infection (Garcia-Garcia et al. 2009a; Borjesson et al. 2005) (Fig. 6.3b). Overall, these findings highlight that *Anaplasma* infection induces large epigenomic changes as a result from the combined action of diverse mechanisms, including changes in expression of epigenetic factors and cross talk between these factors. Pharmacologic inhibition of histone deacetylases or DNA methyltransferases decreases *Anaplasma* intracellular survival (Garcia-Garcia et al. 2009a; Sinclair et al. 2015b), supporting the notion that broad epigenetic changes contribute to disease.

In summary, the *Listeria* and *Anaplasma* paradigms illustrate the diversity of mechanisms involved in modification of chromatin structure during bacterial infection, both at local and large genomic scales.

6.3.2 Chromatin Modifications Driven by Bacteria: Additional Examples

6.3.2.1 Histone Modifications

As shown for *Listeria,* bacteria in contact with eukaryotic cells have the ability to activate a large repertoire of host signaling pathways (e.g., MAPKs, NF-κB, and PI3K pathways) acting on histone kinases and acetylases (Yamamoto et al. 2003; Baek 2011). This is particularly the case of pro-inflammatory pathways. For instance, *Moraxella catarrhalis*, a saprophytic bacterium of the respiratory tract, and *Bacteroides vulgatus*, a commensal of the intestinal flora, induce inflammatory signaling cascades leading to phosphorylation/acetylation of H3 (Haller et al. 2003; Slevogt et al. 2006). The gastric pathogen *Helicobacter pylori* secretes the peptidyl prolyl *cis-*, *trans-*isomerase HP0175 that activates a TLR4-MAPK-MSK1 pathway leading to H3 phosphorylation and activation of the pro-inflammatory gene *IL-6* in THP-1 monocytes (Pathak et al. 2006). Bacterial lipopolysaccharide (LPS) and flagellin trigger histone acetylation and phosphorylation events at the *IL-8* gene downstream of the NF-κB pathway (Saccani et al. 2002; Schmeck et al. 2008).

However, inflammation is often counteracted by bacteria-driven mechanisms. In the case of *B. vulgatus,* this is performed by induction of the TGF-β1 anti-inflammatory pathway, which in turn induces H3 deacetylation and gene silencing via HDAC recruitment at pro-inflammatory gene promoters (Haller et al. 2003). This mechanism prevents *B. vulgatus* from eliciting a strong inflammatory response in the gut and contributes to its tolerance by the host. Bacterial toxins also dampen the host innate immune responses by inhibiting H3 phosphorylation/acetylation events. As described above for *Listeria* LLO (Fig. 6.3a), the pore-forming toxins PFO of *Clostridium perfringens*, PLY of *Streptococcus pneumoniae*, and aerolysin from *Aeromonas hydrophila* share a common mechanism that modulates histone marks, and subsequent gene expression, by acting on intracellular potassium levels (Hamon et al. 2007). Lethal toxin (LT) from *Bacillus anthracis,* the agent of anthrax, uses another mechanism by cleaving and inactivating MAPKKs, leading to disruption of MAPK signaling (Bardwell et al. 2004). In lung epithelial cells activated by TNF-α, LT-mediated MAPK inhibition promotes a decrease in the levels of H3S10p and H3K14ac at the promoters of *IL-8* and *KC* genes (Raymond et al. 2009). In macrophages exposed to LT, MAPK inhibition induces expression of HDAC8, which results in a decrease of H3K27ac levels at one enhancer of the *IL1-β* gene and the subsequent repression of this gene (Ha et al. 2016).

It is interesting to notice that besides immunity and inflammatory genes, changes in histone PTMs can also enable a pathogen to control expression of host genes involved in cell proliferation and death, as illustrated by the carcinogenic bacterium *H. pylori*. In gastric epithelial cells, this pathogen induces transient dephosphorylation of H3S10 and H3T3, as well as deacetylation of H3K23 (Fehri et al. 2009; Ding et al. 2010a). These modifications impact both the cell cycle (Fehri et al. 2009) and transcription of the oncogene *c-JUN* and heat shock gene *hsp70* (Ding et al. 2010a). In addition, *H. pylori*-mediated pre-mitotic arrest involves dephosphorylation of H3S10 upon deregulation of the mitotic histone kinase VRK1, followed by rephosphorylation of H3S10 by an IKKα-dependent pathway. Furthermore, exposure of *H. pylori* to gastric epithelial cells promotes release of HDAC1 from the promoter of the cell cycle regulator gene $p21^{WAF}$, hyper-acetylation of H4, and increased expression of $p21^{WAF}$ (Xia et al. 2008). These mechanisms may contribute to various *H. pylori*-associated gastric pathologies, including ulcers, mucosa-associated lymphoid tissue lymphoma, and cancer.

IFN responses are also modulated by diverse chromatin-based mechanisms. Like *Listeria*, *Mycobacterium tuberculosis* (Mtb), the causative agent of tuberculosis, controls histone-modifying multiprotein complexes at IFN-responsive genes (Lebreton et al. 2012). In macrophages infected with Mtb or exposed to Mtb components, such as the lipoprotein LpqH, genes induced in response to IFN-γ are partly repressed (Wang et al. 2005; Pennini et al. 2006). These genes include *CIITA,* coding for the master regulator of MHC class II genes, as well as some of its targets (e.g., *HLA-DR*). Activation of the TLR2-MAPK-dependent pathway upon Mtb infection stimulates recruitment of the transcriptional repressor C/EBP and histone deacetylation at the promoter of *CIITA*, antagonizing the nucleosome-remodeling activity of the SWI/SNF complex and downregulating *CIITA* expression (Pennini et al. 2007). Additionally, mycobacterial infection upregulates the expression of *SIN3A,* which encodes a core subunit of a HDAC-associated macromolecular complex (Wang et al. 2005) related to NuRD and BAHD1 complexes. Thus, to counteract IFN-γ-induced pathways, Mtb not only silences *CIITA* but also CIITA-regulated genes, such as *HLA-DR*, upon increased recruitment of SIN3A-HDACs to their promoters.

6.3.2.2 DNA Methylation

The importance of DNA methylation events associated with bacterial infections is also becoming increasingly appreciated. However, as for histone PTMs, alteration of 5mC patterns can result from an amalgam of bacteria-driven and host-driven effects on chromatin. Moreover, it can be difficult to connect gain or loss of this epigenetic mark to transcriptional changes. Indeed, the transcriptional effects of 5mC marks depend on their localization. Gain of DNA methylation is mainly coupled with transcriptional silencing at CpG-rich regions in promoters and enhancers and transcriptional activation at CpG-poor regions in gene bodies (Klose and Bird 2006; Chen and Riggs 2011). Interpreting the effect of bacteria on cytosine methylation is thus complex. This is illustrated by genome-wide studies of *Mycobacterium*-induced DNA methylation changes. On one hand, the response of human dendritic cells to Mtb infection is accompanied by both widespread de novo methylation and active demethylation primarily at enhancer elements (Pacis et al. 2015). On the other hand, the response of human THP-1 macrophages to Mtb infection is mostly accompanied by hypermethylation predominantly at cytosines present in a non-CpG dinucleotide context (Sharma et al. 2016). These differences may be explained by the use of different host cell models and infection times. Several DNA methylome maps have probably to be drawn in order to assess with robustness the dynamics of cytosine methylation during infection.

H. pylori infection also induces aberrant DNA methylation. In the human gastric mucosa, changes in 5mC patterns upon infection have been identified, strikingly at promoters of genes found methylated in gastric cancer cells (Maekita et al. 2006; Ding et al. 2010b; Hattori and Ushijima 2016). *H. pylori*-associated hypermethylation occurs for instance at the E-cadherin gene *CDH1* (Chan et al. 2003), tumor suppressor genes (e.g., *USF1*/*2* and *WWOX* (Bussiere et al. 2010; Yan et al. 2011), DNA repair genes [e.g., *MLH1* (Yao et al. 2006)], as well as in CpG islands of miRNA genes (Ando et al. 2009). The ability of *H. pylori* to induce DNA methylation in the gastric mucosa was confirmed in the gerbil animal model, and, interestingly, this effect was relieved upon treatment with the immunosuppressor cyclosporin A (Niwa et al. 2010). Moreover, *H. pylori*-mediated inflammation triggers lymphocyte and macrophage infiltration, which appears to have a key role in induction of DNA methylation (Hur et al. 2011). It is currently believed that DNA methylation changes upon *H. pylori* infection are mostly the indirect consequence of the associated inflammatory responses (Hattori and Ushijima 2016). Signals from macrophages produced by chronic inflammation, such as IL-1β, TNF-α, or nitric oxide, may affect factors that protect DNA from methylation, such as TET proteins (Hattori and Ushijima 2016). It remains unclear whether *H. pylori* effectors contribute more directly to aberrant epigenetic changes during gastric cancer progression (Valenzuela et al. 2015). There is a growing number of studies showing that epigenetic changes, particularly in DNA methylation, are linked to an increased inflammatory response (Bayarsaihan 2011; Medzhitov and Horng 2009), as well as increased risk of chronic disease development and cancerization. The role of bacteria in shaping patho-epigenetic landscapes will be discussed in detail in Sect. 5.

Epithelia other than that of the stomach can undergo bacteria-induced DNA methylation changes. There is evidence that in the oral cavity bacterial-induced chronic infection and uncontrolled inflammatory response may trigger epigenetic modifications. As an illustration, periodontally inflamed gingival biopsies showed a significant increase in promoter methylation of the gene encoding the pro-inflammatory enzyme COX-2, compared with non-inflamed biopsy samples (Zhang et al. 2010). This would allow a chronic inflammatory stimulus to be tolerated, preventing unrestricted tissue destruction. Whether this is a bacteria-triggered phenomenon is unknown, but it is noteworthy that resident bacteria, such as *Porphyromonas gingivalis,* can induce hypermethylation of specific genes in gingival epithelial cells (Yin and Chung 2011).

In human uroepithelial cells, infection with uropathogenic *Escherichia coli* (UPEC) results in the upregulation of DNMT activity and *DNMT1* expression and induces CpG methylation and downregulation of *CDKN2A*, a G1 cell cycle inhibitor regulator (Tolg et al. 2011). This may increase uroepithelial cell proliferation and pathogen persistence, by counteracting infection-stimulated host cell apoptosis. The placenta can also be targeted by bacteria-mediated epigenetic changes. Indeed, maternal infection with *Campylobacter rectus* induces hypermethylation of the imprinted *IGF2* gene promoter in murine placental tissue (Bobetsis et al. 2007). This finding suggests that bacterial infections during pregnancy might epigenetically affect genes involved in fetal development.

Last but not least, there is evidence that nonpathogenic inhabitants of the gut shape the DNA methylome. The gene *TLR4,* which encodes a LPS-sensing receptor, is downregulated in intestinal epithelial cells, and a role of the commensal bacteria in *TLR4* methylation and silencing is suspected (Takahashi et al. 2011). This is proposed to maintain intestinal homeostasis by preventing an excessive inflammatory reaction to the gut microbiota.

6.3.3 Bacterial Nucleomodulins

6.3.3.1 Nucleomodulins Acting Via Protein–Protein or Protein–DNA Interactions

As discussed above, *L. monocytogenes* LntA is a paradigm for nucleomodulins acting as inhibitor of HDAC-associated complexes, while *A. phagocytophilum* AnkA is a paradigm for nucleomodulins binding DNA and recruiting HDAC-associated complexes. So far, LntA orthologs have not been identified in other bacterial species, at least at the level of the primary protein sequence. In contrast, Ank-containing proteins are present in several human intracellular bacterial pathogens, such as *Ehrlichia, Rickettsia, Orientia, Coxiella*, and *Legionella* species. In particular, the protein Ank200 (or p200) from *Ehrlichia chaffeensis* (Wakeel et al. 2010) binds Alu-Sx elements located in promoters and introns of various human genes (Zhu et al. 2009). Several p200 target genes are strongly upregulated during infection, suggesting that p200 may affect gene transcription at a large genomic scale through mechanisms associated with Alu element gene regulation. Several other tandem-repeat containing proteins (TRPs) from *E. chaffeensis* may also enter into the nucleus (Luo et al. 2011; Luo and McBride 2012). Of those, the *E. chaffeensis* 32-kDa and 120-kDa tandem repeat proteins, TRP32 and TRP120, are nucleomodulins that binds host cell DNA particularly at G-rich motifs (Luo et al. 2011; Farris et al. 2016). Genes targeted by TRP120 are most frequently associated with transcriptional regulation, signal transduction, and apoptosis, whereas those targeted by TRP32 are linked to immune cell differentiation, chromatin remodeling, and RNA transcription. Interestingly, like many host nuclear factors, these nucleomodulins are subjected to post-translational modifications, TRP32 being phosphorylated and TRP120 sumoylated in host cells (Dunphy et al. 2014; Farris et al. 2016). Nucleomodulins may not only bind DNA and chromatin factors but also chromatin-anchoring factors. SinC, a protein secreted by *Chlamydia psittaci* via a type III secretion system (T3SS), exemplifies this potential mechanism. This effector targets the inner membrane of the nucleus in infected cells and may control chromatin interaction with the nuclear lamina (Mojica et al. 2015).

6.3.3.2 Nucleomodulins Acting as Epigenetic Modifiers

Several bacterial pathogens, and particularly those living in intracellular vacuolar compartments, can alter host chromatin structure by producing mimics of chromatin-modifying enzymes (Fig. 6.4). A first of such bacterial mimics, NUE, is a HMT discovered in the human pathogen *Chlamydia trachomatis*, based on its sequence similarities with eukaryotic lysine-specific methyltransferases containing a SET domain. After secretion by a T3SS, NUE enters the nucleus and associates with chromatin (Pennini et al. 2010). However, while NUE methylates mammalian histones in vitro, its target genes in the infected cell remain unknown.

Other SET domain-containing proteins were thereafter identified in *Legionella pneumophila* (Rolando et al. 2013; Li et al. 2013), *Burkholderia thailandensis* (Li et al. 2013), and *Bacillus anthracis* (Mujtaba et al. 2013) (Fig. 6.4). In *L. pneumophila,* LpSET is characterized as a bacterial HMT with dual functions. In *L. pneumophila* strain *Paris*, LpSET (named RomA: "Regulator of methylation A") has been shown to act as a HMT that trimethylates K14 of H3 (H3K14me3), a modification that does not exist in mammals (Rolando et al. 2013). By promoting a burst of H3K14me3 genome wide, including at innate immune gene loci, RomA decreases H3K14 acetylation, which is an activating mark, thus leading to repression of host gene expression and playing an important role in bacterial replication inside macrophages. In a separate study performed with *L. pneumophila* strain *Philadelphia*, LpSET (named LegAS4) was reported to act in the nucleolus on the expression of ribosomal RNA (rRNA) (Fig. 6.4). Human cells contain several hundred rDNA genes organized in tandem repeats that are clustered into nucleolar organizer regions. The chromatin structure of rRNA genes plays a fundamental role in regulating transcription of rDNA loci. LegAS4 binds rDNA at promoter and intergenic-spacer regions, by interaction with the chromatin reader HP1. In vitro studies suggest that LegAS4 catalyzes dimethylation of histone H3 on lysine 4 (H3K4me2). Consistently, ectopic expression of LegAS4 in human cells is associated with increased levels of H3K4me2 at rDNA promoters and activation of the transcription of these genes (Li et al. 2013). *B. thailandensis* secretes a LegAS4-like protein (BtSET) that also activates rDNA transcription in the nucleolus. Stimulation of rDNA expression and increased 45S pre-RNA synthesis seems

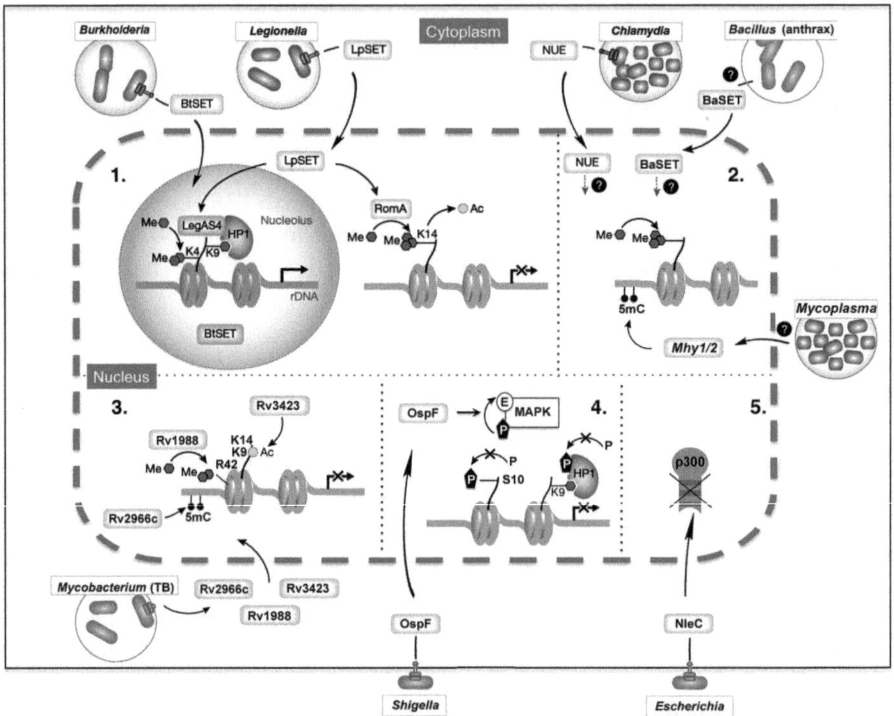

Fig. 6.4 Bacterial nucleomodulins with enzymatic activities. Several bacterial pathogens inject bacterial effector proteins into the host nucleus, and some are enzymes that modify chromatin residues or regulators. **1.** *Legionella pneumophila, Burkholderia thailandensis*, and *Chlamydia trachomatis* secrete SET domain containing effectors via type 3 (T3SS; *Burkholderia; Chlamydia*) or type 4 (T4SS; *Legionella*) secretion systems. *L. pneumophila* secretes a histone methyltransferase (LpSET) that has been assigned two functions: (*i*) in the nucleus, LpSET termed "RomA" trimethylates histone H3 at K14, causing a switch from acetylated to methylated H3K14 at specific gene promoters and thus transcriptional repression, and (*ii*) in the nucleolus, LpSET termed "LegAS4" binds HP1 at rDNA promoters and activates transcription by stimulating H3K4 methylation. *B. thailandensis* secretes a LegAS4-like protein (BtSET) that also activates rDNA transcription in the nucleolus. **2.** *C. trachomatis* secretes the histone methyltransferase NUE that methylates host histones H2B, H3, and H4. *Bacillus anthracis* produces a histone H1 methyltransferase, BaSET. *Mycoplasma hyorhinis* Mhy1, Mhy2, and Mhy3 are nucleomodulins with CG- and GATC-specific cytosine methyltransferase activities. **3.** *Mycobacterium tuberculosis* secretes at least three nucleomodulins: Rv1988 is a histone methyltransferase that methylates H3 on R42 in the nucleosome core; Rv3423 is a histone acetyltransferase; and Rv2966c is a DNA methyltransferases. **4.** The *Shigella flexneri* T3SS effector OspF is a posttranslational modifier with a phosphothreonine lyase activity. OspF eliminylates MAP-kinases in the nucleus, leading to the downregulation of a subset of immunity genes. **5.** The T3SS effector NleC from pathogenic *E. coli* is a metalloproteinase targeting the host histone acetyltransferase p300 for degradation. *Ac* Acetylation, *Me* methylation, *P* phosphorylation, *E* eliminylation. Adapted from (Bierne 2013)

to contribute to bacterial replication, though the mechanism at play is not yet understood. Having different substrates and functions is a property that LpSET shares with eukaryotic HMTs. For instance, the H3K9 HMT G9a preferentially methylates K9 on histone H3 but can also methylate K27 and K56. This HMT predominantly represses genes at euchromatic regions but also acts as a positive activator of rDNA transcription (Yuan et al. 2007). Considering the number of SET domain proteins present in bacterial species that interact with eukaryotes, it is tempting to speculate that several bacteria might employ this strategy.

The agent of tuberculosis also modulates the host epigenetic machinery by secreting an original HMT, here methylating a noncanonical arginine located in the core of histone H3 (Yaseen et al. 2015) (Fig. 6.4). This effector, Rv1988, targets genes involved in defense against pathogens, including genes participating in the generation of reactive oxygen species (ROS). During infection, Rv1988 not only targets gene promoters for H3R42me2 but also putative regulatory regions. Aside from HMTs, MtB also secretes effectors with histone acetyltransferase [Rv3423.1 (Jose et al. 2016)] or DNA methyltransferase [(Rv2966c, (Sharma et al. 2015)] activity. It is interesting to notice that Rv2966c methylates cytosines present in a non-CpG context. Thus, MtB has evolved diverse strategies to directly manipulate chromatin in the nucleus. *Mycoplasma* species also produce nucleomodulins. For instance, three DNA methyltransferases have been identified in *Mycoplasma hyorhinis*, an intracellular commensal that can shift to an opportunist pathogen. *M. hyorhinis* produces Mhy1 and Mhy2, promoting CG methylation, and Mhy3 acting on GATC sites (Chernov et al. 2015). There is evidence that these bacterial DNMTs have the ability to translocate to the human cell nucleus and establish aberrant genome-wide methylation patterns. Yet, it remains to be proven that the host epigenome is reshaped in human cells naturally infected by *M. hyorhinis*.

Bacteria-induced epigenetic effects can also occur by specific modifications of epigenetic factors. The *Shigella flexneri* T3SS effector OspF nicely illustrates this mechanism. OspF is a phosphothreonine lyase that irreversibly modifies host MAPKs by eliminylation (Li et al. 2007; Brennan and Barford 2009). This enzymatic reaction converts a phosphothreonine residue into a dehydrobutyrine residue that can no longer be phosphorylated and hence locks the substrate in an inactive form. Inhibition of MAPK phosphorylation in the nucleus enables OspF to abrogate phosphorylation of histone H3 at a set of NF-κB-regulated promoters, thus impairing expression of a pool of pro-inflammatory genes (Arbibe et al. 2007). In addition, this effector alters the phosphorylation at S83 of the heterochromatin protein HP1-γ, demonstrating that in addition to histones, bacteria can control chromatin regulator PTMs (Harouz et al. 2014) (Fig. 6.4). Furthermore, OspF and another nuclear-targeted effector, OspB, interact with the human retinoblastoma protein Rb, which is known to bind several chromatin-remodeling factors (Zurawski et al. 2009). *Shigella* likely uses OspF–OspB synergy to downregulate host innate immunity via alteration of the chromatin structure at specific genes.

S. flexneri also secretes an effector acting as an E3 ubiquitin ligase, IpaH9.8, which targets several host cytosolic or nuclear proteins for proteasome-dependent degradation (Rohde et al. 2007). In the nucleus, IpaH9.8 disrupts the activity of a mRNA splicing factor, $U2AF^{35}$, thus interfering with $U2AF^{35}$-dependent splicing (Toyotome 2001; Seyedarabi et al. 2011) and impairing host inflammatory responses (Okuda et al. 2005). IpaH9.8 defines a family of bacterial effectors characterized by an N-terminal domain containing leucine-rich repeats (LRR) involved in substrate recognition and a C-terminal E3 ligase domain (Hicks and Galan 2010). The ortholog of IpaH9.8 in *Salmonella enterica* is SspH1, a nucleomodulin that targets for instance the host kinase PKN1 (Haraga and Miller 2003; Rohde et al. 2007). *Yersinia pestis* also encodes a LRR-containing nucleomodulin targeting host kinases, YopM (Benabdillah et al. 2004; Soundararajan et al. 2011). However, YopM is not itself a modifier but rather acts as a scaffolding protein that facilitates the formation of a complex between serine/threonine kinases RSK1 and PKN2 (McDonald et al. 2003; McCoy et al. 2010). Recent data suggest that the YopM causes enhanced phosphorylation of RSK1 in the nucleus, leading to enhanced transcription of immunosuppressive cytokines, such as IL-10. YopM intranuclear levels are dependent on its interaction with the DEAD-box helicase 3 (DDX3) (Berneking et al. 2016).

Other types of bacterial modifiers acting in the nucleus include proteases and phosphatases. The T3SS effector NleC from enteropathogenic (EPEC) and enterohemorrhagic (EHEC) *E. coli* is a zinc metalloprotease that targets the HAT p300/CBP and decreases the abundance of this epigenetic factor in the nucleus (Fig. 6.4). Overexpression or knockdown of NleC impacts IL-8 secretion by EPEC, indicating that NleC contributes to dampening of inflammatory signaling during infection (Shames et al. 2011). The Gram-positive bacterium *Streptococcus pyogenes* expresses a serine/threonine phosphatase, SP-STP, which is secreted into host cells and targets the nucleus (Agarwal et al. 2012). There, it acts as a pro-apoptotic factor that induces apoptosis of pharyngeal cells, a hallmark of streptococcal infections, by influencing transcription of apoptotic genes and preventing the transcription of other genes, such as cytochrome p450.

6.3.4 Change in Expression and/or Activity of Epigenetic Regulators

As illustrated above with upregulation of HDAC1 in *A. phagocytophilum*-infected cells (Garcia-Garcia et al. 2009a) and SIN3a in Mtb-infected cells (Wang et al. 2005), some bacterial species positively or negatively modulate expression of epigenetic factors. Mtb infection induces HDAC1 expression in macrophages (Chandran et al. 2015). In contrast, the levels of HDAC1 and DNMT1 transcripts decrease in gingival epithelial cells treated with the oral pathogen *Porphyromonas gingivalis* (Yin and Chung 2011), and LPS from this bacterial species downregulates DNMT1, DNMT3a, and JMJD3 gene expression levels (de Camargo Pereira et al. 2013). *M. catarrhalis* also reduce HDAC1/2 expression in bronchial epithelial cells (Slevogt et al. 2006). In human urothelial cells, infection with UPEC results in the upregulation DNMT1 expression (Tolg et al. 2011), as well as of EZH2, encoding the H3K27 HMT of the Polycomb chromatin-repressive complex PRC2 (Ting et al. 2016). EZH2 plays a role in early host cell proliferative responses to infection. Bacteria can also produce metabolites, acting as inhibitors of chromatin-modifying enzymes, such as lactate and butyrate, proven to be potent inhibitors of HDACs (Latham et al. 2012).

6.3.5 Bacterial Molecular Patterns Acting on the Epigenetic Machinery

Certain bacterial-derived metabolites can modify the epigenome of host cells and in turn alter the development and function of the cell, either by acting on precursors of enzymatic reactions involved in chromatin modifications or by modulating the activity of epigenetic regulators (Alenghat and Artis 2014). In particular, commensals of the microbiota produce diet-dependent molecules that influence DNA methylation and histone acetylation. One such product is butyrate, a potent inhibitor of HDACs (Riggs et al. 1977). Butyrate exerts beneficial anti-inflammatory effects on the host, particularly on immune cells (Segain et al. 2000; Arpaia et al. 2013; Chang et al. 2014), possibly via epigenetic upregulation of anti-inflammatory genes (see Sect. 4.1). Such observations open the interesting possibility to use butyrate-producing probiotic bacteria as immunosuppressors (Licciardi et al. 2010). Interestingly, a recent study has shown that a bacterial quorum-sensing molecule can also act on HDACs. Prolonged exposure to 2-aminoacetophenone, which is excreted by the pathogen *Pseudomonas aeruginosa*, increases the expression and activity of HDAC1, resulting in hypoacetylation of H3K18 and attenuated expression of pro-inflammatory cytokine genes. This is proposed to promote host tolerance to infection (Bandyopadhaya et al. 2016).

6.4 Epigenetic Factors Engaged in Host Responses to Bacteria and Bacterial Imprinting

6.4.1 Lessons from BAHD1-, SIRT2-, and HDAC3-Deficient Mice

Considering the crucial role of epigenetic factors in embryonic development and cell differentiation, it is not surprising that knocking out their coding genes in the mouse often causes embryonic or perinatal lethality. Thus, so far only a few studies have addressed the role of epigenetic regulators in bacterial infections at the level of a mammalian organism. The use of *Bahd1* haplo-deficient (heterozygous) mice (Lebreton et al. 2011) and *Sirt2* knockout mice (Eskandarian et al. 2013) has confirmed the involvement of BAHD1 and SIRT2 in murine listeriosis. Following intravenous inoculation with *L. monocytogenes,* the spleens of these deficient mice were significantly less infected than those of wild-type littermates after 72 hours of infection. It was not possible to further refine the role of BAHD1 by infecting *Bahd1* knockout mice, because the total ablation of *Bahd1* induces a high neonatal mortality rate (Lakisic et al. 2016). However, it is worthy to note that it also causes restriction of placental growth, indicating a key role of BAHD1 in placental development. This finding opens the possibility that manipulation of the BAHD1 complex by *Listeria* contributes to the fetoplacental step of listeriosis. *Bahd1* deficiency also leads to deregulation of carbohydrate and lipid metabolism, which may play roles in infection (Lakisic et al. 2016). Together these results support the notion that mutations in epigenetic regulatory genes influence the outcome of bacterial infections.

The use of HDAC conditional knockout mice permitted to investigate the functional roles of chromatin regulation in intestinal homeostasis and its cross talk with the microbiota. Intestinal epithelial cells (IECs) integrate numerous microbial signals from the intestinal microenvironment and respond to these stimuli by changing their transcriptional program. Interestingly, IEC-specific *Hdac3*-deficient mice show increased susceptibility to intestinal damage and inflammation and a change in microbial communities (i.e., dysbiosis) (Alenghat et al. 2013). Strikingly, when rendered germ free, *Hdac3*-conditional KO mice recover a normal intestinal barrier function, as observed in wild-type germ-free mice. These data indicate that HDAC3 has an important role in maintaining intestinal homeostasis and establishing normal host–commensal relationships. Ablating both *Hdac1* and *Hdac2* in murine IECs also alters the structure and functions of the gut, via a defect in cell differentiation and chronic intestinal inflammation (Turgeon et al. 2013). While a possible dysbiosis induced by this double mutation has not yet been addressed, it is possible that major alterations of the gut induced by the double *Hdac1–Hdac2* mutation impacts the composition of the population of commensals. Thus, HDACs are likely to be key epigenetic programmers of the host in response to signals from the gut microbiota.

It is important to note that while loss of *Hdac3* expression in IECs impairs microbiota-dependent intestinal barrier function, inhibition of HDACs by commensal bacteria-derived SCFAs, such as butyrate, generally protects from pathologic intestinal inflammation. However, rather than acting on epithelial cells, butyrate is described to inhibit HDAC function in intestinal immune cells, such as peripheral blood mononuclear cells (PBMCs) (Segain et al. 2000), Tregs (Arpaia et al. 2013), and macrophages (Chang et al. 2014). These results suggest that HDACs may have opposite effects in different intestinal cell populations, leading to either protective or pathologic immunity (Alenghat and Artis 2014).

6.4.2 Immune Tolerance and Toxin-Induced Resistance

Host–commensal mutual relationships require that commensal bacteria do not trigger an uncontrolled immune response, and thus become tolerated by the host immune system, while the latter efficiently eliminates invading pathogens. HDACs and other epigenetic regulators are involved in this tolerance. For instance, intestinal commensal bacteria induce DNA methylation at the gene encoding the main sensor of LPS, TLR4, leading to its downregulation in the large intestine (Takahashi et al. 2011). This is believed to maintain intestinal homeostasis by preventing an excessive inflammatory reaction to the gut microbiota. The commensal bacterium *Bacteroides vulgatus* triggers an anti-inflammatory response via recruitment of HDACs at pro-inflammatory gene promoters (Haller et al. 2003). Likewise, *Bifidobacterium breve* and *Lactobacillus rhamnosus GG* inhibit transcriptional activation of inflammatory bowel disease-causing factors through inhibition of histone acetylation and enhancement of DNA methylation (Ghadimi et al. 2012).

In the presence of opportunist pathogens, similar mechanisms may dampen uncontrolled inflammatory responses triggered by bacteria-induced chronic infections. For instance, in the oral cavity, periodontally inflamed gingival biopsies show a significant increase in promoter methylation of the gene encoding the pro-inflammatory enzyme COX-2, when compared with non-inflamed biopsy samples (Zhang et al. 2010). This would allow a chronic inflammatory stimulus to be tolerated, preventing unrestricted tissue destruction. Whether this is a bacteria-triggered phenomenon remains unknown, but it is remarkable that resident bacteria, such as *P. gingivalis,* can induce DNA hypermethylation of specific genes in gingival epithelial cells (Yin and Chung 2011).

When pathogenic species manage to cross epithelial barriers and to multiply in the blood, sustained exposure to microbial inflammatory products, such as LPS, leads to tissue damage, multi-organ dysfunction, septic shock, and death. To compensate these adverse effects, the immune system has developed post-septic immunosuppression (PSI) mechanisms that enable hematopoietic cells to become hypo-responsive to repeated stimulation by microbial insults. This compensatory anti-inflammatory response counteracts the harmful effects of sepsis but leaves individuals more susceptible to opportunistic infections for extended periods of time (weeks to years). Although PSI is a complex multifactorial process, the contribution of epigenetic regulation is recognized (McCall et al. 2010; Carson et al. 2011). One of the facets of PSI is LPS tolerance, in which LPS-elicited TLR4 responses are reprogrammed towards silencing of pro-inflammatory cytokine genes and expression of anti-inflammatory or antimicrobial mediators. LPS activation of TLR4 first elicits transcription of poised pro-inflammatory genes, which are rapidly derepressed and then returned to basal state within hours. Opening the chromatin at target genes during this acute phase involves histone phosphorylation and acetylation. However, sustained exposure to LPS or subsequent LPS challenge activates a pathway leading to permanent gene repression (Fig. 6.5). One mechanism was studied at the proximal promoters of *TNF*-α and *IL1*-β genes, where a change in the composition of the NF-κB transcription factor occurs (El Gazzar et al. 2008; Chen et al. 2009). LPS-mediated upregulation of *RelB* expression induces a shift from activating p65-p50 TF to repressive RelB-p50 TF. RelB interacts with H3K9 HMT G9a, leading to H3K9me2 and subsequent recruitment of HP1. The

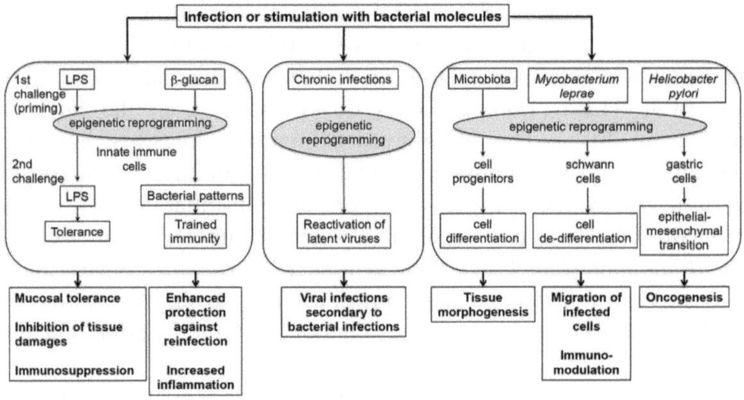

Fig. 6.5 Bacterial imprints. Scheme of mechanisms supporting an epigenetic memory of bacterial infection or bacterial colonization

repressive complex formed by G9a and HP1 recruits DNMT3A/B, which induces de novo CpG methylation and assembly of silent, facultative heterochromatin (McCall et al. 2010). Interestingly, LPS stimulation has also been shown to leave epigenetic footprints on enhancers (Ostuni et al. 2013). As discussed in Sect. 2.4, the epigenetic topography of distal regulatory elements, such as enhancers and insulators, plays a key role in maintaining epigenetic memory. Thus, immune tolerance is likely to involve reshaping of the epigenetic landscape at both proximal and distal regions of pro-inflammatory genes.

LPS tolerance can last for weeks in humans. Tissue-resident macrophages, which appear to persist in the long term, may be the cells that support this memory (Perdiguero and Geissmann 2016). However, whether these cells keep an epigenetic memory along cell divisions is not yet proven. Furthermore, even if imprinted cells divide, why new cells from progenitors in the bone marrow do not restore an efficient immune system is an open question. A tempting hypothesis would be that epigenetic imprinting also occurs at the level of stem cells. This hypothesis needs to be investigated by analyzing the epigenome of stem cells isolated from animal models of sepsis. The reversal of heterochromatin to euchromatin at genes targeted for LPS-mediated repression is also a key issue to understand how "imprinted" immune cells return to homeostasis.

The reduced responsiveness to an effect of a bacterial product caused by prior exposure to this product has also been observed in the case of anthrax Lethal toxin. Some studies indicate that epigenetic modifications contribute to this toxin-induced tolerance (TIR) (Salles et al. 2003). Macrophages exposed to a sublethal dose of anthrax LT become refractory to subsequent cytolytic doses of toxin, and a subset of them retains this phenotype for up to six weeks. The histone deacetylase HDAC8 promotes TIR by changing the chromatin acetylation state of promoter regions of mitochondrial death genes, producing tolerance to a next intoxication (Ha et al. 2014).

6.4.3 Trained Innate Immunity

The priming of innate immune cells by a bacterial challenge can trigger effects opposite to that of tolerance, with cells mounting a faster and longer response than the initial response. This adaptive feature of innate immunity has been defined as "trained immunity" (Netea et al. 2016). A first observation of this phenomenon in humans has been reported upon vaccination with bacilli Calmette-Guérin (BCG). When compared to unvaccinated patients, vaccinated healthy volunteers mounted a more robust response to subsequent infection with unrelated pathogens, and this effect persisted three months after vaccination. The phenomenon was dependent on NOD2 but independent from T- and B-lymphocyte protection (Kleinnijenhuis et al. 2012). The mechanism at play involves BCG-induced epigenetic reprogramming of monocytes through the activating mark H3K4me3 (Kleinnijenhuis et al. 2012). The effect could be downregulated by vitamin A, which increases the levels of the silencing mark H3K9me3 (Arts et al. 2015). Likewise, training of monocytes with a fungal component, β-glucan, induces stable change in histone marks (Quintin et al. 2012) (Fig. 6.5). The list of genes whose expression is induced by this epigenetic reprogramming includes genes involved in glucose metabolism (Cheng et al. 2014). Interestingly, a study supports the notion that the functional programming of monocytes towards either enhanced (training) or decreased (tolerance) cytokine production depends on the nature of the bacterial ligand and of the subsequent activated pattern-recognition receptor (Ifrim et al. 2014).

6.4.4 Polymicrobial Infections and Viral Reactivation

A bacterial infection can also influence viral infections by reactivating latent viruses (Fig. 6.5). This kind of adverse effect is suspected to be associated, for instance, with periodontal pathogens colonizing the oral cavity, such as *P. gingivalis*. This opportunistic bacterium is proposed to be a risk factor for AIDS or Herpes, by reactivating latent Kaposi's sarcoma-associated herpesvirus (KSHV) (Morris et al. 2007), human immunodeficiency virus (HIV), and Epstein–Barr virus (EBV) (Imai et al. 2009, 2011). It is proposed that the high production of butyrate by this bacterial species in gingival pockets reactivates viral genes maintained silent by HDAC-containing complexes. One effect of butyrate-mediated EBV reactivation is the increase of the expression of ZEBRA, a lytic gene transactivator (Imai et al. 2012). Butyrate and other SCFAs also downregulate expression of the HDAC SIRT1 and of the HMT EZH2 and SUV39H1, leading to histone hyperacetylation and reduction in the levels of H3K27me3 and H3K9me3, respectively (Ye and Karn 2015). *P. gingivalis* also elicits changes in the expression of gene encoding chromatin modifiers (as discussed in Sect. 3.4).

6.5 Bacterial Reprogramming of Cell and Tissue Fate

The ultimate effect of bacteria on the epigenome could manifest by a change of identity in the host cell itself. Several recent lines of evidence strongly support the existence of such a drastic change, which would occur when bacterial signals target stem cells or they are potent enough to disrupt the epigenetic barriers that maintain the differentiated cells in their locked state (Fig. 6.5).

6.5.1 Bacteria-Induced Cell Differentiation

The observation that the gut of germ-free mice is altered by defects in the maturation of the intestinal epithelium and of the immune and vascular systems of the gastrointestinal tract is a strong indication that the microbiota contributes to tissue morphogenesis (Sommer and Backhed 2013). The mechanisms are complex and the contribution of epigenetic regulation is not clearly established. Nevertheless, there is evidence that bacteria manipulate the stem cells that generate the different cell types of the intestinal epithelium at the bottom of the crypts. First, intestinal crypts of germ-free mice exhibit a slower turnover of the epithelial cells than conventional mice. Second, stem cells respond to bacterial patterns, such as muramyl-dipeptide (MDP) of the peptidoglycan (Nigro et al. 2014). The commensal microbiota also shapes the intestinal immune system by regulating T helper (TH) cell lineage differentiation (Furusawa et al. 2015). One mechanism involves butyrate secretion by the anaerobic commensal class of bacteria, Clostridia. In particular, butyrate enhances histone acetylation at the promoter of the master regulator of regulatory T cells, FOXP3 (Furusawa et al. 2013).

There is also evidence for pathogen-mediated targeting of cell differentiation pathways in the gut. For instance, the enteroinvasive species *Salmonella enterica* serovar Typhimurium can convert lymphoid follicle-associated enterocytes into intestinal epithelial microfold (M) cells, though activation of the Wnt/β-catenin signaling pathway mediated by the T3SS effector SopB (Tahoun et al. 2012). This is proposed to promote intestinal invasion by this pathogen.

6.5.2 Bacteria-Induced Cell Dedifferentiation

The study of the behavior of human dermal fibroblasts (HDF) artificially infected with nonpathogenic lactic acid bacteria (LAD) led to an intriguing observation: these LAB-treated HDFs became clustered like embryoid spheres and lost their self-renewal ability (Ohta et al. 2012). In addition, LAB-incorporated cell clusters expressed a subset of pluripotent stem cell marker genes, such as NANOG, OCT3/4, and SOX2, while expression HOX genes, which control the body plan of an embryo, was decreased. Furthermore, LAB-incorporated cell clusters could transform into any of the derivatives of the three germ layers. The mechanism involved in this artificial cell reprogramming by LAD is unclear, but supports the concept that there is a potential for bacterial molecules to revert the host transcriptional program of differentiated cells towards pluripotency.

The study of *Mycobacterium leprae* (ML), the causative agent of human leprosy, supports this hypothesis. During infection of the peripheral nervous system, this pathogenic bacterium promotes an amazing reprogramming process on adult Schwann cells, by triggering their dedifferentiation into progenitor/stem cell-like cells (Masaki et al. 2013). By using this sophisticated strategy, bacteria disseminate to other niches without being detected by immune cells (Fig. 6.5). ML not only migrates within reprogrammed cells but also spreads the infection to skeletal and smooth muscles by re-differentiating stem cells into these tissues. Moreover, infected stem cells display immunomodulatory properties that promote recruitment of macrophages and formation of granuloma-like structures (Masaki et al. 2014). The mechanism of dedifferentiation of Schwann cells involves activation of differentiation/myelination and lineage-associated genes as well as silencing of numerous developmental genes. ML-induced cellular reprogramming also correlates with changes in DNA methylation supporting a key role of epigenetic regulation in this phenomenon (Masaki et al. 2013).

6.5.3 Bacteria-Induced Epithelial–Mesenchymal Transition and Oncogenesis

The existence of a link between bacteria-mediated aberrant somatic cell reprogramming and cancer is supported by the example of *H. pylori*, an important acquired risk factor for gastric cancer. Besides *H. pylori*-mediated effects on cell proliferation, DNA integrity, and DNA methylation (Ushijima and Hattori 2012), infection by this bacterium may also induce dedifferentiation of mature epithelial cells by changing the expression program of the stem cell signaling network. Recent evidence supports a role for *H. pylori* in inducing the so-called “intestinal metaplasia,” which transforms stomach cells to intestine-like cells via epithelial–mesenchymal transition (EMT) (Fig. 6.5) (Bessede et al. 2014). EMT is known to participate in different carcinogenesis processes and is involved in the generation of cancer stem cells. The bacterial secreted effector CagA promotes the EMT phenotype by activating the expression of master transcription factor genes regulating intestinal differentiation and maintenance.

The “Helicobacter paradigm” may be transposable to bacteria targeting other tissues. It is speculated that *E. coli* infection may be linked with bladder carcinoma risk (Tolg et al. 2011) and a set of intestinal bacteria might predispose to colon cancer (Sun 2010). More generally, deregulation of tumor-suppressor and/or stem cell-associated pathways (e.g., WNT, JAK-STAT, JNK, and NOTCH) upon genetic alteration and epigenetic reprogramming induced by bacteria is a possible cause of cancer development in epithelial niches.

6.6 Conclusions

The "patho-epigenetics" research field has been defined by Janos Minarovits as the elucidation of the epigenetic consequences of microbe–host interactions leading to pathogenesis (Minarovits 2009). We now propose to add the field of "probio-epigenetics" to cover studies addressing the beneficial effects of bacterial interactions with epigenetic factors. Both themes are developing rapidly as important emerging subtopics at the frontier of Epigenetics and Microbiology sciences. Deciphering the mechanisms underlying the plasticity of gene expression under the action of endogenous (from the microbiota) or exogenous (from food-associated or aerosol-transported bacteria), stimuli may have important impacts on health and disease. Patho-epigenetics may lead to new treatments against bacterial infectious diseases, at a time when pathogenic bacteria became a serious concern due to the emergence of drug resistances. In addition, epigenetic marks may be used as biomarkers to monitor latent infection, disease reactivation, or responses to treatment. Probio-epigenetics can help to characterize the benefits of commensal bacteria on health and to understand the deleterious effects of dysbiosis, which may promote pathological epigenetic signals. Indeed, it is recognized that alterations of microbial communities can cause immune dysregulation, leading to autoimmune disorders, and may contribute to metabolic diseases, neuropathies, and behavioral problems. Restoring the composition of altered intestinal microbiota with fecal flora transfer is now tested as a way to counteract these adverse effects. The interplay of bacteria and nutrients is also an important emerging field of research, due to the key role of metabolites on epigenetic regulation. Western food may change epigenetic patterns in the gut, favoring pathogens associated with intestinal inflammatory diseases. There is evidence that changes in nutritional habits, such as low intake in methyl donor molecules, promote abnormal epigenetic marks in a mouse model mimicking susceptibility to *E. coli*-mediated gut inflammation and Crohn's disease (Denizot et al. 2015).

At the level of basic science, these investigations highlight the amazing molecular tools used by intracellular pathogens to manipulate chromatin and to fine-tune host gene expression. The number of nucleomodulins discovered in the bacterial world is rising steadily, in particular thanks to bioinformatic analysis. For instance, five of over 80 T4SS substrates of *Coxiella burnetii,* the agent of Q fever, are predicted to carry a nuclear localization signal. One of them, Cbu1314, has recently been confirmed as a protein binding to chromatin and controlling transcription of host cell genes (Weber et al. 2016). In the toolbox of bacterial molecules, bacterial metabolic by-products, such as SCFAs, are also prone to induce epimutations. Several of such molecules may emerge as novel epigenetic drugs. It is also striking that the study of bacterial factors promotes innovation in the epigenetics field. For instance, studies carried out in *L. monocytogenes* have led to the discovery of the BAHD1 chromatin-repressive complex (Bierne et al. 2009; Lebreton et al. 2011; Libertini et al. 2015; Lakisic et al. 2016) and of a nuclear function for the histone deacetylase SIRT2 (Eskandarian et al. 2013).

How bacterial modulation of the epigenetic information is spatio-temporally coordinated, at specific genome loci, according to time, cell type, and stimuli, is not fully understood. In addition, the impact of bacterial signals on the epigenetic profile of structural elements, such as enhancers, insulators, and DNA repeats, has been poorly addressed. Moreover, how expression and secretion of bacterial factors is controlled in the host and how chromatin writers and erasers become engaged in response to a cocktail of bacterial stimuli are important questions to be solved. In this regard, deciphering how chromatin modifications are spread throughout the genome (the "patho-epigenome" or the "probio-epigenome") is likely to provide important clues. Technological advances in human genome-wide mapping of DNA methylation and histone modifications, as well as in systems biology, will help to make significant progress. Investigations will have to be performed at both the tissue and single cell levels, with the objective of analyzing precursors of cell lineages or differentiated cells with long life spans, in order to determine the "chromatin signature" of bacterial cues. As epigenetic processes can be reverted, elimination of patho-epigenetic changes induced by microbes may prevent chronic or latent infections, as well as some cancers and autoimmune diseases. This opens avenues for future research.

Acknowledgements I apologize to colleagues whose work was not cited here. I acknowledge support from INRA, French Ligue Nationale Contre le Cancer (LNCC RS10/75-76 Bierne), French National Research Agency (ANR, grant EPILIS), and the iXcore Research Foundation.

References

Agarwal S, Agarwal S, Jin H, Pancholi P, Pancholi V (2012) Serine/threonine phosphatase (SP-STP), secreted from Streptococcus pyogenes, is a pro-apoptotic protein. J Biol Chem 287(12):9147–9167. doi:10.1074/jbc.M111.316554

Aguilera C, Nakagawa K, Sancho R, Chakraborty A, Hendrich B, Behrens A (2011) c-Jun N-terminal phosphorylation antagonises recruitment of the Mbd3/NuRD repressor complex. Nature 469 (7329):231–235. doi:10.1038/nature09607

Alenghat T, Artis D (2014) Epigenomic regulation of host-microbiota interactions. Trends in immunology 35(11):518–525. doi:10.1016/j.it.2014.09.007

Alenghat T, Osborne LC, Saenz SA, Kobuley D, Ziegler CG, Mullican SE, Choi I, Grunberg S, Sinha R, Wynosky-Dolfi M, Snyder A, Giacomin PR, Joyce KL, Hoang TB, Bewtra M, Brodsky IE, Sonnenberg GF, Bushman FD, Won KJ, Lazar MA, Artis D (2013) Histone deacetylase 3 coordinates commensal-bacteria-dependent intestinal homeostasis. Nature 504 (7478):153–157. doi:10.1038/nature12687

Allemand E, Batsche E, Muchardt C (2008) Splicing, transcription, and chromatin: a menage a trois. Curr Opin Genet Dev 18(2):145–151. doi:10.1016/j.gde.2008.01.006

Alvarez-Errico D, Vento-Tormo R, Sieweke M, Ballestar E (2015) Epigenetic control of myeloid cell differentiation, identity and function. Nat Rev Immunol 15(1):7–17. doi:10.1038/nri3777

Ando T, Yoshida T, Enomoto S, Asada K, Tatematsu M, Ichinose M, Sugiyama T, Ushijima T (2009) DNA methylation of microRNA genes in gastric mucosae of gastric cancer patients: its possible involvement in the formation of epigenetic field defect. Int J Cancer 124 (10):2367–2374. doi:10.1002/ijc.24219

Arbibe L, Kim DW, Batsche E, Pedron T, Mateescu B, Muchardt C, Parsot C, Sansonetti PJ (2007) An injected bacterial effector targets chromatin access for transcription factor NF-kappaB to alter transcription of host genes involved in immune responses. Nat Immunol 8(1):47–56. doi:10.1038/ni1423

Arpaia N, Campbell C, Fan X, Dikiy S, van der Veeken J, deRoos P, Liu H, Cross JR, Pfeffer K, Coffer PJ, Rudensky AY (2013) Metabolites produced by commensal bacteria promote peripheral regulatory T-cell generation. Nature 504(7480):451–455. doi:10.1038/nature12726

Arts RJ, Blok BA, van Crevel R, Joosten LA, Aaby P, Benn CS, Netea MG (2015) Vitamin A induces inhibitory histone methylation modifications and down-regulates trained immunity in human monocytes. J Leukoc Biol 98(1):129–136. doi:10.1189/jlb.6AB0914-416R

Arzate-Mejia RG, Valle-Garcia D, Recillas-Targa F (2011) Signaling epigenetics: novel insights on cell signaling and epigenetic regulation. IUBMB Life 63(10):881–895. doi:10.1002/iub.557

Baek SH (2011) When signaling kinases meet histones and histone modifiers in the nucleus. Mol Cell 42(3):274–284. doi:10.1016/j.molcel.2011.03.022

Bandyopadhaya A, Tsurumi A, Maura D, Jeffrey KL, Rahme LG (2016) A quorum-sensing signal promotes host tolerance training through HDAC1-mediated epigenetic reprogramming. Nat Microbiol 1:16174. doi:10.1371/journal.ppat.1003024

Bardwell AJ, Abdollahi M, Bardwell L (2004) Anthrax lethal factor-cleavage products of MAPK (mitogen-activated protein kinase) kinases exhibit reduced binding to their cognate MAPKs. Biochem J 378(Pt 2):569–577. doi:10.1042/BJ20031382

Bayarsaihan D (2011) Epigenetic mechanisms in inflammation. J Dent Res 90(1):9–17. doi:10.1177/0022034510378683

Benabdillah R, Ls JM, Lützelschwab S, Demoinet E, Cornelis GR (2004) Identification of a nuclear targeting signal in YopM from Yersinia spp. Microbial pathogenesis 36(5):247–261. doi:10.1016/j.micpath.2003.12.006

Berneking L, Schnapp M, Rumm A, Trasak C, Ruckdeschel K, Alawi M, Grundhoff A, Kikhney AG, Koch-Nolte F, Buck F, Perbandt M, Betzel C, Svergun DI, Hentschke M, Aepfelbacher M (2016) Immunosuppressive Yersinia effector YopM binds DEAD box helicase DDX3 to control ribosomal S6 kinase in the nucleus of host cells. PLoS Pathog 12(6):e1005660. doi:10.1371/journal.ppat.1005660

Bessede E, Staedel C, Acuna Amador LA, Nguyen PH, Chambonnier L, Hatakeyama M, Belleannee G, Megraud F, Varon C (2014) Helicobacter pylori generates cells with cancer stem cell properties via epithelial-mesenchymal transition-like changes. Oncogene 33 (32):4123–4131. doi:10.1038/onc.2013.380

Bhavsar AP, Guttman JA, Finlay BB (2007) Manipulation of host-cell pathways by bacterial pathogens. Nature 449(7164):827–834. doi:10.1038/nature06247

Bhutani N, Burns DM, Blau HM (2011) DNA demethylation dynamics. Cell 146(6):866–872. doi:10.1016/j.cell.2011.08.042

Bierne H (2013) Nuclear microbiology—bacterial assault on the nucleolus. EMBO Rep 14 (8):663–664. doi:10.1038/embor.2013.105

Bierne H, Cossart P (2012) When bacteria target the nucleus: the emerging family of nucleomodulins. Cell Microbiol. doi:10.1111/j.1462-5822.2012.01758.x

Bierne H, Hamon M, Cossart P (2012) Epigenetics and bacterial infections. Cold Spring Harb Perspect Med 2(12):a010272. doi:10.1101/cshperspect.a010272

Bierne H, Tham TN, Batsche E, Dumay A, Leguillou M, Kerneis-Golsteyn S, Regnault B, Seeler JS, Muchardt C, Feunteun J, Cossart P (2009) Human BAHD1 promotes heterochromatic gene silencing. Proc Natl Acad Sci U S A 106(33):13826–13831. doi:10.1073/pnas.0901259106

Bird A (2007) Perceptions of epigenetics. Nature 447(7143):396–398. doi:10.1038/nature05913

Bobetsis YA, Barros SP, Lin DM, Weidman JR, Dolinoy DC, Jirtle RL, Boggess KA, Beck JD, Offenbacher S (2007) Bacterial infection promotes DNA hypermethylation. J Dent Res 86 (2):169–174. doi:10.1177/154405910708600212

Borjesson DL, Kobayashi SD, Whitney AR, Voyich JM, Argue CM, Deleo FR (2005) Insights into pathogen immune evasion mechanisms: Anaplasma phagocytophilum fails to induce an apoptosis differentiation program in human neutrophils. J Immunol 174(10):6364–6372

Brennan DF, Barford D (2009) Eliminylation: a post-translational modification catalyzed by phosphothreonine lyases. Trends Biochem Sci 34(3):108–114. doi:10.1016/j.tibs.2008.11.005

Brestoff JR, Artis D (2013) Commensal bacteria at the interface of host metabolism and the immune system. Nat Immunol 14(7):676–684. doi:10.1038/ni.2640

Buck-Koehntop BA, Defossez PA (2013) On how mammalian transcription factors recognize methylated DNA. Epigenetics 8(2):131–137. doi:10.4161/epi.23632

Bussiere FI, Michel V, Memet S, Ave P, Vivas JR, Huerre M, Touati E (2010) H. pylori-induced promoter hypermethylation downregulates USF1 and USF2 transcription factor gene expression. Cell Microbiol 12(8):1124–1133. doi:10.1111/j.1462-5822.2010.01457.x

Cai Y, Geutjes EJ, de Lint K, Roepman P, Bruurs L, Yu LR, Wang W, van Blijswijk J, Mohammad H, de Rink I, Bernards R, Baylin SB (2014) The NuRD complex cooperates with DNMTs to maintain silencing of key colorectal tumor suppressor genes. Oncogene 33 (17):2157–2168. doi:10.1038/Onc.2013.178

Canonne J, Rivas S (2012) Bacterial effectors target the plant cell nucleus to subvert host transcription. Plant Signal Behav 7(2):217–221. doi:10.4161/psb.18885

Cao J (2014) The functional role of long non-coding RNAs and epigenetics. Biol Proced Online 16:11. doi:10.1186/1480-9222-16-11

Carson WF, Cavassani KA, Dou Y, Kunkel SL (2011) Epigenetic regulation of immune cell functions during post-septic immunosuppression. Epigenetics 6(3):273–283

Cedar H, Bergman Y (2009) Linking DNA methylation and histone modification: patterns and paradigms. Nature Rev Genetics 10(5):295–304. doi:10.1038/nrg2540

Chan AO, Lam SK, Wong BC, Wong WM, Yuen MF, Yeung YH, Hui WM, Rashid A, Kwong YL (2003) Promoter methylation of E-cadherin gene in gastric mucosa associated with Helicobacter pylori infection and in gastric cancer. Gut 52(4):502–506

Chandran A, Antony C, Jose L, Mundayoor S, Natarajan K, Kumar RA (2015) Mycobacterium tuberculosis infection induces HDAC1-mediated suppression of IL-12B gene expression in macrophages. Front Cell Infect Microbiol 5:90. doi:10.3389/fcimb.2015.00090

Chang PV, Hao L, Offermanns S, Medzhitov R (2014) The microbial metabolite butyrate regulates intestinal macrophage function via histone deacetylase inhibition. Proc Natl Acad Sci U S A 111(6):2247–2252. doi:10.1073/pnas.1322269111

Chen ZX, Riggs AD (2011) DNA methylation and demethylation in mammals. J Biol Chem 286 (21):18347–18353. doi:10.1074/jbc.R110.205286

Chen X, Yoza BK, El Gazzar M, Hu JY, Cousart SL, McCall CE (2009) RelB sustains IkappaBalpha expression during endotoxin tolerance. Clin Vaccine Immunol 16(1):104–110. doi:10.1128/CVI.00320-08

Cheng SC, Quintin J, Cramer RA, Shepardson KM, Saeed S, Kumar V, Giamarellos-Bourboulis EJ, Martens JH, Rao NA, Aghajanirefah A, Manjeri GR, Li Y, Ifrim DC, Arts RJ, van der Veer BM, Deen PM, Logie C, O'Neill LA, Willems P, van de Veerdonk FL, van der Meer JW, Ng A, Joosten LA, Wijmenga C, Stunnenberg HG, Xavier RJ, Netea MG (2014) mTOR- and HIF-1alpha-mediated aerobic glycolysis as metabolic basis for trained immunity. Science 345(6204):1250684. doi:10.1126/science.1250684

Chernov AV, Reyes L, Xu Z, Gonzalez B, Golovko G, Peterson S, Perucho M, Fofanov Y, Strongin AY (2015) Mycoplasma CG- and GATC-specific DNA methyltransferases selectively and efficiently methylate the host genome and alter the epigenetic landscape in human cells. Epigenetics 10(4):303–318. doi:10.1080/15592294.2015.1020000

Cuddapah S, Barski A, Zhao K (2010) Epigenomics of T cell activation, differentiation, and memory. Curr Opin Immunol 22(3):341–347. doi:10.1016/j.coi.2010.02.007

de Camargo Pereira G, Guimaraes GN, Planello AC, Santamaria MP, de Souza AP, Line SR, Marques MR (2013) Porphyromonas gingivalis LPS stimulation downregulates DNMT1, DNMT3a, and JMJD3 gene expression levels in human HaCaT keratinocytes. Clin Oral Investig 17(4):1279–1285. doi:10.1007/s00784-012-0816-z

Denizot J, Desrichard A, Agus A, Uhrhammer N, Dreux N, Vouret-Craviari V, Hofman P, Darfeuille-Michaud A, Barnich N (2015) Diet-induced hypoxia responsive element demethylation increases CEACAM6 expression, favouring Crohn's disease-associated Escherichia coli colonisation. Gut 64(3):428–437. doi:10.1136/gutjnl-2014-306944

Diacovich L, Gorvel JP (2010) Bacterial manipulation of innate immunity to promote infection. Nat Rev Microbiol 8(2):117–128. doi:10.1038/nrmicro2295

Ding SZ, Fischer W, Kaparakis-Liaskos M, Liechti G, Merrell DS, Grant PA, Ferrero RL, Crowe SE, Haas R, Hatakeyama M, Goldberg JB (2010a) Helicobacter pylori-induced histone modification, associated gene expression in gastric epithelial cells, and its implication in pathogenesis. PLoS One 5(4):e9875. doi:10.1371/journal.pone.0009875

Ding SZ, Goldberg JB, Hatakeyama M (2010b) Helicobacter pylori infection, oncogenic pathways and epigenetic mechanisms in gastric carcinogenesis. Future Oncol 6(5):851–862. doi:10.2217/fon.10.37

Dominissini D, Moshitch-Moshkovitz S, Schwartz S, Salmon-Divon M, Ungar L, Osenberg S, Cesarkas K, Jacob-Hirsch J, Amariglio N, Kupiec M, Sorek R, Rechavi G (2012) Topology of the human and mouse m6A RNA methylomes revealed by m6A-seq. Nature 485 (7397):201–206. doi:10.1038/nature11112

Dominissini D, Nachtergaele S, Moshitch-Moshkovitz S, Peer E, Kol N, Ben-Haim MS, Dai Q, Di Segni A, Salmon-Divon M, Clark WC, Zheng GQ, Pan T, Solomon O, Eyal E, Hershkovitz V, Han D, Dore LC, Amariglio N, Rechavi G, He C (2016) The dynamic N-1-methyladenosine methylome in eukaryotic messenger RNA. Nature 530(7591):441–446. doi:10.1038/nature16998

Du J, Johnson LM, Jacobsen SE, Patel DJ (2015) DNA methylation pathways and their crosstalk with histone methylation. Nat Rev Mol Cell Biol 16(9):519–532. doi:10.1038/nrm4043

Dumler JS, Sinclair SH, Pappas-Brown V, Shetty AC (2016) Genome-wide *Anaplasma phagocytophilum* AnkA-DNA interactions are enriched in intergenic regions and gene promoters and correlate with infection-induced differential gene expression. Front Cell Infect Microbiol 20(6):97. doi:10.3389/fcimb.2016.00097

Dunphy PS, Luo T, McBride JW (2014) Ehrlichia chaffeensis exploits host SUMOylation pathways to mediate effector-host interactions and promote intracellular survival. Infect Immun 82 (10):4154–4168. doi:10.1128/IAI.01984-14

Dussurget O, Bierne H, Cossart P (2014) The bacterial pathogen Listeria monocytogenes and the interferon family: type I, type II and type III interferons. Front Cell Infect Microbiol 4:50. doi:10.3389/fcimb.2014.00050

El Gazzar M, Yoza BK, Chen X, Hu J, Hawkins GA, McCall CE (2008) G9a and HP1 couple histone and DNA methylation to TNFalpha transcription silencing during endotoxin tolerance. J Biol Chem 283(47):32198–32208. doi:10.1074/jbc.M803446200

Eskandarian HA, Impens F, Nahori MA, Soubigou G, Coppee JY, Cossart P, Hamon MA (2013) A role for SIRT2-dependent histone H3K18 deacetylation in bacterial infection. Science 341 (6145):1238858. doi:10.1126/science.1238858

Faghihi MA, Wahlestedt C (2009) Regulatory roles of natural antisense transcripts. Nat Rev Mol Cell Biol 10(9):637–643. doi:10.1038/nrm2738

Farris TR, Dunphy PS, Zhu B, Kibler CE, McBride JW (2016) *Ehrlichia chaffeensis* TRP32 is a nucleomodulin that directly regulates expression of host genes governing differentiation and proliferation. Infect Immun 84:3182–3194. doi:10.1128/IAI.00657-16

Fehri LF, Rechner C, Janssen S, Mak TN, Holland C, Bartfeld S, Bruggemann H, Meyer TF (2009) Helicobacter pylori-induced modification of the histone H3 phosphorylation status in gastric epithelial cells reflects its impact on cell cycle regulation. Epigenetics 4(8):577–586

Fujita N, Watanabe S, Ichimura T, Tsuruzoe S, Shinkai Y, Tachibana M, Chiba T, Nakao M (2003) Methyl-CpG binding domain 1 (MBD1) interacts with the Suv39h1-HP1 heterochromatic complex for DNA methylation-based transcriptional repression. J Biol Chem 278 (26):24132–24138. doi:10.1074/jbc.M302283200

Furusawa Y, Obata Y, Fukuda S, Endo TA, Nakato G, Takahashi D, Nakanishi Y, Uetake C, Kato K, Kato T, Takahashi M, Fukuda NN, Murakami S, Miyauchi E, Hino S, Atarashi K, Onawa S, Fujimura Y, Lockett T, Clarke JM, Topping DL, Tomita M, Hori S, Ohara O, Morita T, Koseki H, Kikuchi J, Honda K, Hase K, Ohno H (2013) Commensal microbe-derived butyrate induces the differentiation of colonic regulatory T cells. Nature 504(7480):446–450. doi:10.1038/nature12721

Furusawa Y, Obata Y, Hase K (2015) Commensal microbiota regulates T cell fate decision in the gut. Semin Immunopathol 37(1):17–25. doi:10.1007/s00281-014-0455-3

Garcia-Dominguez M, Reyes JC (2009) SUMO association with repressor complexes, emerging routes for transcriptional control. Biochim Biophys Acta 1789(6–8):451–459. doi:10.1016/j.bbagrm.2009.07.001

Garcia-Garcia JC, Barat NC, Trembley SJ, Dumler JS (2009a) Epigenetic silencing of host cell defense genes enhances intracellular survival of the rickettsial pathogen *Anaplasma phagocytophilum*. PLoS Pathog 5(6):e1000488. doi:10.1371/journal.ppat.1000488

Garcia-Garcia JC, Rennoll-Bankert KE, Pelly S, Milstone AM, Dumler JS (2009b) Silencing of host cell CYBB gene expression by the nuclear effector AnkA of the intracellular pathogen *Anaplasma phagocytophilum*. Infect Immun 77(6):2385–2391. doi:10.1128/IAI.00023-09

Ghadimi D, Helwig U, Schrezenmeir J, Heller KJ, de Vrese M (2012) Epigenetic imprinting by commensal probiotics inhibits the IL-23/IL-17 axis in an in vitro model of the intestinal mucosal immune system. J Leukoc Biol 92(4):895–911. doi:10.1189/jlb.0611286

Guelen L, Pagie L, Brasset E, Meuleman W, Faza MB, Talhout W, Eussen BH, de Klein A, Wessels L, de Laat W, van Steensel B (2008) Domain organization of human chromosomes revealed by mapping of nuclear lamina interactions. Nature 453(7197):948–951. doi:10.1038/nature06947

Ha SD, Han CY, Reid C, Kim SO (2014) HDAC8-mediated epigenetic reprogramming plays a key role in resistance to anthrax lethal toxin-induced pyroptosis in macrophages. J Immunol 193 (3):1333–1343. doi:10.4049/jimmunol.1400420

Ha SD, Reid C, Meshkibaf S, Kim SO (2016) Inhibition of interleukin 1beta (IL-1beta) expression by anthrax Lethal Toxin (LeTx) is reversed by Histone Deacetylase 8 (HDAC8) inhibition in murine macrophages. J Biol Chem 291(16):8745–8755. doi:10.1074/jbc.M115.695809

Haller D, Holt L, Kim SC, Schwabe RF, Sartor RB, Jobin C (2003) Transforming growth factor-beta 1 inhibits non-pathogenic Gram negative bacteria-induced NF-kappa B recruitment to the interleukin-6 gene promoter in intestinal epithelial cells through modulation of histone acetylation. J Biol Chem 278(26):23851–23860. doi:10.1074/jbc.M300075200

Hamon MA, Batsche E, Regnault B, Tham TN, Seveau S, Muchardt C, Cossart P (2007) Histone modifications induced by a family of bacterial toxins. Proc Natl Acad Sci U S A 104 (33):13467–13472. doi:10.1073/pnas.0702729104

Hamon M, Bierne H, Cossart P (2006) *Listeria monocytogenes*: a multifaceted model. Nat Rev Microbiol 4(6):423–434. doi:10.1038/nrmicro1413

Hamon MA, Cossart P (2011) K+ efflux is required for histone H3 dephosphorylation by Listeria monocytogenes listeriolysin O and other pore-forming toxins. Infect Immun 79(7):2839–2846. doi:10.1128/IAI.01243-10

Haraga A, Miller SI (2003) A Salmonella enterica serovar typhimurium translocated leucine-rich repeat effector protein inhibits NF-kappa B-dependent gene expression. Infect Immun 71 (7):4052–4058

Harouz H, Rachez C, Meijer BM, Marteyn B, Donnadieu F, Cammas F, Muchardt C, Sansonetti P, Arbibe L (2014) Shigella flexneri targets the HP1gamma subcode through the phosphothreonine lyase OspF. EMBO J 33 (22):2606–2622. doi:10.15252/embj.201489244

Hattori N, Ushijima T (2016) Epigenetic impact of infection on carcinogenesis: mechanisms and applications. Genome Med 8(1):10. doi:10.1186/s13073-016-0267-2

Hicks SW, Galan JE (2010) Hijacking the host ubiquitin pathway: structural strategies of bacterial E3 ubiquitin ligases. Curr Opin Microbiol 13(1):41–46. doi:10.1016/j.mib.2009.11.008

Hiragami-Hamada K, Shinmyozu K, Hamada D, Tatsu Y, Uegaki K, Fujiwara S, Nakayama J (2011) N-terminal phosphorylation of HP1{alpha} promotes its chromatin binding. Mol Cell Biol 31(6):1186–1200. doi:10.1128/MCB.01012-10

Hnilicova J, Stanek D (2011) Where splicing joins chromatin. Nucleus 2(3):182–188. doi:10.4161/nucl.2.3.15876

Holeski LM, Jander G, Agrawal AA (2012) Transgenerational defense induction and epigenetic inheritance in plants. Trends Ecol Evol 27(11):618–626. doi:10.1016/j.tree.2012.07.011

Hur K, Niwa T, Toyoda T, Tsukamoto T, Tatematsu M, Yang HK, Ushijima T (2011) Insufficient role of cell proliferation in aberrant DNA methylation induction and involvement of specific types of inflammation. Carcinogenesis 32(1):35–41. doi:10.1093/carcin/bgq219

Ichimura T, Watanabe S, Sakamoto Y, Aoto T, Fujita N, Nakao M (2005) Transcriptional repression and heterochromatin formation by MBD1 and MCAF/AM family proteins. J Biol Chem 280(14):13928–13935. doi:10.1074/jbc.M413654200

Ifrim DC, Quintin J, Joosten LA, Jacobs C, Jansen T, Jacobs L, Gow NA, Williams DL, van der Meer JW, Netea MG (2014) Trained immunity or tolerance: opposing functional programs induced in human monocytes after engagement of various pattern recognition receptors. Clin Vaccine Immunol 21(4):534–545. doi:10.1128/CVI.00688-13

Imai K, Inoue H, Tamura M, Cueno ME, Inoue H, Takeichi O, Kusama K, Saito I, Ochiai K (2012) The periodontal pathogen Porphyromonas gingivalis induces the Epstein-Barr virus lytic switch transactivator ZEBRA by histone modification. Biochimie 94(3):839–846. doi:10.1016/j.biochi.2011.12.001

Imai K, Inoue H, Tamura M, Cueno ME, Takeichi O, Kusama K, Saito I, Ochiai K (2011) The periodontal pathogen Porphyromonas gingivalis induces the Epstein-Barr virus lytic switch transactivator ZEBRA by histone modification. Biochimie. doi:10.1016/j.biochi.2011.12.001

Imai K, Ochiai K, Okamoto T (2009) Reactivation of latent HIV-1 infection by the periodontopathic bacterium Porphyromonas gingivalis involves histone modification. J Immunol 182(6):3688–3695. doi:10.4049/jimmunol.0802906

Jenner RG, Young RA (2005) Insights into host responses against pathogens from transcriptional profiling. Nat Rev Microbiol 3(4):281–294. doi:10.1038/nrmicro1126

Jimenez A, Chen D, Alto NM (2016) How bacteria subvert animal cell structure and function. Annu Rev Cell Dev Biol. doi:10.1146/annurev-cellbio-100814-125227

Jose L, Ramachandran R, Bhagavat R, Gomez RL, Chandran A, Raghunandanan S, Omkumar RV, Chandra N, Mundayoor S, Kumar RA (2016) Hypothetical protein Rv3423.1 of Mycobacterium tuberculosis is a histone acetyltransferase. FEBS J 283(2):265–281. doi:10.1111/febs.13566

Kaikkonen MU, Lam MT, Glass CK (2011) Non-coding RNAs as regulators of gene expression and epigenetics. Cardiovasc Res 90(3):430–440. doi:10.1093/cvr/cvr097

Kleinnijenhuis J, Quintin J, Preijers F, Joosten LA, Ifrim DC, Saeed S, Jacobs C, van Loenhout J, de Jong D, Stunnenberg HG, Xavier RJ, van der Meer JW, van Crevel R, Netea MG (2012) Bacille Calmette-Guerin induces NOD2-dependent nonspecific protection from reinfection via epigenetic reprogramming of monocytes. Proc Natl Acad Sci U S A 109(43):17537–17542. doi:10.1073/pnas.1202870109

Klose RJ, Bird AP (2006) Genomic DNA methylation: the mark and its mediators. Trends Biochem Sci 31 (2):89–97. doi:1016/j.tibs.2005.12.008

Kohwi-Shigematsu T, Kohwi Y, Takahashi K, Richards HW, Ayers SD, Han HJ, Cai S (2012) SATB1-mediated functional packaging of chromatin into loops. Methods 58(3):243–254. doi:10.1016/j.ymeth.2012.06.019

Kouzarides T (2007) Chromatin modifications and their function. Cell 128(4):693–705. doi:10.1016/j.cell.2007.02.005

Lai AY, Wade PA (2011) Cancer biology and NuRD: a multifaceted chromatin remodelling complex. Nat Rev Cancer 11(8):588–596. doi:10.1038/nrc3091

Lakisic G, Lebreton A, Pourpre R, Wendling O, Libertini E, Radford EJ, Le Guillou M, Champy MF, Wattenhofer-Donze M, Soubigou G, Ait-Si-Ali S, Feunteun J, Sorg T, Coppee JY, Ferguson-Smith AC, Cossart P, Bierne H (2016) Role of the BAHD1 chromatin-repressive complex in placental development and regulation of steroid metabolism. PLoS Genet 12(3): e1005898. doi:10.1371/journal.pgen.1005898

Langst G, Manelyte L (2015) Chromatin remodelers: from function to dysfunction. Genes (Basel) 6(2):299–324. doi:10.3390/genes6020299

Latham JA, Dent SY (2007) Cross-regulation of histone modifications. Nat Struct Mol Biol 14 (11):1017–1024. doi:10.1038/nsmb1307

Latham T, Mackay L, Sproul D, Karim M, Culley J, Harrison DJ, Hayward L, Langridge-Smith P, Gilbert N, Ramsahoye BH (2012) Lactate, a product of glycolytic metabolism, inhibits histone deacetylase activity and promotes changes in gene expression. Nucleic Acids Res 40 (11):4794–4803. doi:10.1093/nar/gks066

Lebreton A, Cossart P, Bierne H (2012) Bacteria tune interferon responses by playing with chromatin. Virulence 3(1):87–91

Lebreton A, Job V, Ragon M, Le Monnier A, Dessen A, Cossart P, Bierne H (2014) Structural basis for the inhibition of the chromatin repressor BAHD1 by the bacterial nucleomodulin LntA. MBio 5(1):e00775–e00713. doi:10.1128/mBio.00775-13

Lebreton A, Lakisic G, Job V, Fritsch L, Tham TN, Camejo A, Mattei P-J, Regnault B, Nahori M-A, Cabanes D, Gautreau A, Ait-Si-Ali S, Dessen A, Cossart P, Bierne H (2011) A bacterial protein targets the BAHD1 chromatin complex to stimulate type III interferon response. Science 331(6022):1319–1321

Lee HC, Kioi M, Han J, Puri RK, Goodman JL (2008) Anaplasma phagocytophilum-induced gene expression in both human neutrophils and HL-60 cells. Genomics 92(3):144–151. doi:10.1016/j.ygeno.2008.05.005

Lee JS, Smith E, Shilatifard A (2010) The language of histone crosstalk. Cell 142(5):682–685. doi:10.1016/j.cell.2010.08.011

Li T, Lu Q, Wang G, Xu H, Huang H, Cai T, Kan B, Ge J, Shao F (2013) SET-domain bacterial effectors target heterochromatin protein 1 to activate host rDNA transcription. EMBO Rep 14 (8):733–740. doi:10.1038/embor.2013.86

Li H, Rauch T, Chen ZX, Szabo PE, Riggs AD, Pfeifer GP (2006) The histone methyltransferase SETDB1 and the DNA methyltransferase DNMT3A interact directly and localize to promoters silenced in cancer cells. J Biol Chem 281(28):19489–19500. doi:10.1074/jbc.M513249200

Li H, Xu H, Zhou Y, Zhang J, Long C, Li S, Chen S, Zhou J-M, Shao F (2007) The phosphothreonine lyase activity of a bacterial type III effector family. Science 315 (5814):1000–1003. doi:10.1126/science.1138960

Libertini E, Lebreton A, Lakisic G, Dillies MA, Beck S, Coppee JY, Cossart P, Bierne H (2015) Overexpression of the heterochromatinization factor BAHD1 in HEK293 cells differentially reshapes the DNA methylome on autosomes and X chromosome. Front Genet 6:339. doi:10.3389/fgene.2015.00339

Licciardi PV, Wong SS, Tang ML, Karagiannis TC (2010) Epigenome targeting by probiotic metabolites. Gut Pathog 2(1):24. doi:10.1186/1757-4749-2-24

Liu Y, Zhang X, Blumenthal RM, Cheng X (2013) A common mode of recognition for methylated CpG. Trends Biochem Sci 38(4):177–183. doi:10.1016/j.tibs.2012.12.005

Luo T, Kuriakose JA, Zhu B, Wakeel A, McBride JW (2011) Ehrlichia chaffeensis TRP120 interacts with a diverse array of eukaryotic proteins involved in transcription, signaling, and cytoskeleton organization. Infect Immun 79(11):4382–4391. doi:10.1128/IAI.05608-11

Luo T, McBride JW (2012) Ehrlichia chaffeensis TRP32 interacts with host cell targets that influence intracellular survival. Infect Immun 80(7):2297–2306. doi:10.1128/IAI.00154-12

Ma KW, Flores C, Ma W (2011) Chromatin configuration as a battlefield in plant-bacteria interactions. Plant Physiol. doi:10.1104/pp.111.182295

Maekita T, Nakazawa K, Mihara M, Nakajima T, Yanaoka K, Iguchi M, Arii K, Kaneda A, Tsukamoto T, Tatematsu M, Tamura G, Saito D, Sugimura T, Ichinose M, Ushijima T (2006) High levels of aberrant DNA methylation in *Helicobacter pylori*-infected gastric mucosae and its possible association with gastric cancer risk. Clin Cancer Res 12(3 Pt 1):989–995. doi:10.1158/1078-0432.CCR-05-2096

Margueron R, Reinberg D (2011) The Polycomb complex PRC2 and its mark in life. Nature 469 (7330):343–349. doi:10.1038/nature09784

Masaki T, McGlinchey A, Cholewa-Waclaw J, Qu J, Tomlinson SR, Rambukkana A (2014) Innate immune response precedes Mycobacterium leprae-induced reprogramming of adult Schwann cells. Cell Reprogram 16(1):9–17. doi:10.1089/cell.2013.0064

Masaki T, Qu J, Cholewa-Waclaw J, Burr K, Raaum R, Rambukkana A (2013) Reprogramming adult Schwann cells to stem cell-like cells by leprosy bacilli promotes dissemination of infection. Cell 152(1–2):51–67. doi:10.1016/j.cell.2012.12.014

Mattick JS, Makunin IV (2005) Small regulatory RNAs in mammals. Hum Mol Genet 14(1): R121–R132. doi:10.1093/hmg/ddi101

Mattout A, Cabianca DS, Gasser SM (2015) Chromatin states and nuclear organization in development—a view from the nuclear lamina. Genome Biol 16:174. doi:10.1186/s13059-015-0747-5

McCall CE, Yoza B, Liu T, El Gazzar M (2010) Gene-specific epigenetic regulation in serious infections with systemic inflammation. J Innate Immun 2(5):395–405. doi:10.1159/000314077

McCoy MW, Marre ML, Lesser CF, Mecsas J (2010) The C-terminal tail of Yersinia pseudotuberculosis YopM is critical for interacting with RSK1 and for virulence. Infect Immun 78 (6):2584–2598. doi:10.1128/IAI.00141-10

McDonald C, Vacratsis PO, Bliska JB, Dixon JE (2003) The yersinia virulence factor YopM forms a novel protein complex with two cellular kinases. J Biol Chem 278(20):18514–18523. doi:10.1074/jbc.M301226200

Medvedeva YA, Lennartsson A, Ehsani R, Kulakovskiy IV, Vorontsov IE, Panahandeh P, Khimulya G, Kasukawa T, Consortium F, Drablos F (2015) EpiFactors: a comprehensive database of human epigenetic factors and complexes. Database (Oxford) 2015:bav067. doi:10.1093/database/bav067

Medzhitov R, Horng T (2009) Transcriptional control of the inflammatory response. Nature Rev Immunol 9(10):692–703. doi:10.1038/nri2634

Meyer KD, Jaffrey SR (2014) The dynamic epitranscriptome: N6-methyladenosine and gene expression control. Nat Rev Mol Cell Biol 15(5):313–326. doi:10.1038/nrm3785

Mikkelsen TS, Ku M, Jaffe DB, Issac B, Lieberman E, Giannoukos G, Alvarez P, Brockman W, Kim TK, Koche RP, Lee W, Mendenhall E, O'Donovan A, Presser A, Russ C, Xie X, Meissner A, Wernig M, Jaenisch R, Nusbaum C, Lander ES, Bernstein BE (2007) Genome-wide maps of chromatin state in pluripotent and lineage-committed cells. Nature 448 (7153):553–560. doi:10.1038/nature06008

Minarovits J (2009) Microbe-induced epigenetic alterations in host cells: the coming era of patho-epigenetics of microbial infections. A review. Acta Microbiol Immunol Hung 56(1):1–19. doi:10.1556/AMicr.56.2009.1.1

Mohammad HP, Baylin SB (2010) Linking cell signaling and the epigenetic machinery. Nat Biotechnol 28(10):1033–1038. doi:10.1038/nbt1010-1033

Mojica SA, Hovis KM, Frieman MB, Tran B, Hsia RC, Ravel J, Jenkins-Houk C, Wilson KL, Bavoil PM (2015) SINC, a type III secreted protein of Chlamydia psittaci, targets the inner nuclear membrane of infected cells and uninfected neighbors. Mol Biol Cell 26 (10):1918–1934. doi:10.1091/mbc.E14-11-1530

Morris TL, Arnold RR, Webster-Cyriaque J (2007) Signaling cascades triggered by bacterial metabolic end products during reactivation of Kaposi's sarcoma-associated herpesvirus. J Virol 81(11):6032–6042. doi:10.1128/JVI.02504-06

Mujtaba S, Winer BY, Jaganathan A, Patel J, Sgobba M, Schuch R, Gupta YK, Haider S, Wang R, Fischetti VA (2013) Anthrax SET protein: a potential virulence determinant that epigenetically represses NF-kappaB activation in infected macrophages. J Biol Chem 288(32):23458–23472. doi:10.1074/jbc.M113.467696

Netea MG, Joosten LA, Latz E, Mills KH, Natoli G, Stunnenberg HG, O'Neill LA, Xavier RJ (2016) Trained immunity: a program of innate immune memory in health and disease. Science 352(6284):aaf1098. doi:10.1126/science.aaf1098

Nigro G, Rossi R, Commere PH, Jay P, Sansonetti PJ (2014) The cytosolic bacterial peptidoglycan sensor Nod2 affords stem cell protection and links microbes to gut epithelial regeneration. Cell Host Microbe 15(6):792–798. doi:10.1016/j.chom.2014.05.003

Niwa T, Tsukamoto T, Toyoda T, Mori A, Tanaka H, Maekita T, Ichinose M, Tatematsu M, Ushijima T (2010) Inflammatory processes triggered by Helicobacter pylori infection cause aberrant DNA methylation in gastric epithelial cells. Cancer Res 70(4):1430–1440. doi:10.1158/0008-5472.CAN-09-2755

Noordermeer D, Duboule D (2013) Chromatin looping and organization at developmentally regulated gene loci. Wiley Interdiscip Rev Dev Biol 2(5):615–630. doi:10.1002/wdev.103

Ohta K, Kawano R, Ito N (2012) Lactic acid bacteria convert human fibroblasts to multipotent cells. PloS one 7(12):e51866. doi:10.1371/journal.pone.0051866

Okuda J, Toyotome T, Kataoka N, Ohno M, Abe H, Shimura Y, Seyedarabi A, Pickersgill R, Sasakawa C (2005) Shigella effector IpaH9.8 binds to a splicing factor U2AF35 to modulate host immune responses. Biochem Biophys Res Commun 333 (2):531–539. doi:10.1016/j.bbrc.2005.05.145

Opitz B, Puschel A, Beermann W, Hocke AC, Forster S, Schmeck B, van Laak V, Chakraborty T, Suttorp N, Hippenstiel S (2006) Listeria monocytogenes activated p38 MAPK and induced IL-8 secretion in a nucleotide-binding oligomerization domain 1-dependent manner in endothelial cells. J Immunol 176(1):484–490. doi:10.4049/jimmunol.176.1.484

Ostuni R, Piccolo V, Barozzi I, Polletti S, Termanini A, Bonifacio S, Curina A, Prosperini E, Ghisletti S, Natoli G (2013) Latent enhancers activated by stimulation in differentiated cells. Cell 152 (1–2):157–171. doi:10.1016/j.cell.2012.12.018

Pacis A, Tailleux L, Morin AM, Lambourne J, MacIsaac JL, Yotova V, Dumaine A, Danckaert A, Luca F, Grenier JC, Hansen KD, Gicquel B, Yu M, Pai A, He C, Tung J, Pastinen T, Kobor MS, Pique-Regi R, Gilad Y, Barreiro LB (2015) Bacterial infection remodels the DNA methylation landscape of human dendritic cells. Genome Res 25(12):1801–1811. doi:10.1101/gr.192005.115

Padeken J, Heun P (2014) Nucleolus and nuclear periphery: velcro for heterochromatin. Curr Opin Cell Biol 28:54–60. doi:10.1016/j.ceb.2014.03.001

Pamer EG (2004) Immune responses to Listeria monocytogenes. Nat Rev Immunol 4 (10):812–823. doi:10.1038/nri1461

Park J, Kim KJ, K-S C, Grab DJ, Dumler JS (2004) Anaplasma phagocytophilum AnkA binds to granulocyte DNA and nuclear proteins. Cell Microbiol 6(8):743–751. doi:10.1111/j.1462-5822.2004.00400.x

Pathak SK, Basu S, Bhattacharyya A, Pathak S, Banerjee A, Basu J, Kundu M (2006) TLR4-dependent NF-kappaB activation and mitogen- and stress-activated protein kinase 1-triggered phosphorylation events are central to *Helicobacter pylori* peptidyl prolyl cis-, trans-isomerase (HP0175)-mediated induction of IL-6 release from macrophages. J Immunol 177 (11):7950–7958. doi:10.4049/jimmunol.177.11.7950

Pennini ME, Liu Y, Yang J, Croniger CM, Boom WH, Harding CV (2007) CCAAT/enhancer-binding protein beta and delta binding to CIITA promoters is associated with the inhibition of CIITA expression in response to Mycobacterium tuberculosis 19-kDa lipoprotein. J Immunol 179 (10):6910–6918. doi:179/10/6910 [pii]

Pennini ME, Pai RK, Schultz DC, Boom WH, Harding CV (2006) Mycobacterium tuberculosis 19-kDa lipoprotein inhibits IFN-gamma-induced chromatin remodeling of MHC2TA by TLR2 and MAPK signaling. J Immunol 176(7):4323–4330. doi:10.4049/jimmunol.176.7.4323

Pennini M, Perrinet S, Dautry-Varsat A, Subtil A (2010) Histone methylation by NUE, a novel nuclear effector of the intracellular pathogen *Chlamydia trachomatis*. PLoS Pathog 6(7): e1000995. doi:10.1371/journal.ppat.1000995

Perdiguero EG, Geissmann F (2016) The development and maintenance of resident macrophages. Nat Immunol 17(1):2–8. doi:10.1038/ni.3341

Quintin J, Saeed S, Martens JH, Giamarellos-Bourboulis EJ, Ifrim DC, Logie C, Jacobs L, Jansen T, Kullberg BJ, Wijmenga C, Joosten LA, Xavier RJ, van der Meer JW, Stunnenberg HG, Netea MG (2012) Candida albicans infection affords protection against reinfection via functional reprogramming of monocytes. Cell Host Microbe 12(2):223–232. doi:10.1016/j.chom.2012.06.006

Rais Y, Zviran A, Geula S, Gafni O, Chomsky E, Viukov S, Mansour AA, Caspi I, Krupalnik V, Zerbib M, Maza I, Mor N, Baran D, Weinberger L, Jaitin DA, Lara-Astiaso D, Blecher-Gonen R, Shipony Z, Mukamel Z, Hagai T, Gilad S, Amann-Zalcenstein D, Tanay A, Amit I, Novershtern N, Hanna JH (2013) Deterministic direct reprogramming of somatic cells to pluripotency. Nature 502(7469):65–70. doi:10.1038/nature12587

Ram O, Goren A, Amit I, Shoresh N, Yosef N, Ernst J, Kellis M, Gymrek M, Issner R, Coyne M, Durham T, Zhang X, Donaghey J, Epstein CB, Regev A, Bernstein BE (2011) Combinatorial patterning of chromatin regulators uncovered by genome-wide location analysis in human cells. Cell 147(7):1628–1639. doi:10.1016/j.cell.2011.09.057

Raymond B, Batsche E, Boutillon F, Wu YZ, Leduc D, Balloy V, Raoust E, Muchardt C, Goossens PL, Touqui L (2009) Anthrax lethal toxin impairs IL-8 expression in epithelial cells through inhibition of histone H3 modification. PLoS Pathog 5(4):e1000359. doi:10.1371/journal.ppat.1000359

Rennoll-Bankert KE, Dumler JS (2012) Lessons from Anaplasma phagocytophilum: chromatin remodeling by bacterial effectors. Infect Disord Drug Targets 12(5):380–387

Rennoll-Bankert KE, Garcia-Garcia JC, Sinclair SH, Dumler JS (2015) Chromatin-bound bacterial effector ankyrin A recruits histone deacetylase 1 and modifies host gene expression. Cell Microbiol 17(11):1640–1652. doi:10.1111/cmi.12461

Riggs AD, Martienssen RA, Russo VEA (1996) Introduction. In Russo, VEA et al (ed) Epigenetic mechanisms of gene regulation. Cold Spring Harbor Laboratory Press, New York, p 1

Riggs MG, Whittaker RG, Neumann JR, Ingram VM (1977) n-Butyrate causes histone modification in HeLa and Friend erythroleukaemia cells. Nature 268(5619):462–464

Roadmap Epigenomics C, Kundaje A, Meuleman W, Ernst J, Bilenky M, Yen A, Heravi-Moussavi A, Kheradpour P, Zhang Z, Wang J, Ziller MJ, Amin V, Whitaker JW, Schultz MD, Ward LD, Sarkar A, Quon G, Sandstrom RS, Eaton ML, Wu YC, Pfenning AR, Wang X, Claussnitzer M, Liu Y, Coarfa C, Harris RA, Shoresh N, Epstein CB, Gjoneska E, Leung D, Xie W, Hawkins RD, Lister R, Hong C, Gascard P, Mungall AJ, Moore R, Chuah E, Tam A, Canfield TK, Hansen RS, Kaul R, Sabo PJ, Bansal MS, Carles A, Dixon JR, Farh KH, Feizi S, Karlic R, Kim AR, Kulkarni A, Li D, Lowdon R, Elliott G, Mercer TR, Neph SJ, Onuchic V, Polak P, Rajagopal N, Ray P, Sallari RC, Siebenthall KT, Sinnott-Armstrong NA, Stevens M, Thurman RE, Wu J, Zhang B, Zhou X, Beaudet AE, Boyer LA, De Jager PL, Farnham PJ, Fisher SJ, Haussler D, Jones SJ, Li W, Marra MA, McManus MT, Sunyaev S, Thomson JA, Tlsty TD, Tsai LH, Wang W, Waterland RA, Zhang MQ, Chadwick LH, Bernstein BE, Costello JF, Ecker JR, Hirst M, Meissner A, Milosavljevic A, Ren B, Stamatoyannopoulos JA, Wang T, Kellis M (2015) Integrative analysis of 111 reference human epigenomes. Nature 518(7539):317–330. doi:10.1038/nature14248

Rohde JR, Breitkreutz A, Chenal A, Sansonetti PJ, Parsot C (2007) Type III secretion effectors of the IpaH family are E3 ubiquitin ligases. Cell Host Microbe 1(1):77–83. doi:10.1016/j.chom.2007.02.002

Rolando M, Sanulli S, Rusniok C, Gomez-Valero L, Bertholet C, Sahr T, Margueron R, Buchrieser C (2013) Legionella pneumophila effector RomA uniquely modifies host chromatin to repress gene expression and promote intracellular bacterial replication. Cell Host Microbe 13(4):395–405. doi:10.1016/j.chom.2013.03.004

Romanoski CE, Glass CK, Stunnenberg HG, Wilson L, Almouzni G (2015) Epigenomics: roadmap for regulation. Nature 518(7539):314–316. doi:10.1038/518314a

Rye M, Sandve GK, Daub CO, Kawaji H, Carninci P, Forrest AR, Drablos F, Fantom Consortium (2014) Chromatin states reveal functional associations for globally defined transcription start sites in four human cell lines. BMC Genomics 15:120. doi:10.1186/1471-2164-15-120

Saccani S, Pantano S, Natoli G (2002) p38-Dependent marking of inflammatory genes for increased NF-kappa B recruitment. Nat Immunol 3(1):69–75. doi:10.1038/ni748

Sadakierska-Chudy A, Filip M (2015) A comprehensive view of the epigenetic landscape. Part II: Histone post-translational modification, nucleosome level, and chromatin regulation by ncRNAs. Neurotox Res 27(2):172–197. doi:10.1007/s12640-014-9508-6

Salles, II, Tucker AE, Voth DE, Ballard JD (2003) Toxin-induced resistance in Bacillus anthracis lethal toxin-treated macrophages. Proc Natl Acad Sci U S A 100 (21):12426–12431. doi:10.1073/pnas.2134042100

Samba-Louaka A, Pereira JM, Nahori MA, Villiers V, Deriano L, Hamon MA, Cossart P (2014) Listeria monocytogenes dampens the DNA damage response. PLoS Pathog 10(10):e1004470. doi:10.1371/journal.ppat.1004470

Sasai N, Defossez PA (2009) Many paths to one goal? The proteins that recognize methylated DNA in eukaryotes. Int J Dev Biol 53(2–3):323–334. doi:10.1387/ijdb.082652ns

Schmeck B, Lorenz J, N'Guessan PD, Opitz B, van Laak V, Zahlten J, Slevogt H, Witzenrath M, Flieger A, Suttorp N, Hippenstiel S (2008) Histone acetylation and flagellin are essential for Legionella pneumophila-induced cytokine expression. J Immunol 181(2):940–947. doi:10.4049/jimmunol.181.2.940

Schmitz KM, Mayer C, Postepska A, Grummt I (2010) Interaction of noncoding RNA with the rDNA promoter mediates recruitment of DNMT3b and silencing of rRNA genes. Genes Dev 24(20):2264–2269. doi:10.1101/gad.590910

Segain JP, Raingeard de la Bletiere D, Bourreille A, Leray V, Gervois N, Rosales C, Ferrier L, Bonnet C, Blottiere HM, Galmiche JP (2000) Butyrate inhibits inflammatory responses through NFkappaB inhibition: implications for Crohn's disease. Gut 47(3):397–403

Seyedarabi A, Sullivan JA, Sasakawa C, Pickersgill RW (2011) A disulfide driven domain swap switches off the activity of Shigella IpaH9.8 E3 ligase. FEBS Lett 584(19):4163–4168. doi:10.1016/j.febslet.2010.09.006

Shames SR, Bhavsar AP, Croxen MA, Law RJ, Mak SH, Deng W, Li Y, Bidshari R, de Hoog CL, Foster LJ, Finlay BB (2011) The pathogenic Escherichia coli type III secreted protease NleC degrades the host acetyltransferase p300. Cell Microbiol 13(10):1542–1557. doi:10.1111/j.1462-5822.2011.01640.x

Sharma G, Sowpati DT, Singh P, Khan MZ, Ganji R, Upadhyay S, Banerjee S, Nandicoori VK, Khosla S (2016) Genome-wide non-CpG methylation of the host genome during M. tuberculosis infection. Sci Rep 6:25006. doi:10.1038/srep25006

Sharma G, Upadhyay S, Srilalitha M, Nandicoori VK, Khosla S (2015) The interaction of mycobacterial protein Rv2966c with host chromatin is mediated through non-CpG methylation and histone H3/H4 binding. Nucleic Acids Res 43(8):3922–3937. doi:10.1093/nar/gkv261

Sinclair SH, Garcia-Garcia JC, Dumler JS (2015a) Bioinformatic and mass spectrometry identification of Anaplasma phagocytophilum proteins translocated into host cell nuclei. Front Microbiol 6:55. doi:10.3389/fmicb.2015.00055

Sinclair SH, Rennoll-Bankert KE, Dumler JS (2014) Effector bottleneck: microbial reprogramming of parasitized host cell transcription by epigenetic remodeling of chromatin structure. Front Genet 5:274. doi:10.3389/fgene.2014.00274

Sinclair SH, Yegnasubramanian S, Dumler JS (2015b) Global DNA methylation changes and differential gene expression in Anaplasma phagocytophilum-infected human neutrophils. Clin Epigenetics 7(1):77. doi:10.1186/s13148-015-0105-1

Slevogt H, Schmeck B, Jonatat C, Zahlten J, Beermann W, van Laak V, Opitz B, Dietel S, N'Guessan PD, Hippenstiel S, Suttorp N, Seybold J (2006) Moraxella catarrhalis induces inflammatory response of bronchial epithelial cells via MAPK and NF-kappaB activation

and histone deacetylase activity reduction. Am J Physiol Lung Cell Mol Physiol 290(5):L818–L826. doi:10.1152/ajplung.00428.2005
Sommer F, Backhed F (2013) The gut microbiota—masters of host development and physiology. Nat Rev Microbiol 11(4):227–238. doi:10.1038/nrmicro2974
Soundararajan V, Patel N, Subramanian V, Sasisekharan V, Sasisekharan R (2011) The many faces of the YopM effector from plague causative bacterium Yersinia pestis and its implications for host immune modulation. Innate Immun 17(6):548–557. doi:10.1177/1753425910377099
Suganuma T, Workman JL (2011) Signals and combinatorial functions of histone modifications. Annu Rev Biochem 80:473–499. doi:10.1146/annurev-biochem-061809-175347
Sun J (2010) Enteric bacteria and cancer stem cells. Cancers 3(1):285–297. doi:10.3390/cancers3010285
Tahoun A, Mahajan S, Paxton E, Malterer G, Donaldson DS, Wang D, Tan A, Gillespie TL, O'Shea M, Roe AJ, Shaw DJ, Gally DL, Lengeling A, Mabbott NA, Haas J, Mahajan A (2012) Salmonella transforms follicle-associated epithelial cells into M cells to promote intestinal invasion. Cell Host Microbe 12(5):645–656. doi:10.1016/j.chom.2012.10.009
Takahashi K, Sugi Y, Nakano K, Tsuda M, Kurihara K, Hosono A, Kaminogawa S (2011) Epigenetic control of the host gene by commensal bacteria in large intestinal epithelial cells. J Biol Chem 286(41):35755–35762. doi:10.1074/jbc.M111.271007
Takahashi K, Yamanaka S (2006) Induction of pluripotent stem cells from mouse embryonic and adult fibroblast cultures by defined factors. Cell 126(4):663–676. doi:10.1016/j.cell.2006.07.024
Taverna SD, Li H, Ruthenburg AJ, Allis CD, Patel DJ (2007) How chromatin-binding modules interpret histone modifications: lessons from professional pocket pickers. Nat Struct Mol Biol 14(11):1025–1040. doi:10.1038/nsmb1338
Ting K, Aitken KJ, Penna F, Samiei AN, Sidler M, Jiang JX, Ibrahim F, Tolg C, Delgado-Olguin P, Rosenblum N, Bagli DJ (2016) Uropathogenic E.coli (UPEC) infection induces proliferation through Enhancer of Zeste Homologue 2 (EZH2). PLoS One 11(3):e0149118. doi:10.1371/journal.pone.0149118
Tisseur M, Kwapisz M, Morillon A (2011) Pervasive transcription – Lessons from yeast. Biochimie 93(11):1889–1896. doi:10.1016/j.biochi.2011.07.001
Tiwari VK, Stadler MB, Wirbelauer C, Paro R, Schubeler D, Beisel C (2011) A chromatin-modifying function of JNK during stem cell differentiation. Nat Genet 44(1):94–100. doi:10.1038/ng.1036
Tolg C, Sabha N, Cortese R, Panchal T, Ahsan A, Soliman A, Aitken KJ, Petronis A, Bägli DJ (2011) Uropathogenic E. coli infection provokes epigenetic downregulation of CDKN2A (p16INK4A) in uroepithelial cells. Lab Invest 91:825–836
Toyotome T (2001) Shigella protein IpaH9.8 Is secreted from bacteria within mammalian cells and transported to the nucleus. J Biol Chem 276(34):32071–32079. doi:10.1074/jbc.M101882200
Turgeon N, Blais M, Gagne JM, Tardif V, Boudreau F, Perreault N, Asselin C (2013) HDAC1 and HDAC2 restrain the intestinal inflammatory response by regulating intestinal epithelial cell differentiation. PLoS One 8(9):e73785. doi:10.1371/journal.pone.0073785
Turner AM, Morris KV (2010) Controlling transcription with noncoding RNAs in mammalian cells. Biotechniques 48(6):ix–xvi. doi:10.2144/000113442
Ushijima T, Hattori N (2012) Molecular Pathways: involvement of helicobacter pylori -triggered inflammation in the formation of an epigenetic field defect, and its usefulness as cancer risk and exposure markers. Clin Cancer Res. doi:10.1158/1078-0432.CCR-11-2011
Valenzuela MA, Canales J, Corvalan AH, Quest AF (2015) Helicobacter pylori-induced inflammation and epigenetic changes during gastric carcinogenesis. World J Gastroenterol 21(45):12742–12756. doi:10.3748/wjg.v21.i45.12742
Vire E, Brenner C, Deplus R, Blanchon L, Fraga M, Didelot C, Morey L, Van Eynde A, Bernard D, Vanderwinden JM, Bollen M, Esteller M, Di Croce L, de Launoit Y, Fuks F (2006) The

Polycomb group protein EZH2 directly controls DNA methylation. Nature 439 (7078):871–874. doi:10.1038/nature04431
Wakeel A, Zhu B, Yu XJ, McBride JW (2010) New insights into molecular Ehrlichia chaffeensis-host interactions. Microbes Infect 12(5):337–345. doi:10.1016/j.micinf.2010.01.009
Wang Y, Curry HM, Zwilling BS, Lafuse WP (2005) Mycobacteria inhibition of IFN-gamma induced HLA-DR gene expression by up-regulating histone deacetylation at the promoter region in human THP-1 monocytic cells. J Immunol 174(9):5687–5694. doi:10.4049/jimmunol.174.9.5687
Wang TY, Han ZM, Chai YR, Zhang JH (2010) A mini review of MAR-binding proteins. Mol Biol Rep 37 (7):3553–3560. doi:10.1007/s11033-010-0003-8
Wani AH, Boettiger AN, Schorderet P, Ergun A, Munger C, Sadreyev RI, Zhuang X, Kingston RE, Francis NJ (2016) Chromatin topology is coupled to Polycomb group protein subnuclear organization. Nat Commun 7:10291. doi:10.1038/ncomms10291
Weber MM, Faris R, McLachlan J, Tellez A, Wright WU, Galvan G, Luo ZQ, Samuel JE (2016) Modulation of the host transcriptome by Coxiella burnetii nuclear effector Cbu1314. Microbes Infect 18(5):336–345. doi:10.1016/j.micinf.2016.01.003
Wu H, Zhang Y (2011) Mechanisms and functions of Tet protein-mediated 5-methylcytosine oxidation. Genes Dev 25(23):2436–2452. doi:10.1101/gad.179184.111
Xia G, Schneider-Stock R, Diestel A, Habold C, Krueger S, Roessner A, Naumann M, Lendeckel U (2008) Helicobacter pylori regulates p21(WAF1) by histone H4 acetylation. Biochem Biophys Res Commun 369 (2):526–531. doi:1016/j.bbrc.2008.02.073
Yamamoto Y, Verma UN, Prajapati S, Kwak YT, Gaynor RB (2003) Histone H3 phosphorylation by IKK-alpha is critical for cytokine-induced gene expression. Nature 423(6940):655–659. doi:10.1038/nature01576
Yan J, Zhang M, Zhang J, Chen X, Zhang X (2011) Helicobacter pylori infection promotes methylation of WWOX gene in human gastric cancer. Biochem Biophys Res Commun 408 (1):99–102. doi:10.1016/j.bbrc.2011.03.127
Yang L, Lin C, Liu W, Zhang J, Ohgi KA, Grinstein JD, Dorrestein PC, Rosenfeld MG (2011) ncRNA- and Pc2 methylation-dependent gene relocation between nuclear structures mediates gene activation programs. Cell 147(4):773–788. doi:10.1016/j.cell.2011.08.054
Yao Y, Tao H, Park DI, Sepulveda JL, Sepulveda AR (2006) Demonstration and characterization of mutations induced by Helicobacter pylori organisms in gastric epithelial cells. Helicobacter 11(4):272–286. doi:10.1111/j.1523-5378.2006.00408.x
Yaseen I, Kaur P, Nandicoori VK, Khosla S (2015) Mycobacteria modulate host epigenetic machinery by Rv1988 methylation of a non-tail arginine of histone H3. Nat Commun 6:8922. doi:10.1038/ncomms9922
Ye F, Karn J (2015) Bacterial short chain fatty acids push all the buttons needed to reactivate latent viruses. Stem Cell Epigenetics 2(1). doi:10.14800/sce.532
Yin L, Chung WO (2011) Epigenetic regulation of human beta-defensin 2 and CC chemokine ligand 20 expression in gingival epithelial cells in response to oral bacteria. Mucosal Immunol 4(4):409–419. doi:10.1038/mi.2010.83
Yuan X, Feng W, Imhof A, Grummt I, Zhou Y (2007) Activation of RNA polymerase I transcription by cockayne syndrome group B protein and histone methyltransferase G9a. Mol Cell 27(4):585–595. doi:10.1016/j.molcel.2007.06.021
Zhang S, Barros SP, Niculescu MD, Moretti AJ, Preisser JS, Offenbacher S (2010) Alteration of PTGS2 promoter methylation in chronic periodontitis. J Dent Res 89(2):133–137. doi:10.1177/0022034509356512
Zhou Y, Kim J, Yuan X, Braun T (2011) Epigenetic modifications of stem cells: a paradigm for the control of cardiac progenitor cells. Circ Res 109(9):1067–1081. doi:10.1161/CIRCRESAHA.111.243709

Zhu B, Nethery KA, Kuriakose JA, Wakeel A, Zhang X, McBride JW (2009) Nuclear translocated ehrlichia chaffeensis ankyrin protein interacts with a specific adenine-rich motif of host promoter and intronic alu elements. Infect Immun 77(10):4243–4255. doi:10.1128/IAI.00376-09

Zurawski DV, Mumy KL, Faherty CS, Mccormick BA, Maurelli AT (2009) Shigella flexneritype III secretion system effectors OspB and OspF target the nucleus to downregulate the host inflammatory response via interactions with retinoblastoma protein. Mol Microbiol 71(2):350–368. doi:10.1111/j.1365-2958.2008.06524.x

Chapter 7
Epigenetic Phase Variation in Bacterial Pathogens

Marjan W. van der Woude

Abstract Epigenetics is defined as heritable but reversible phenotype that does not involve a change in the DNA sequence. Thus, in a bacterial population epigenetic regulation of gene expression can define the phenotype of an individual cell. For a pathogen, the relevance lies in the phenotype of the infecting population and how this impacts on temporal changes in phenotype during an infection. The focus of this chapter is bacterial epigenetic regulation by DNA methylation and in particular phase variation that is controlled by the maintenance methyltransferase in *Escherichia coli* and *Salmonella*. The occurrence, the relevance for virulence, and key concepts of the mechanism are discussed. Our understanding of the relevance is presented based on the roles of the genes that are regulated and by considering the impact of population heterogeneity that occurs as a result of phase variation. More recently, the significance for the virulence of a different range of bacterial pathogens of DNA methyltransferases other than Dam has been identified. This leads to new understanding of possible origins and occurrence of heterogeneity and virulent phenotypes in clonal populations. The contribution of new technologies to explore bacterial methylomes and the challenges in identifying actual epigenetic regulation based on that is discussed. Overall, the current state of knowledge suggests that more examples of epigenetic control in bacteria are yet to be discovered and that this could enhance our understanding of virulence strategies in bacteria.

Keywords Bacterial pathogens • Salmonella • *E. coli* • Phase variation • Phasevarion • DNA methylation • Deoxyadenosine methyltransferase

7.1 Introduction

The outcome of a bacterial infection is a result of a complex set of interactions between the host and the pathogen. Epigenetics will affect the host response to an infection, but pathogen epigenetics also may affect the outcome of infection.

M.W. van der Woude (✉)
Department of Biology, Centre of Immunology and Infection, Hull York Medical School, University of York, York, UK
e-mail: marjan.vanderwoude@york.ac.uk

W. Doerfler, J. Casadesús (eds.), *Epigenetics of Infectious Diseases*, Epigenetics and Human Health, DOI 10.1007/978-3-319-55021-3_7

Bacterial epigenetics specifically helps define a variable phenotype within an infecting, clonal population, and this in turn will impact upon the outcome of infection and disease progression.

In the late 1980s and early 1990s, seminal experiments showed that a mouse embryo was not viable in the absence of a cytosine DNA methyltransferase. This started a field of investigation that firmly established that epigenetic regulation by DNA methylation plays a critical role in many eukaryotic organisms. It is now also well established, but perhaps not as widely known, that DNA methylation can also be crucial to the success of bacteria. This is the focus of this chapter: epigenetic regulation in bacterial pathogens by DNA methylation. The significance of this for the success of pathogen population will be discussed and the main molecular mechanism outlined.

Bacterial epigenetics is not a new field. Studies on the *lac* operon in *Escherichia coli* in the 1950s showed that inheriting elevated levels of a key protein was sufficient to convey a phenotype that was as if the cells were induced by lactose, even if this inducer was absent [reviewed in (Casadesus and Low 2013)]. This describes "epigenetics" in its broadest sense: a heritable but reversible phenotype that does not involve a change in the DNA sequence. In the 1980s, the epigenetics concept was expanded to include regulation by DNA methylation. Initially, bacterial adenine DNA methylation in *E. coli* by the enzyme deoxyadenosine methyltransferase (Dam) was identified as a key feature in controlling a variety of cellular processes, including activity of a transposase promoter and initiation of chromosome replication. These processes involve a transient hemi-methylated state that occurs after DNA replication, when the newly synthesized strand has not been fully methylated yet [reviewed in (Marinus and Casadesus 2009)]. Then, a different type of regulation involving Dam in *E. coli* was identified and led to elucidating DNA methylation-dependent epigenetic regulation in bacteria (Blyn et al. 1989; Hernday et al. 2004).

In a seminal body of work by Low and colleagues that started in the late 1980s, the mechanism of regulation of the pyelonephritis-associated pilus (*pap)* operon in *E. coli pap* operon was unraveled. This showed that specifically Dam-dependent adenine methylation was a critical element and that the characteristic feature is that specific methyltransferase target sequences in the chromosome remain non-methylated throughout the cell cycle establishing "methylation patterns." The gene expression state is reversible and heritable, and there is no associated change in DNA sequence, and thus, this mechanism meets the definition of epigenetic regulation. This regulation is also referred to as phase variation and results in a mixed expression state in a clonal population with cells with the gene either in the "on" or "off" expression state. The discovery by Low and colleagues added an epigenetic mechanism to the known genetic mechanisms of phase variation. This and other DNA methylation-dependent phase variation is the main focus of this chapter.

7.2 The Contribution of Epigenetic Phase Variation to Bacterial Pathogenesis

Dam methylates the adenine in context of GATC sequences and can be essential for viability and affect virulence. For example, *dam* mutants of Salmonella are avirulent, and *dam* is essential for *Vibrio cholera* and *Yersinia pseudotuberculosis* viability [reviewed in (Marinus and Casadesus 2009; Casadesus and Low 2013)]. However, a *dam* mutation has pleiotropic effects that relates to the multiple roles of DNA methylation, and an attenuated phenotype may not be due to disrupted epigenetic phase variation (Low and Casadesus 2008; Marinus and Casadesus 2009). However, no studies have addressed whether the *epigenetic mechanism* of phase variation itself is a virulence factor. To assess the significance of DNA methylation-dependent epigenetic regulation, here the general role of phase variation and the impact of the phenotypic heterogeneity of the clonal population is considered, as well as the significance of the proteins that are controlled by epigenetic phase variation.

Epigenetic phase variation mainly has been identified in controlling expression of surface structures. A significant category of epigenetic controlled structures are fimbriae in *E. coli* and Salmonella. Fimbriae are bacterial adhesins that mediate attachment to surfaces including host tissue and can contribute to tissue tropism of the pathogen through receptor specificity. Strains may encode for multiple fimbrial operons that may allow for redundancy in function. In *E. coli*, a regulatory network coordinates the expression between many fimbrial operons allowing for temporal and spatial coordination of expression, and epigenetic regulatory mechanism is integrated in this fimbrial network (van der Woude 2006, 2011). Phase variation of fimbriae thus may facilitate a "stick or swim" strategy and dispersal through the body. Fimbriae are also highly antigenic, and since phase variation results in heterogeneity of gene expression state, this should facilitate immune evasion of a subpopulation. Indeed, population heterogeneity generated by other means is also implicated as a *Salmonella* virulence strategy, by allowing initiation of an immune response that benefits the pathogen, and escape of this same response by a different subpopulation (Diard et al. 2013). It is not inconceivable that some phase variation contributes to a strain's virulence in this way.

These aspects may contribute for example to the virulence of uropathogenic *E. coli* that encode the epigenetic controlled *pap* operon. Pap fimbriae have an adhesive tip that facilitates interaction of the bacteria with receptors on uroepithelial cells. Pap fimbrial phase variation thus may facilitate the bacterial migration from the bladder to the kidney (Westerlund-Wikstrom and Korhonen 2005). Pap-mediated adhesion also may be involved specifically in initiating an inflammatory response, indicating a complex contribution to virulence by Pap fimbriae but also identifying a possible further significance for a heterogeneous population with cells both expressing Pap and not expressing Pap as a result of the regulatory mechanism (Lane and Mobley 2007).

The significance of phase variation as a virulence factor has been directly addressed for two fimbriae, specifically the Long Polar fimbriae (LPF) in Salmonella and type 1 fimbriae in *E. coli*. Neither, however, is controlled by a Dam-dependent regulatory mechanism (Norris et al. 1998; Blomfield and van der Woude 2007). Studies on LPF suggested that phase variation allowed evasion of cross-immunity between Salmonella serovars (van der Woude and Baumler 2004). In a pyelonephritis model infection system, phase variation of type 1 fimbriae itself, however, was not an evident virulence factor (Snyder et al. 2006). In the absence of direct studies, further significance for specifically epigenetic phase variation can only be hypothesized based on the known features of the mechanism (see below). The links to bacterial DNA replication and thus bacterial growth, possibly conferring benefits to the population, may not occur with genetic mechanisms.

In addition to fimbrial regulation, epigenetic mechanisms in *Salmonella* also control the O-antigen of the LPS, which is a main antigenic surface carbohydrate of Gram-negative bacteria. First of these is the *gtr* system that controls the expression of a family of operons that mediate a variety of biochemical modifications of the O-antigen, including the well-studied modification genes encoded on the phage P22 genome (Davies et al. 2013; Makela 1973). This mechanism was identified several decades after the elucidation of the enzymatic pathway and built upon knowledge of the key regulatory concepts of epigenetic phase variation of the *pap* operon. In direct relevance to virulence, phase variation of Gtr-mediated changes to the O-antigen will contribute immune evasion due to the antigenic nature of the O-antigen. The variation in O-antigen directly affects the classification by the serology-based Kaufmann–White scheme and several of the *gtr*-mediated modifications were known to exist based on that even though the heterogeneity and mechanism of regulation were not (Davies et al. 2013).

Epigenetic control of the *opvAB* operon in *Salmonella* in contrast determines the length of the O-antigen (Cota et al. 2015a, b). Expression of *opv* results in a short O-antigen, which is a phenotype with decreased virulence. This is consistent with a general protective nature of the O-antigen. Indeed, the bias of phase variation frequency results in a population that mainly consists of cells that do not express *opvAB* and have the normal O-antigen length.

The advantage to the cell of phase variation of *opvAB* and the resulting short O-antigen lies in the interaction with bacteriophage. The short O-antigen is associated with the ability of enhanced interactions with phage (Cota et al. 2015b). This in turn is important for pathogen evolution and adaptation, as phage facilitate horizontal gene transfer and thus may allow beneficial traits to be acquired. Similarly, the population heterogeneity generated by *gtr* epigenetic regulation can directly affect the susceptibility of the population for (specific) bacteriophage. For two *gtr*-dependent modifications, it has been directly shown that *gtr*-dependent modification provides protection to the lysogen for superinfection by other similar phage. Due to phase variation, a subpopulation, however, remains susceptible to infection allowing evolution of the genome through introducing novel genes (Kim and Ryu 2012; Kintz et al. 2015; Broadbent et al. 2010). Furthermore, the repertoire of the epigenetically controlled O-antigen modification genes varies per *Salmonella* strain, reflecting in part that *gtr* is introduced into the genome as cargo genes by lysogenic phage. Thus, changes to the O-antigen as a result of phase variation of *gtr* may allow a balance to be struck between phage-mediated cell lysis, genome disruption, and genome evolution. The change in phenotype due to *opvAB* phase variation on the other hand is a direct trade-off between phage susceptibility and thus phage-mediated cell lysis, genome disruption and genome evolution, and the population's virulence.

In addition to fimbriae in *E. coli,* the surface protein Ag43 and its antigenic variants are controlled by epigenetic phase variation. This protein mediates autoaggregation and as such can contribute to biofilm formation (van der Woude and Henderson 2008; Heras et al. 2014). This in turn may assist persistence of colonization by the Ag43 expressing strain (Ulett et al. 2007). The final example of epigenetic regulation that is known is that of a type VI secretion system in *E. coli.* This *sci1* type VI secretion system in enteroaggregative *E. coli* is also required for efficient biofilm formation (Brunet et al. 2011; Aschtgen et al. 2010).

The significance, if any, of epigenetic phase variation as opposed to genetic phase variation can only be hypothesized based on our understanding of the mechanism (see below). There may be significance in the fact that a change in gene expression state requires DNA synthesis and thus links tightly to bacterial growth. The phase variation switch frequency is highly sensitive to just a few base pair changes, which may allow evolutionary adaptation to different niches. Furthermore, when considering horizontal transfer, the regulation is linked to a common enzyme, Dam, and a few global regulators. Together, these features may allow pathogens to respond to selective pressure more rapidly and flexibly using epigenetic than some genetic phase variation regulatory systems.

7.3 Core Principles of the Dam-Dependent Epigenetic Phase Variation

The epigenetic regulation that controls expression of the genes described above is associated with methylation of the adenine in GATC sequences that is carried out by the DNA maintenance methyltransferase Dam. Maintenance DNA methyltransferases are not associated with restriction enzymes, and as a result, a non-methylated GATC sequence is not lethal. As evident from the examples above, the occurrence of Dam-dependent phase variation has only been identified in *E. coli* and *Salmonella* spp., even though Dam is present in other species as well.

When mutants are generated that lack Dam in either *E. coli* or *Salmonella* spp., pleiotropic effects on bacterial phenotype and physiology are evident, in part as a direct result of changes in DNA supercoiling and indirect effects of changes in the initiation of chromosome replication (Marinus and Casadesus 2009). In contrast, the changes in methylation that are important for epigenetic phase variation are localized to the specific regulatory regions, where in each case the Dam methylation target sequence GATC and the regulatory protein DNA binding sequence overlap. There is one key common principle for this regulation, and this requires that two prerequisites are fulfilled. First is that the interaction of a DNA binding protein is affected by the methylation state of the DNA. Secondly, and conversely, the methylation state of the target sequence is affected by the binding of this same protein. In other words, the binding of the regulatory protein will protect the target sequence from methylation by Dam. As a result, epigenetic Dam-dependent regulation will be localized and limited to sequences where these prerequisites are met.

Fully methylated DNA describes DNA that is methylated at the methyltransferase target sequences on both strands. This is the state of the Dam target sequences in a cell expressing DNA methyltransferase, in the absence of interfering factors or DNA replication. In contrast, when the sequence on only one of the two strands is methylated, this is referred to as hemi-methylated DNA. This latter state occurs temporarily after the DNA replication fork has passed (Fig. 7.1). In contrast, the regulation under consideration here focuses on "methylation patterns" that represent the relatively few target sequences in the genome that can have stable, aberrant DNA methylation state: a non-methylated sequence in the context of a fully methylated genome.

DNA methylation is an elegant way to link regulation to DNA synthesis and thus DNA replication and growth. DNA is synthesized as non-methylated DNA, and to date no DNA de-methylating enzymes have been identified in *E. coli* or *Salmonella*. Thus, the methylated state of DNA can only be changed to a non-methylated state passively as a result of new DNA synthesis by way of a generation with a hemi-methylated intermediate state. The converse change from non-methylated to methylated requires access of Dam to the site, and thus, the DNA binding regulatory protein must be dislodged from the DNA at the right time and for a sufficiently long period of time. The regulatory proteins must at least be released from the DNA every time the replication form passes. Thus, based on current understanding, changes in expression state of epigenetically controlled operons or genes will be prevalent in growing cells with active DNA synthesis, and the expression state at most will change once per cell cycle (Fig. 7.1).

Three commonly occurring DNA binding regulatory proteins have been identified as contributors to epigenetic regulation: Lrp (leucine responsive regulatory protein), OxyR, and Fur. Interestingly, the apparent complexity of the systems does not align with the chronology of discovery and understanding: *pap* and similar Lrp- and Dam-dependent systems established the paradigm, but many layers of complexity have been discovered for this system. In contrast, the apparently basic system of OxyR and Dam-dependent control of Ag43 outer membrane protein expression in *E. coli* was only the second system that was described, but this was more than a decade after the first report of *pap* epigenetic regulation.

The most simple form of Dam-dependent regulation consists of a system involving only one DNA binding protein that can bind to its recognition sequence when the Dam target sequences contained in it are non-methylated DNA, but cannot bind when the sequences are methylated. Such a system controls expression of the Ag43 outer membrane protein in *Escherichia coli* that encodes for an autotransporter that

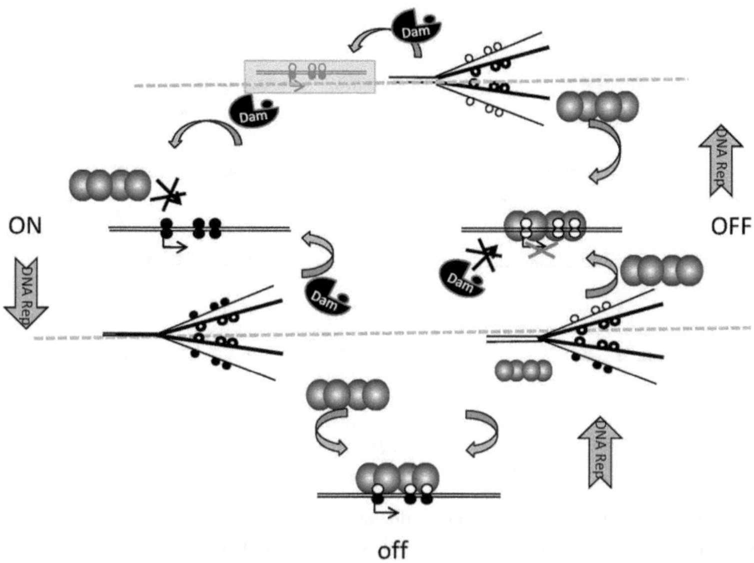

Fig. 7.1 Schematic of the DNA and protein interactions during DNA replication depicted for the *agn43* regulatory region, leading to a model of epigenetic regulation, is shown. Dam and DNA replication are identified; methyl groups are presented as *small circles* (*closed*, methylated; *open*, non-methylated), OxyR is represented as *gray* tetramer, and DNA is represented as double stranded by *straight lines*. V-shaped DNA is the transient state after passage of the replication fork, with newly synthesized strands depicted with *thicker lines*. *Curved arrows* identify the outcome between two states dependent on interaction of the DNA with the relevant protein. *Dark cross* means protein cannot interact with the specific DNA (occlusion), and *gray cross* identifies non-active promoter in OFF state. The process from OFF to ON is related to two DNA replication events, where the lower case "off" state depicts a state with lower OxyR binding affinity for the depicted hemi-methylated state. The process from OFF to ON may not occur in one generation, but this is not clear. DNA replication is required to facilitate removal of highaffinity bound OxyR. Dam is processive, and one scenario is that, as depicted, methylation of one strand occurs first. The ON and OFF state, the occurrence occlusion, the OxyR binding affinities, and a role for DNA replication are derived from experimental data. Other processes are hypothesized

mediates bacterial autoaggregation (van der Woude and Henderson 2008). The DNA binding protein OxyR and Dam are the only known regulators for expression of the Ag43 encoding gene *agn43*. OxyR is a DNA binding protein that binds DNA as a dimer of dimers, and will become oxidized under certain oxidative stress conditions, which affects the DNA–protein contact sites (Zheng and Storz 2000). In the regulatory region for the Ag43 coding sequence, the *agn43* (originally referred to as the *flu*) gene, the OxyR binding site encompasses three Dam target sequences, GATC (Figs. 7.1 and 7.2). Methylation of the three GATC sequences affects the affinity of OxyR (van der Woude and Henderson 2008). In full agreement with the core principle that protein binding affinity is important in the output, the frequency of switching between expression states varies with the number of Dam target sequences in the binding site (Haagmans and van der Woude 2000; Wallecha et al. 2002). The reduced form of OxyR is sufficient to obtain Ag43 phase

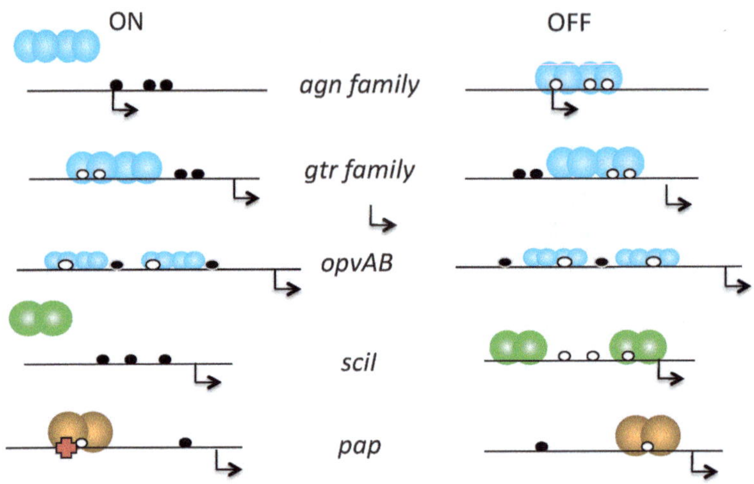

Fig. 7.2 Signature sequences of epigenetic phase variation. A cartoon of the organization of the signature sequence elements for a few examples of epigenetically controlled operons is shown with DNA binding proteins and the relation to the expression state. Features of the signature sequences are identified as the Dam target sequences (*open circles* = non-methylated, *closed circles* = methylated); the DNA binding proteins (*blue spheres* = OxyR monomer; *brown spheres* = a tetramer of Lrp dimers, *Green sphere* = Fur monomer). An *arrow* identifies the +1 transcription start site. No additional regulators are depicted except the PapI (*brown cross*) for *pap*; the main *pap* promoter is shown, but the divergent promoter for PapI that is also controlled by the epigenetic mechanism is not shown. Most features shown have been confirmed experimentally

variation. Even though a hypothesis is appealing that oxidative stress induces expression and thus Ag43-dependent aggregation to enhance survival for individual cells, there is no evidence of oxidative stress affecting this epigenetic regulation (Wallecha et al. 2003; Tree et al. 2007).

A similar "basic" epigenetic principle as above controls expression of the *sci1* type VI secretion gene cluster in enteroaggregative *E. coli* (EAEC) (Brunet et al. 2011) (Fig. 7.2). Here, the iron responsive regulatory protein Fur is a key regulator of expression. Iron is a key element for bacterial pathogens, and biofilm formation is an important virulence trait. However, the rationale specifically of linking biofilm formation to iron dependency for this pathovar has been not directly addressed and can only be speculated upon (Brunet et al. 2011). Similarly, the relevance of a switch in expression is speculative, and would appear to relate to other "stick or swim" strategies, and possibly immune evasion due to the antigenicity of the structure. Regardless, this is an elegant system that allows a random switch to incorporate information from the environment as relevant for virulence. For this and the previously described "simple" one protein-one binding site systems, it cannot be ruled out that additional layers of complexity exist that have not yet been identified.

Even though OxyR and Dam are present in many species, OxyR and Dam-dependent regulation has to date only been identified for additional examples in *Salmonella*. OxyR and Dam control phase variation of the *gtr* family of operons that modify the O-antigen and the *opvAB* operon that affects the O-antigen length in *Salmonella* spp. In these regulatory regions, the organization of the Dam target sequences, the OxyR binding sites, and how these align in relation to the promoter vary (Fig. 7.2). The main difference is whether, as in the *agn43* gene, it is an "all-or-none" binding event, or whether two alternate bindings sites are involved that are differentially occupied in the two expression states (on or off). It is unclear what the precise contribution of the non-repressing binding site is for control of phase variation, nor if there are additional, less obvious complexities when there are two OxyR binding sites.

Based on our understanding of the detailed mechanism, the most complex epigenetic phase variation system known to date was the first one identified and the one that is best characterized: the mechanism controlling expression of the *pap* fimbrial operon and similarly regulated fimbrial operons in *E. coli* and *Salmonella*. This mechanism relies on the global regulator Lrp (leucine responsive regulatory protein), and its interaction with the regulatory sequence furthermore requires a small protein (PapI for *pap*). In the *pap* regulatory region, there are two Lrp binding sites that do not overlap. Regulation of phase variation involves mutually exclusive occupation of these two sites (Fig. 7.2). Transcription of *pap* can additionally be modified by integration of environmental signals into the phase variation mechanisms mediated by global regulators including cAMP-CRP, CpxAR and HNS. This leads to a "random but regulated" mechanism: randomness or stochasticity determines whether the Dam- and Lrp-dependent methylation pattern in a cell will change or be maintained (and thus the expression state), but there is also epistatic control. The interaction between the regulatory region of the DNA and additional proteins in this complex regulatory mechanism is discussed elsewhere in detail (Hernday et al. 2002, 2004; Blomfield and van der Woude 2007; Casadesus and Low 2013; Low and Casadesus 2008).

A different, complex system controls expression of the Std fimbrial operon in *Salmonella*. This involves Dam, and the regulator HfdR, but also SeqA that is associated with binding with highest affinity to specifically hemi-methylated DNA (Jakomin et al. 2008). In contrast to this system, in the other examples described above, additional factors can affect the Dam-dependent epigenetic regulation, which include those that influence chromosome organization, e.g., SeqA and HNS. Mutations in these genes will result in global, pleiotropic effects, including indirectly affecting the overall DNA methylation level of the chromosome that in turn may be responsible for the effect on the epigenetic regulation (Correnti et al. 2002).

7.4 Dam-Dependent Epigenetic Regulation in Natural Isolates

Many of the examples above illustrate the value of exploring pathogenic strains for the occurrence of epigenetic regulation. The cumulative findings suggest that epigenetic regulation may be especially prevalent for surface structures encoded on pathogenicity islands. The examples of Pap and Sci-1 illustrate that *E. coli* pathovars have acquired or evolved specific strategies of virulence involving pathovar-specific epigenetic regulation.

By comparing natural isolates, it also becomes evident that allelic variants of the coding sequences of epigenetically controlled surface structures exist, for example, for the Ag43 encoding gene and *pap*. Similarly, there are many variants of the regulatory sequences either associated with allelic variants or controlling functionally related but different structures or proteins, as in fimbriae or *gtr* operons (Totsika et al. 2008; Davies et al. 2013; van der Woude and Henderson 2008; Broadbent et al. 2010; Blomfield and van der Woude 2007). In each case, essential sequence elements for phase variation are conserved, which has been referred to as the "signature sequence". Indeed, in the S. Typhi glucosyltransferase gtr operon a naturally acquired two nucleotide deviation from the signature sequence was shown to be necessary and sufficient to abrogate phase variation, affecting the strain's serum resistance (Kintz et al 2017). Outside of the signature sequence, variations in the binding sites for the DNA binding proteins as well as the intervening sequences exist and can affect expression. The variation is mostly evident as variable rates at which the expression state changes ("switch frequency") (Broadbent et al. 2010; Totsika et al. 2008; Kintz et al. 2015). In the case of OxyR-dependent regulation of Ag43, the variation furthermore raises the possibility that some sequences may facilitate input of oxidative stress into the mechanism if the binding affinity of the oxidized and reduced forms of OxyR is differentially affected, despite this not being evident from the sequence derived from *E. coli* strain MG1655 (Wallecha et al. 2003).

Taking together our understanding of the shared principle underpinning epigenetic regulation, we can speculate what the main advantage of specifically epigenetic phase variation is. The requirement of DNA synthesis and thus indirectly a link to growth to facilitate change may be an advantage. The ability to superimpose and integrate environmental factors on a stochastic system as seen for *pap* clearly can be an advantage, but this is not unique to epigenetic variation. An aspect that can be considered is that perhaps the epigenetic mechanism can evolve more readily from established regulatory systems to generate population heterogeneity. As long as the host cell encodes for a DNA methyltransferase, a single binding site for a regulatory protein may evolve to allow epigenetic phase variation without the need for additional proteins or factors. Selective pressure then may allow retention of this regulatory feature.

7.5 Epigenetic Regulation in Pathogens by DNA Methyltransferases Other than Dam

DNA methylation is also involved in controlling expression of "phasevarions" (Srikhanta et al. 2010). The term phasevarion reflects a regulon that phase varies as a result of being controlled by a (single) initial phase variation event controlling the expression of a DNA methyltransferase associated with a restriction-modification system. The DNA methyltransferase itself is controlled by a (genetic) phase variation mechanism, and it is the variable DNA methylation that directly or indirectly leads to global changes in gene expression (i.e., the phasevarion). Phasevarions associated with single DNA modification genes have been described for the pathogens *Haemophilus influenzae, Helicobacter pylori,* and *Neisseria meningitidis* and *N. gonorrhoeae* (Srikhanta et al. 2010, 2011). The mechanism (s) that links DNA methylation to gene regulation have not been fully elucidated and may involve epigenetic regulation similar to that described for Dam, global changes due to, for example, changes in the chromosome structure, or a combination of both. In *H. influenzae,* the virulence phenotype is significantly affected, as is virulence in an infection model system (VanWagoner et al. 2016). Similar strong correlations between DNA modification gene expression and expression of virulence traits were observed for *N. meningitidis* (Seib et al. 2011).

A similar concept of variable expression of DNA modifying genes also underpins a complex and elaborate use of DNA methylation to control virulence in *Streptococcus pneumoniae* (Manso et al. 2014). The virulence of this pathogen is directly affected by the DNA specificity of the DNA modification protein, SpnD39III. Like for the phasevarions described above, but in contrast to the epigenetic Dam-dependent phase variation, SpnD39III does not appear to be a maintenance methyltransferase. The complexity arises from the fact that six variants of SpnD39III, each modifying a different DNA sequence, can be expressed as a result of genetic recombination. Each variant gives rise to a population with a unique gene expression profile and virulence traits. Clonal populations are heterogeneous regarding the expressed variant. This strategy not only provides further evidence of the flexible use of DNA methyltransferases to control virulence but also suggests that further occurrence of epigenetic regulation may yet be discovered.

7.6 What Lies Ahead?

Considering it was almost 30 years before the publication of this book that the first study was published on bacterial epigenetic phase variation, it could be considered surprising that not many more examples are known. However, when considering mechanism of regulation, it is clear that identifying the occurrence from genome sequences is a significant challenge. The accurate prediction of DNA binding of regulatory proteins is still a challenge, and whether the binding is affected by DNA methylation requires sufficient knowledge of the protein–DNA contacts.

We know of the requirements and possibilities for Lrp, OxyR, and Fur, and these, and Dam, are present in a range of bacterial species. This raises an important question in context of understanding virulence: is epigenetic phase variation occurring in other species? If it is, which regulatory proteins are involved and which genes are affected? Conversely, if there is no epigenetic regulation even when the required players are present, is this because it does not confer an advantage? Understanding the rationale behind the presence or absence may provide insight into the significance of specific mechanisms that result in population heterogeneity.

Exploring the methylome of bacterial pathogens on a larger scale than was previously achievable has become feasible with the advent of SMRT sequencing (Flusberg et al. 2010). As described above, this can be used to identify the specificity of DNA methyltransferases (Manso et al. 2014), but also allow identification of the degree of methylation of individual sites (Cota et al. 2015a; Sanchez-Romero et al. 2015), even though this approach is not suitable for identification of a minor fraction of the population with a different methylated state, and some phase variation events would still not be identifiable therefore. Additionally, a non-methylated state does not always reflect methylation-dependent epigenetic regulation: this state may fulfill one of the two prerequisites, specifically that the regulatory protein blocks methylation, but if the second prerequisite is not also fulfilled, specifically that the methylation fails to alter the binding affinity of this same protein (Hale et al. 1994), then the non-methylated state will not reflect epigenetic regulation. Furthermore, non-methylated sequences may occur as a result of non-regulatory DNA structures or various cellular processes, including rapid growth. Thus, whether non-methylated sites affect gene expression will need to be validated gene by gene, but SMRT sequencing is now used instead of methylation-sensitive restriction analysis to provide evidence of the correlation between expression state and methylation state of a specific site (Cota et al. 2015a). Significant value would be had in improving in silico analysis of detailed protein–DNA interactions to facilitate interpretation of methylome data. Furthermore, the methylome may be variable and, much like transcriptome analyses, results need to be considered in context of the growth conditions of the cells.

Since the first description of Dam-dependent phase variation in 1989 (Blyn et al. 1989), the understanding of the mechanisms and impact of DNA methylation and epigenetics in bacteria has progressed significantly, and it is likely that there is much left to be discovered. Indeed, as is evident from the growing body of work on phasevarions, the role of DNA methylation for bacterial virulence may be more diverse and widespread than could have been envisioned then. Targeting DNA methylation for drug development may not be feasible since methyltransferases are ubiquitous and share key structural features, but this could be further explored. However, there does seem to be scope to combine the old and new insights to design better, more stable vaccine strains. The application of new technologies will provide the support that is needed to identify the full contribution of epigenetic regulation in bacteria, to understand how that contributes to the success of both pathogens and commensals and perhaps to use that knowledge to our benefit in controlling bacterial infectious disease.

References

Aschtgen MS, Gavioli M, Dessen A, Lloubes R, Cascales E (2010) The SciZ protein anchors the enteroaggregative *Escherichia coli* Type VI secretion system to the cell wall. Mol Microbiol 75 (4):886–899

Blomfield I, van der Woude M (2007) Regulation of fimbrial expression. EcoSal Plus 2(2). doi:10.1128/ecosal.2.4.2.2

Blyn LB, Braaten BA, White-Ziegler CA, Rolfson DH, Low DA (1989) Phase-variation of pyelonephritis-associated pili in *Escherichia coli*: evidence for transcriptional regulation. EMBO J 8(2):613–620

Broadbent SE, Davies MR, van der Woude MW (2010) Phase variation controls expression of *Salmonella* lipopolysaccharide modification genes by a DNA methylation-dependent mechanism. Mol Microbiol 77(2):337–353

Brunet YR, Bernard CS, Gavioli M, Lloubes R, Cascales E (2011) An epigenetic switch involving overlapping fur and DNA methylation optimizes expression of a type VI secretion gene cluster. PLoS Genet 7(7):e1002205

Casadesus J, Low DA (2013) Programmed heterogeneity: epigenetic mechanisms in bacteria. J Biol Chem 288(20):13929–13935. doi:10.1074/jbc.R113.472274

Correnti J, Munster V, Chan T, van der Woude M (2002) Dam-dependent phase variation of Ag43 in *E. coli* is altered in a seqA mutant. Mol Microbiol 44:521–532

Cota I, Bunk B, Sproer C, Overmann J, Konig C, Casadesus J (2015a) OxyR-dependent formation of DNA methylation patterns in OpvABOFF and OpvABON cell lineages of *Salmonella enterica*. Nucleic Acids Res 44(8):3595–3609

Cota I, Sanchez-Romero MA, Hernandez SB, Pucciarelli MG, Garcia-Del Portillo F, Casadesus J (2015b) Epigenetic control of salmonella enterica O-antigen chain length: a tradeoff between virulence and bacteriophage resistance. PLoS Genet 11(11):e1005667

Davies MR, Broadbent SE, Harris SR, Thomson NR, van der Woude MW (2013) Horizontally acquired glycosyltransferase operons drive salmonellae lipopolysaccharide diversity. PLoS Genet 9(6):e1003568

Diard M, Garcia V, Maier L, Remus-Emsermann MN, Regoes RR, Ackermann M, Hardt WD (2013) Stabilization of cooperative virulence by the expression of an avirulent phenotype. Nature 494(7437):353–356

Flusberg BA, Webster DR, Lee JH, Travers KJ, Olivares EC, Clark TA, Korlach J, Turner SW (2010) Direct detection of DNA methylation during single-molecule, real-time sequencing. Nat Methods 7(6):461–465

Haagmans W, van der Woude M (2000) Phase variation of Ag43 in E. coli: Dam-dependent methylation abrogates OxyR binding and OxyR-mediated repression of transcription. Mol Microbiol 35(4):877–887

Hale WB, Van der Woude MW, Low DA (1994) Analysis of nonmethylated GATC sites in the *Escherichia coli* chromosome and identification of sites that are differentially methylated in response to environmental stimuli. J Bacteriol 176(11):3438–3441

Heras B, Totsika M, Peters KM, Paxman JJ, Gee CL, Jarrott RJ, Perugini MA, Whitten AE, Schembri MA (2014) The antigen 43 structure reveals a molecular Velcro-like mechanism of autotransporter-mediated bacterial clumping. Proc Natl Acad Sci U S A 111:457–462

Hernday A, Braaten B, Low D (2004) The intricate workings of a bacterial epigenetic switch. Adv Exp Med Biol 547:83–89

Hernday A, Krabbe M, Braaten B, Low D (2002) Self-perpetuating epigenetic pili switches in bacteria. Proc Natl Acad Sci U S A 99(Suppl 4):16470–16476. doi:10.1073/pnas.182427199

Jakomin M, Chessa D, Baumler AJ, Casadesus J (2008) Regulation of the *Salmonella enterica std* fimbrial operon by DNA adenine methylation, SeqA, and HdfR. J Bacteriol 190 (22):7406–7413

Kim M, Ryu S (2012) Spontaneous and transient defence against bacteriophage by phase-variable glucosylation of O-antigen in *Salmonella enterica* serovar Typhimurium. Mol Microbiol 86 (2):411–425

Kintz E, Davies MR, Hammarlof DL, Canals R, Hinton JC, van der Woude MW (2015) A BTP1 prophage gene present in invasive non-typhoidal *Salmonella* determines composition and length of the O-antigen of the lipopolysaccharide. Mol Microbiol 96(2):263–275

Kintz E, Heiss C, Black I, Donohue N, Brown N, Davies MR, Azadi P, Baker S, Kaye PM, van der Woude M (2017) *Salmonella enterica* serovar typhi lipopolysaccharide O-antigen modification impact on serum resistance and antibody recognition. Infect Immun 85(4):e01021-16

Lane MC, Mobley HL (2007) Role of P-fimbrial-mediated adherence in pyelonephritis and persistence of uropathogenic *Escherichia coli* (UPEC) in the mammalian kidney. Kidney Int 72(1):19–25

Low DA, Casadesus J (2008) Clocks and switches: bacterial gene regulation by DNA adenine methylation. Curr Opin Microbiol 11(2):106–112

Makela PH (1973) Glucosylation of lipopolysaccharide in *Salmonella*: mutants negative for O antigen factor 1221. J Bacteriol 116(2):847–856

Manso AS, Chai MH, Atack JM, Furi L, De Ste CM, Haigh R, Trappetti C, Ogunniyi AD, Shewell LK, Boitano M, Clark TA, Korlach J, Blades M, Mirkes E, Gorban AN, Paton JC, Jennings MP, Oggioni MR (2014) A random six-phase switch regulates pneumococcal virulence via global epigenetic changes. Nat Commun 5:5055

Marinus MG, Casadesus J (2009) Roles of DNA adenine methylation in host pathogen interactions: mismatch repair, transcriptional regulation, and more. FEMS Microbiol Rev 33 (3):488–503

Norris TL, Kingsley RA, Bumler AJ (1998) Expression and transcriptional control of the Salmonella typhimurium lpf fimbrial operon by phase variation. Mol Microbiol 29(1):311–320

Sanchez-Romero MA, Cota I, Casadesus J (2015) DNA methylation in bacteria: from the methyl group to the methylome. Curr Opin Microbiol 25:9–16

Seib KL, Pigozzi E, Muzzi A, Gawthorne JA, Delany I, Jennings MP, Rappuoli R (2011) A novel epigenetic regulator associated with the hypervirulent Neisseria meningitidis clonal complex 41/44. FASEB J 25(10):3622–3633

Snyder JA, Lloyd AL, Lockatell CV, Johnson DE, Mobley HL (2006) Role of phase variation of type 1 fimbriae in a uropathogenic Escherichia coli cystitis isolate during urinary tract infection. Infect Immun 74(2):1387–1393

Srikhanta YN, Fox KL, Jennings MP (2010) The phasevarion: phase variation of type III DNA methyltransferases controls coordinated switching in multiple genes. Nat Rev Microbiol 8 (3):196–206

Srikhanta YN, Gorrell RJ, Steen JA, Gawthorne JA, Kwok T, Grimmond SM, Robins-Browne RM, Jennings MP (2011) Phasevarion mediated epigenetic gene regulation in Helicobacter pylori. PLoS ONE 6(12):e27569

Totsika M, Beatson SA, Holden N, Gally DL (2008) Regulatory interplay between pap operons in uropathogenic Escherichia coli. Mol Microbiol 67(5):996–1011

Tree JJ, Ulett GC, Hobman JL, Constantinidou C, Brown NL, Kershaw C, Schembri MA, Jennings MP, McEwan AG (2007) The multicopper oxidase (CueO) and cell aggregation in Escherichia coli. Environ Microbiol 9(8):2110–2116

Ulett G, Valle J, Beloin C, Sherlock O, Ghigo J, Schembri M (2007) Functional analysis of antigen 43 in uropathogenic Escherichia coli reveals a role in long-term persistence in the urinary tract. Infect Immun 75(7):3233–3244 van der Woude MW (2006) Re-examining the role and random nature of phase variation. FEMS Microbiol Lett 254(2):190–197

van der Woude MW (2011) Phase variation: how to create and coordinate population diversity. Curr Opin Microbiol 14(2):205–211

van der Woude MW, Baumler AJ (2004) Phase and antigenic variation in bacteria. Clin Microbiol Rev 17(3):581–611

van der Woude MW, Henderson IR (2008) Regulation and function of Ag43 (flu). Annu Rev Microbiol 62:153–169

VanWagoner TM, Atack JM, Nelson KL, Smith HK, Fox KL, Jennings MP, Stull TL, Smith AL (2016) The modA10 phasevarion of nontypeable Haemophilus influenzae R2866 regulates multiple virulence-associated traits. Microb Pathog 92:60–67

Wallecha A, Correnti J, Munster V, van der Woude M (2003) Phase variation of Ag43 is independent of the oxidation state of OxyR. J Bacteriol 185(7):2203–2209

Wallecha A, Munster V, Correnti J, Chan T, van der Woude MW (2002) Dam- and OxyR-dependent phase variation of agn43: essential elements and evidence for a new role of DNA methylation. J Bacteriol 184(12):3338–3347

Westerlund-Wikstrom B, Korhonen TK (2005) Molecular structure of adhesin domains in Escherichia coli fimbriae. Int J Med Microbiol 295(6–7):479–486

Zheng M, Storz G (2000) Redox sensing by prokaryotic transcription factors. Biochem Pharmacol 59(1):1–6

Chapter 8
Epigenetic Switching in the Human Fungal Pathogen *Candida albicans*

Han Du and Guanghua Huang

Abstract The human fungal pathogen *Candida albicans* can grow in a number of distinct morphologies: yeast (including white, gray, opaque, and GUT forms), hyphae, and pseudohyphae. White and opaque cells are two heritable cell types. Despite the differences in several biological aspects including cellular and colony appearances, global gene profiles, virulence, filamentation, mating competency, and susceptibility to antifungals, white and opaque cells contain the same set of genomic DNA. The regulation of white–opaque switching is at the epigenetic level. Although the switch between the white and opaque phenotypes can occur spontaneously at low frequency, it can also be induced by certain environmental cues such as host-related signaling molecules like CO_2 and N-acetylglucosamine (GlcNAc). In this chapter, we will review the environmental and genetic regulatory mechanisms of this epigenetic switch in *C. albicans*. The relationships between white–opaque switching and pathogenesis, sexual reproduction, and other biological processes will be discussed. The newly discovered white–gray–opaque tristable switching system and white–opaque switching in two *C. albicans*-related species, *Candida dubliniensis* and *Candida tropicalis*, will be included in the last two sections. These morphological switches may provide *Candida* species with a better ability to adapt to the changing host environment and survive under diverse niches.

Keywords *Candida albicans* • Epigenetic switching • White–opaque transitions • Sexual mating • Wor1 • Environmental factors

8.1 Introduction

Candida albicans is an important human fungal pathogen that causes superficial infections in healthy people and life-threatening disease in immunocompromised individuals (Berman 2012; Brown et al. 2012). A striking feature of this fungus is its ability to grow in several morphological forms: yeast (including white, gray,

H. Du • G. Huang (✉)
State Key Laboratory of Mycology, Institute of Microbiology, Chinese Academy of Sciences, Beijing 100101, China
e-mail: huanggh@im.ac.cn

W. Doerfler, J. Casadesús (eds.), *Epigenetics of Infectious Diseases*, Epigenetics and Human Health, DOI 10.1007/978-3-319-55021-3_8

opaque, and GUT forms), hyphae, and pseudohyphae (Huang 2012; Sudbery et al. 2004; Tao et al. 2014b). *C. albicans* can switch its growth form between different morphological phenotypes under certain environmental conditions. For example, the yeast-filament (including hyphal and psuedohyphal forms) transition is highly dependent on the presence of environmental stimuli such as serum, high temperature (37 °C), and neutral pH [reviewed by (Biswas et al. 2007; Huang 2012; Sudbery et al. 2004; Whiteway and Bachewich 2007)]. Despite the fact that certain environmental factors favor certain phenotypes, white, gray, and opaque cell types are heritable (Anderson and Soll 1987; Slutsky et al. 1987; Tao et al. 2014b). The switches between the three phenotypes are reversible and can occur spontaneously. The cells, however, can maintain their original phenotype for many generations. In this chapter, we will focus on the genetic and epigenetic regulatory mechanisms of white–opaque switching and its association with sexual mating and pathogenesis.

8.2 General Features of White–Opaque Switching

The white–opaque switching system was first discovered in 1987 in WO-1, a clinical isolate of *C. albicans* by Soll and his colleagues (Slutsky et al. 1987). White and opaque cells differ in a number of biological aspects including cellular and colony morphologies, global gene expression profiles, mating competence, and virulence in different infection systems (Anderson and Soll 1987; Kvaal et al. 1999; Lan et al. 2002; Miller and Johnson 2002; Tsong et al. 2003; Xie et al. 2013). The switch between the two cell types is an epigenetic phenomenon with no changes of the genomic DNA sequence (Srikantha et al. 2001; Zordan et al. 2006). White cells are relatively round and small and form “white” and hemispherical colonies. Opaque cells are elongated, or bean shaped, and form flat, opaque colonies. The cell surface of white cells is smooth while that of mature opaque cells is “pimpled.” The biological function of “pimples” remains to be investigated. On nutrient agar containing phloxine B (a red dye which exclusively stains opaque cells red), opaque colonies appear pink or red, while white colonies maintain in white (Anderson and Soll 1987).

White and opaque cells express a set of cell type-specific genes. Global gene expression profile assays demonstrate that hundreds of protein-encoding genes and noncoding RNAs are differentially expressed in white and opaque cells (Lan et al. 2002; Tuch et al. 2010). The 5′ and 3′ untranslated regions (UTR) of some genes also differ in length between the two cell types (Tuch et al. 2010), suggesting that posttranscriptional regulations may be involved in this phenotypic switching. Differentially expressed genes are involved in the regulation of metabolism, mating, stress response, and virulence. Of them, a large number of genes are related to metabolic pathways. For example, the expression of fermentative metabolism-related genes is upregulated in white cells, while that of oxidative metabolism-related genes is upregulated in opaque cells (Lan et al. 2002). The difference of gene expression profiles provides the two cell types unique features and may confer them different abilities to survive and propagate in diverse natural niches.

8.3 Genetic and Epigenetic Regulation of White–Opaque Switching in *C. albicans*

The Wor1 Master Regulator In 2006, three labs reported the discovery of the master regulator of white–opaque switching in *C. albicans* (Huang et al. 2006; Srikantha et al. 2006; Zordan et al. 2006). It has been named as Wor1 (white-opaque regulator 1) by Huang et al. (2006) and Zordan et al. (2006) and referred to as Tos9 in the Srikantha paper (Srikantha et al. 2006) since it is homologous to the *Saccharomyces cerevisiae* Tos9 protein. The Wor1 transcription factor is a member of the WORP family, which is a fungal-specific protein family with critical biological functions in a range of diverse species (Jonkers et al. 2012; Lohse et al. 2010; Michielse et al. 2011; Mirzadi Gohari et al. 2014; Nguyen and Sil 2008). Wor1 contains two conserved domains (WORPa and WORPb) which are dissimilar to each other and form a WOPR box with DNA sequence-specific binding ability (Lohse et al. 2010). The WORPa domain also contains a conserved protein kinase A (PKA) phosphorylation site (Huang et al. 2006), suggesting that it may function downstream of the cAMP-PKA signaling pathway. The expression of *WOR1* in white and opaque cells exhibits an "all-or-none" feature. Its transcription can hardly be detected in white cells, but it is highly upregulated in opaque cells (Huang et al. 2006). Wor1, together with Efg1, Wor2, Wor3, and Ahr1 (or named as Zcf37), forms interlocking transcriptional feedback loops and controls the transition between the two distinct cell types (Hernday et al. 2013; Zordan et al. 2007). Despite the critical role of the other transcription factors in the control of the feedback loop, only Wor1 is essential for the development of the opaque cell type. Wor2 is required for the white-to-opaque switching when cultured under regular laboratory conditions (like using glucose as the carbon source), but is not required for this process in the presence of GlcNAc, a potent inducer of the opaque phenotype (Tong et al. 2014). The transcriptional and environmental regulatory mechanisms are summarized in Fig. 8.1.

Mating Type Locus (MTL) Under standard laboratory culture conditions (with glucose as the carbon source and an aerobic environment), only a small set of clinical isolates of *C. albicans* were found to be able to undergo white-to-opaque switching. These strains were proved to be homozygous at the *MTL* locus (*C. albicans* is an "obligate" diploid organism) (Lockhart et al. 2002; Miller and Johnson 2002). The **a**1/α2 heterodimer, a transcriptional repressor produced by the *MTL* locus, binds to the promoter region of *WOR1* and represses its expression and thus locks cells in the white phase (Huang et al. 2006; Miller and Johnson 2002; Zordan et al. 2006). We have recently discovered that white–opaque switching is not restricted to *MTL* homozygous isolates of *C. albicans*. In the presence of CO_2

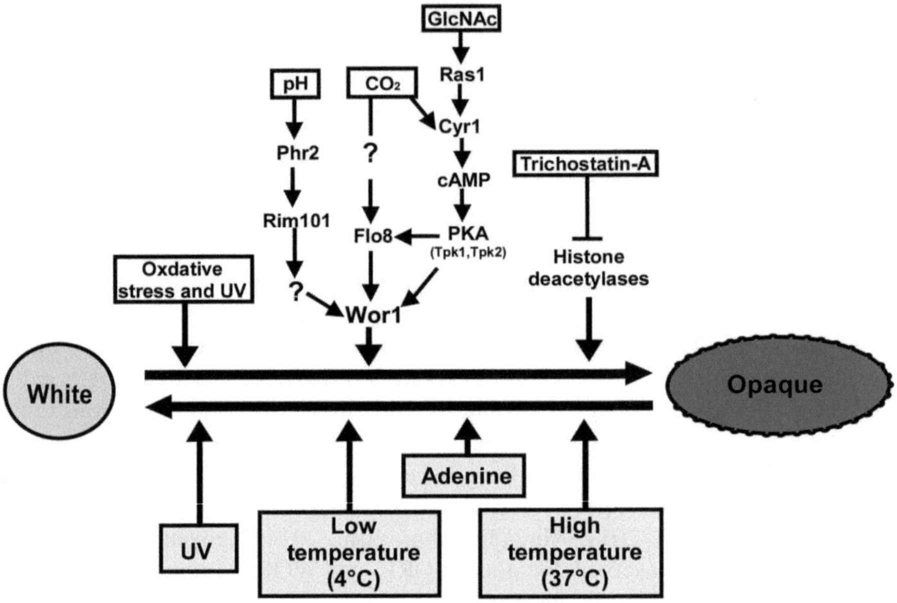

Fig. 8.1 Environmental and genetic regulation of white–opaque switching in *C. albicans*. UV, ultraviolet. Worl plays a central role in the regulation of white–opaque switching. All the signaling pathways could converge on the regulation of Worl

and GlcNAc, a small set of clinical isolated *MTL* heterozygous strains can undergo white-to-opaque switching (Xie et al. 2013). As mentioned earlier, CO_2 and GlcNAc are abundant in the human host, especially in the lower gut. Therefore, to some extent, this culture condition mimics the host environment. Both white and opaque cells of the *MTL* heterozygous strains exhibit a similar gene expression profile and virulence characteristic to their counterparts of the *MTL* homozygous strains. Given the human host is the major reservoir of *C. albicans*, our study implies that white–opaque switching could be a general feature of natural isolates.

Epigenetic Regulation of White–Opaque Switching in *C. albicans* As mentioned earlier, white and opaque cells contain the same genome. It has been proposed by Klar et al. that white–opaque switching involves changes in the structure of chromosomes (Klar et al. 2001). They found that cells of *C. albicans* treated with trichostatin-A (TSA), a histone deacetylase inhibitor, showed a dramatic increased frequency of white-to-opaque switching. Consistently, deletion of the histone deacetylase genes *HDA1* and *RTT109* promotes the opaque phenotype, while deletion of the histone deacetylase gene *RPD3* leads to an increase in the frequency of switching in both white-to-opaque and opaque-to-white transitions (Srikantha et al. 2001). Deletion of one copy of the essential gene *HST3*, which encodes an H3K56 deacetylase, also results in an increase in the white-to-opaque switching frequency (Stevenson and Liu 2011). Hnisz et al. have further demonstrated that the Set3/Hos2 complex plays a critical role in the regulation of white–

opaque switching (Hnisz et al. 2009). Together with the Wor1-Efg1-Wor2-Czf1 transcriptional circuitry, these chromatin modifiers form a dual layer network and control this epigenetic switch in *C. albicans*.

8.4 Environmental Regulation of White–Opaque Switching in *C. albicans*

Although white–opaque switching can occur spontaneously, a plethora of environmental factors have been found to affect the frequency of switching. Host-related environmental factors, such as the physiological temperature (37 °C), pH, GlcNAc, hypoxic conditions, and CO_2, regulate white–opaque switching through distinct pathways (Huang et al. 2009, 2010; Srikantha and Soll 1993; Sun et al. 2015). When cultured in air, both low (lower than 4 °C) and high (higher than 27 °C) temperatures promote opaque-to-white switching (Anderson and Soll 1987; Huang et al. 2009, 2010; Srikantha and Soll 1993; Sun et al. 2015). This phenomenon raises a question: if the physiological temperature of the human body is about 37 °C, how can *C. albicans* switch to and maintain the opaque phase in humans? GlcNAc and CO_2 are abundantly present in the mammalian host (e.g., in the gut and blood) (Chang et al. 2004; Levitt 1971). Huang et al. have found that both GlcNAc and high levels of CO_2 promote white-to-opaque switching and repress opaque-to-white switching at 37 °C. Similar to the promoting effect on filamentation, GlcNAc activates white-to-opaque switching primarily through the Ras1-cAMP-PKA signaling pathway (Huang et al. 2010). The master regulator Wor1, which contains a conserved cAMP-dependent protein kinase A phosphorylation site (T67), is downstream of this pathway. The PKA phosphorylation site is critical for GlcNAc-induced opaque cell formation (Huang et al. 2010). CO_2 regulates white-to-opaque switching primarily via two pathways: the cAMP signaling and an unidentified pathway. The two pathways converge on the Flo8 transcription factor, which is upstream of Wor1. Flo8 is essential for the maintenance of the opaque phenotype both in air and in the presence of high levels of CO_2. Overexpression of *WOR1* promotes the opaque phenotype in the *flo8/flo8* mutant (Du et al. 2012). We have recently reported that acidic pH induces the opaque phenotype in several clinical strains of *C. albicans* (Sun et al. 2015). This induction is dependent on the Rim101-mediated pH sensing pathway. Since the hydrolysis of CO_2 results in acidification of the medium, the induction of the opaque phenotype could be partially due to the intracellular acidification. Interestingly, acidic pH suppresses sexual mating in *C. albicans* through repressing the induction of the expression of *MFA1*, a gene encoding the precursor of **a**-pheromone, and *STE2*, a gene encoding the α-pheromone receptor (Sun et al. 2015). Moreover, environmental stresses (such as oxidative and genotoxic stresses) promote the development of the opaque phenotype and thus facilitate sexual mating in *C. albicans* (Alby and Bennett

2009), suggesting that phenotypic switching and mating could be an adaptive response under stressful environments.

8.5 Relationship Between White–Opaque Switching and Mating in *C. albicans*

As a diploid organism, *C. albicans* was long thought to be asexual. The discovery of the *MTL* locus of *C. albicans*, which is similar to the mating type locus (*MAT*) in *S. cerevisiae*, implies that there could be a cryptic sexual life cycle in this pathogenic fungus (Hull and Johnson 1999). In 2000, two independent groups observed mating in an in vivo and an in vitro system in several genome-engineered *C. albicans* strains (Hull et al. 2000; Magee and Magee 2000). Miller et al. further found that only opaque cells can mate efficiently (Miller and Johnson 2002). Opaque cells mate approximately one million times more efficiently than white cells. Given the spontaneous switching feature of the two cell types, mating observed in white cells could be due to mating of the rare "switched" opaque cells. Despite the poor mating efficiency, when treated with pheromones or exposed to opaque cells with an opposite *MTL* type, white cells can secrete pheromones and facilitate opaque cell mating (Tao et al. 2014a). Soll and his colleagues have reported that pheromone secreted by opaque cells can also signal white cells forming a "sexual" biofilm, which in turn facilitates opaque cell mating (Daniels et al. 2006). This white cell pheromone response is a general feature of clinical isolates of *C. albicans* (Sahni et al. 2009). Given the predominant feature of white cells in natural conditions, this coordinative behavior between white and opaque cells could play an important balancing role in the regulation of sexual and asexual lifestyles in *C. albicans*.

8.6 White–Opaque Switching Regulates Filamentation and Biofilm Development

The yeast-filament switch is another type of morphological transition which plays a central role in the regulation of virulence in *C. albicans* (Lo et al. 1997; Sudbery 2011; Whiteway and Bachewich 2007). The environmental and genetic regulatory mechanisms of filamentation are different in white and opaque cells. Typical filamentation-inducing conditions of white cells such as high temperature (37 °C), YPD + serum, Spider, and Lee's medium do not promote filamentous growth in opaque cells (Anderson et al. 1989; Guan et al. 2013; Si et al. 2013). However, low phosphate medium and sorbitol medium can efficiently induce opaque cell filamentation (Si et al. 2013). High temperature (37 °C) facilitates white cell filamentation but represses opaque cell filamentation (Si et al. 2013). The conserved cAMP signaling pathway and transcription factors including Tup1, Nrg1, and Efg1 control the development of filaments in both white and opaque cells (Guan et al. 2013; Si et al. 2013). The Bcr1 transcription factor, which was previously identified as a biofilm regulator, plays a unique role in the regulation of opaque cell filamentation. Inactivation of *BCR1* in *C. albicans* does not obviously affect white cell filamentation but has a dramatic promoting effect on the induction of opaque cell filamentation (Guan et al. 2013).

White and opaque cells also differ in the ability of biofilm development. Under standard laboratory incubation conditions, only white cells can form biofilms (Daniels et al. 2006; Yi et al. 2011). This could be due to the different expression profiles of cell wall and adhesion-related proteins in the two cell types (Lan et al. 2002; Tsong et al. 2003) which are critical for the development of biofilms in *C. albicans*. Interestingly, in a mixed culture system, opaque cells signal white cells to form "sexual" biofilms through activating the white cell pheromone response pathway. "Sexual" biofilms differ from conventional "pathogenic" biofilms in a number of aspects including the permeability to peptides and small organic molecules, as well as antifungal susceptibility (Daniels et al. 2006; Park et al. 2013; Yi et al. 2011). Park et al. proposed that "sexual" biofilms could facilitate opaque cell mating in *C. albicans* (Park et al. 2013). This white cell response to pheromone is a general feature of clinical isolates of *C. albicans* (Sahni et al. 2009). Another possible reason for the promoting effect of white cells on opaque cell mating in biofilms could be the fact that white cells secrete pheromone and increase the general level of pheromone in the mixed culture system (Tao et al. 2014a).

8.7 White–Opaque Switching and Pathogenesis

White and opaque cells exhibit distinct virulence in different infection models. White cells are more virulent in the mouse systemic infection models, while opaque cells are better at skin colonization in cutaneous infection models. Consistent with these observations, opaque cells express high levels of secreted aspartic proteinases (Saps) (Tao et al. 2014b; Xie et al. 2013), which could degrade the animal tissue and facilitate opaque cell growth. The two cell types also differ in the susceptibility to host immune cells. For example, mouse macrophage-derived RAW264.7 cells and human polymorphonuclear neutrophils preferentially phagocytose white cells (Lohse and Johnson 2008; Sasse et al. 2013). Recently, Pande et al. reported a novel cell type in *C. albicans*, named as the GUT (gastrointestinally induced transition) cell type (Pande et al. 2013). GUT cells are morphologically similar to but functionally different from opaque cells. The GUT form promotes *C. albicans* commensalism, while opaque cells decrease commensalism. The Wor1 transcription factor is essential for both the GUT and opaque cell types. These intriguing findings suggest that phenotypic switching could be used as an escaping strategy for the avoidance of the attack of the host defense system.

8.8 White–Gray–Opaque Tristable Switching

We recently discovered a novel morphological phenotype, referred to as the gray phenotype, in *C. albicans* (Fig. 8.2) (Tao et al. 2014b). This phenotype, together with the white and opaque phenotypes, forms a tristable morphological switching system. Compared with the white and opaque phenotypes, the gray phenotype has several unique features including distinct cellular morphology, high Sap activity in response to proteins or animal tissues, distinct global gene expression profile, and an increased ability to colonize animal cutaneous surfaces. Compared with white and opaque cells, gray cells exhibit an intermediate level of mating competency. The Wor1 and Efg1 transcription factors play a central role in the regulation of white–gray–opaque tristable switching in *C. albicans*. Wor1 is essential for the development of the opaque cell type, while Efg1 is required for the development of the white cell type. Deletion of both *WOR1* and *EFG1* locks cell in the gray phenotype since the mutant cannot form either white or opaque cells. Although two regulators are enough to control a tristable switching system, it is unknown whether or not there is a master regulator controlling the gray phenotype. Gray cells are different from the GUT cell type in a number of aspects. First, gray cells are smaller than GUT cells in cell size. Second, gray cells are stable when cultured in laboratory media, while GUT cells only exist in the gut and cannot grow in vitro.

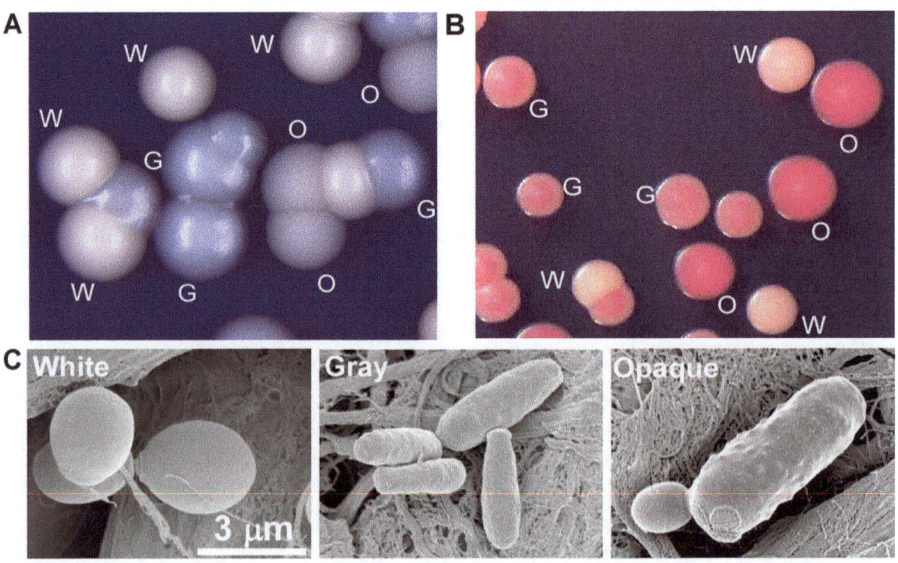

Fig. 8.2 Colony and cellular morphologies of white, gray, and opaque phenotypes of *C. albicans* [adapted from the reference (Tao et al., 2014b)]. W, white; G, gray; O, opaque. (**a**) Colony morphologies on nutrient agar (YPD medium without phloxine B, a red dye). (**b**) Colony morphologies on nutrient agar (Lee's medium with phloxine B). (**c**) Scanning Electron Microscopy images of white, gray, and opaque cells

Third, the formation of gray cells is independent of Wor1, which is essential for the maintenance of the GUT phenotype (Pande et al. 2013).

8.9 White–Opaque Switching in *Candida dubliniensis* and *Candida tropicalis*

C. dubliniensis and *C. tropicalis* are two *C. albicans*-related species and also frequently isolated from clinical settings (Pfaller 1995; Pfaller and Diekema 2007). Similar to *C. albicans*, these two *Candida* species can also undergo white–opaque switching under certain conditions (Porman et al. 2011; Pujol et al. 2004; Xie et al. 2012). The cellular and colony morphologies of white and opaque phenotypes of the two species are similar to their counterparts of *C. albicans*. For example, the surface of both *C. dubliniensis* and *C. tropicalis* cells is also rough (or "pimpled"), and the white–opaque switch regulates sexual mating in the two species. Opaque cells of *C. dubliniensis* and *C. tropicalis* mate more efficiently than white cells. Wor1 plays an essential role in the regulation of white-to-opaque switching in both *C. dubliniensis* (Yue et al. 2016) and *C. tropicalis*. Efg1 is essential for the white phenotype in *C. tropicalis*. The *efg1/efg1* mutant of *C. tropicalis* cannot form white cells but can switch between the opaque and an intermediate phenotype (Mancera et al. 2015). As in *C. albicans*, GlcNAc promotes the opaque phenotype and sexual mating in *C. tropicalis* (Xie et al. 2012). These findings suggest that white–opaque switching is a highly conserved biological process in *C. albicans*, *C. dubliniensis*, and *C. tropicalis*.

8.10 Conclusion

Heritable phenotypic transitions have been observed in several *Candida* species. The switch between different cell types is epigenetically regulated and involves no genomic DNA changes. In *C. albicans*, the white–opaque switch plays an important role in several biological aspects including sexual reproduction, filamentation, virulence, escaping the attack from the host immune cells, and antifungal resistance. *C. albicans* can colonize almost all of the organs of humans. The morphological plasticity may provide this species with strong ability to adapt to different host niches. The involvement of white–opaque switching in mating adds another tier of regulation to sexual reproduction in *Candida* species. In the past decade, the environmental and genetic regulatory mechanisms have been intensively studied. Host-related environmental cues such as the temperature, GlcNAc, and CO_2 play a critical role in the regulation of white–opaque switching and sexual mating. The master regulator Wor1, together with a number of transcription factors including Efg1, Wor2, Wor3, and Czf1, forms an interlocking feedback loop in the control of white–opaque transitions. This transcriptional regulation may also coordinate with the changes of chromosome structure, which involve a number of epigenetic regulators such as histone acetylases and deacetylases. The epigenetic switching could also provide a nice model system for the study of epigenetic regulations of other biological processes in pathogenic *Candida* species.

Acknowledgements The authors are grateful to Mr. Justin Friedman (University at Buffalo) for reading the manuscript and valuable suggestions. The work of the Huang lab was supported by grants from the Chinese National Natural Science Foundation (31625002, 31170086, and 81322026 to G.H.) and the "100 Talent Program" grant from the Chinese Academy of Sciences (to G.H.).

References

Alby K, Bennett RJ (2009) Stress-induced phenotypic switching in Candida albicans. Mol Biol Cell 20:3178–3191

Anderson J, Cundiff L, Schnars B, Gao MX, Mackenzie I, Soll DR (1989) Hypha formation in the white-opaque transition of Candida albicans. Infect Immun 57:458–467

Anderson JM, Soll DR (1987) Unique phenotype of opaque cells in the white-opaque transition of Candida albicans. J Bacteriol 169:5579–5588

Berman J (2012) Candida albicans. Curr Biol 22:R620–R622

Biswas S, Van Dijck P, Datta A (2007) Environmental sensing and signal transduction pathways regulating morphopathogenic determinants of Candida albicans. Microbiol Mol Biol Rev 71:348–376

Brown GD, Denning DW, Gow NA, Levitz SM, Netea MG, White TC (2012) Hidden killers: human fungal infections. Sci Transl Med 4:165rv113

Chang DE, Smalley DJ, Tucker DL, Leatham MP, Norris WE, Stevenson SJ, Anderson AB, Grissom JE, Laux DC, Cohen PS et al (2004) Carbon nutrition of Escherichia coli in the mouse intestine. Proc Natl Acad Sci U S A 101:7427–7432

Daniels KJ, Srikantha T, Lockhart SR, Pujol C, Soll DR (2006) Opaque cells signal white cells to form biofilms in Candida albicans. EMBO J 25:2240–2252

Du H, Guan G, Xie J, Cottier F, Sun Y, Jia W, Muhlschlegel FA, Huang G (2012) The transcription factor Flo8 mediates CO2 sensing in the human fungal pathogen Candida albicans. Mol Biol Cell 23:2692–2701

Guan G, Xie J, Tao L, Nobile CJ, Sun Y, Cao C, Tong Y, Huang G (2013) Bcr1 plays a central role in the regulation of opaque cell filamentation in Candida albicans. Mol Microbiol 89:732–750

Hernday AD, Lohse MB, Fordyce PM, Nobile CJ, DeRisi JL, Johnson AD (2013) Structure of the transcriptional network controlling white-opaque switching in Candida albicans. Mol Microbiol 90:22–35

Hnisz D, Schwarzmuller T, Kuchler K (2009) Transcriptional loops meet chromatin: a dual-layer network controls white-opaque switching in Candida albicans. Mol Microbiol 74:1–15

Huang G (2012) Regulation of phenotypic transitions in the fungal pathogen Candida albicans. Virulence 3:251–261

Huang G, Srikantha T, Sahni N, Yi S, Soll DR (2009) CO(2) regulates white-to-opaque switching in Candida albicans. Curr Biol 19:330–334

Huang G, Wang H, Chou S, Nie X, Chen J, Liu H (2006) Bistable expression of WOR1, a master regulator of white-opaque switching in Candida albicans. Proc Natl Acad Sci U S A 103:12813–12818

Huang G, Yi S, Sahni N, Daniels KJ, Srikantha T, Soll DR (2010) N-acetylglucosamine induces white to opaque switching, a mating prerequisite in Candida albicans. PLoS Pathog 6: e1000806
Hull CM, Johnson AD (1999) Identification of a mating type-like locus in the asexual pathogenic yeast Candida albicans. Science 285:1271–1275
Hull CM, Raisner RM, Johnson AD (2000) Evidence for mating of the "asexual" yeast Candida albicans in a mammalian host. Science 289:307–310
Jonkers W, Dong Y, Broz K, Kistler HC (2012) The Wor1-like protein Fgp1 regulates pathogenicity, toxin synthesis and reproduction in the phytopathogenic fungus Fusarium graminearum. PLoS Pathog 8:e1002724
Klar AJ, Srikantha T, Soll DR (2001) A histone deacetylation inhibitor and mutant promote colony-type switching of the human pathogen Candida albicans. Genetics 158:919–924
Kvaal C, Lachke SA, Srikantha T, Daniels K, McCoy J, Soll DR (1999) Misexpression of the opaque-phase-specific gene PEP1 (SAP1) in the white phase of Candida albicans confers increased virulence in a mouse model of cutaneous infection. Infect Immun 67:6652–6662
Lan CY, Newport G, Murillo LA, Jones T, Scherer S, Davis RW, Agabian N (2002) Metabolic specialization associated with phenotypic switching in Candidaalbicans. Proc Natl Acad Sci U S A 99:14907–14912
Levitt MD (1971) Volume and composition of human intestinal gas determined by means of an intestinal washout technic. N Engl J Med 284:1394–1398
Lo HJ, Kohler JR, DiDomenico B, Loebenberg D, Cacciapuoti A, Fink GR (1997) Nonfilamentous C. albicans mutants are avirulent. Cell 90:939–949
Lockhart SR, Pujol C, Daniels KJ, Miller MG, Johnson AD, Pfaller MA, Soll DR (2002) In Candida albicans, white-opaque switchers are homozygous for mating type. Genetics 162:737–745
Lohse MB, Johnson AD (2008) Differential phagocytosis of white versus opaque Candida albicans by Drosophila and mouse phagocytes. PloS ONE 3:e1473
Lohse MB, Zordan RE, Cain CW, Johnson AD (2010) Distinct class of DNA-binding domains is exemplified by a master regulator of phenotypic switching in Candida albicans. Proc Natl Acad Sci U S A 107:14105–14110
Magee BB, Magee PT (2000) Induction of mating in Candida albicans by construction of MTLa and MTLalpha strains. Science 289:310–313
Mancera E, Porman AM, Cuomo CA, Bennett RJ, Johnson AD (2015) Finding a missing gene: EFG1 regulates morphogenesis in Candida tropicalis. G3 (Bethesda) 5:849–856
Michielse CB, Becker M, Heller J, Moraga J, Collado IG, Tudzynski P (2011) The Botrytis cinerea Reg1 protein, a putative transcriptional regulator, is required for pathogenicity, conidiogenesis, and the production of secondary metabolites. Mol Plant Microbe Interact 24:1074–1085
Miller MG, Johnson AD (2002) White-opaque switching in Candida albicans is controlled by mating-type locus homeodomain proteins and allows efficient mating. Cell 110:293–302
Mirzadi Gohari A, Mehrabi R, Robert O, Ince IA, Boeren S, Schuster M, Steinberg G, de Wit PJ, Kema GH (2014) Molecular characterization and functional analyses of ZtWor1, a transcriptional regulator of the fungal wheat pathogen Zymoseptoria tritici. Mol Plant Pathol 15:394–405
Nguyen VQ, Sil A (2008) Temperature-induced switch to the pathogenic yeast form of Histoplasma capsulatum requires Ryp1, a conserved transcriptional regulator. Proc Natl Acad Sci U S A 105:4880–4885
Pande K, Chen C, Noble SM (2013) Passage through the mammalian gut triggers a phenotypic switch that promotes Candida albicans commensalism. Nat Genet 45:1088–1091
Park YN, Daniels KJ, Pujol C, Srikantha T, Soll DR (2013) Candida albicans forms a specialized "sexual" as well as "pathogenic" biofilm. Eukaryot Cell 12:1120–1131
Pfaller MA (1995) Epidemiology of candidiasis. J Hosp Infect 30(Suppl):329–338
Pfaller MA, Diekema DJ (2007) Epidemiology of invasive candidiasis: a persistent public health problem. Clin Microbiol Rev 20:133–163

Porman AM, Alby K, Hirakawa MP, Bennett RJ (2011) Discovery of a phenotypic switch regulating sexual mating in the opportunistic fungal pathogen Candida tropicalis. Proc Natl Acad Sci U S A 108:21158–21163

Pujol C, Daniels KJ, Lockhart SR, Srikantha T, Radke JB, Geiger J, Soll DR (2004) The closely related species Candida albicans and Candida dubliniensis can mate. Eukaryot Cell 3:1015–1027

Sahni N, Yi S, Pujol C, Soll DR (2009) The white cell response to pheromone is a general characteristic of Candida albicans strains. Eukaryot Cell 8:251–256

Sasse C, Hasenberg M, Weyler M, Gunzer M, Morschhauser J (2013) White-opaque switching of Candida albicans allows immune evasion in an environment-dependent fashion. Eukaryot Cell 12:50–58

Si H, Hernday AD, Hirakawa MP, Johnson AD, Bennett RJ (2013) Candida albicans white and opaque cells undergo distinct programs of filamentous growth. PLoS Pathog 9:e1003210

Slutsky B, Staebell M, Anderson J, Risen L, Pfaller M, Soll DR (1987) "White-opaque transition": a second high-frequency switching system in Candida albicans. J Bacteriol 169:189–197

Srikantha T, Borneman AR, Daniels KJ, Pujol C, Wu W, Seringhaus MR, Gerstein M, Yi S, Snyder M, Soll DR (2006) TOS9 regulates white-opaque switching in Candida albicans. Eukaryot Cell 5:1674–1687

Srikantha T, Soll DR (1993) A white-specific gene in the white-opaque switching system of Candida albicans. Gene 131:53–60

Srikantha T, Tsai L, Daniels K, Klar AJ, Soll DR (2001) The histone deacetylase genes HDA1 and RPD3 play distinct roles in regulation of high-frequency phenotypic switching in Candida albicans. J Bacteriol 183:4614–4625

Stevenson JS, Liu H (2011) Regulation of white and opaque cell-type formation in Candida albicans by Rtt109 and Hst3. Mol Microbiol 81:1078–1091

Sudbery PE (2011) Growth of Candida albicans hyphae. Nat Rev Microbiol 9:737–748

Sudbery P, Gow N, Berman J (2004) The distinct morphogenic states of Candida albicans. Trends Microbiol 12:317–324

Sun Y, Cao C, Jia W, Tao L, Guan G, Huang G (2015) pH regulates white-opaque switching and sexual mating in Candida albicans. Eukaryot Cell 14:1127–1134

Tao L, Cao C, Liang W, Guan G, Zhang Q, Nobile CJ, Huang G (2014a) White cells facilitate opposite- and same-sex mating of opaque cells in Candida albicans. PLoS Genet 10:e1004737

Tao L, Du H, Guan G, Dai Y, Nobile CJ, Liang W, Cao C, Zhang Q, Zhong J, Huang G (2014b) Discovery of a "white-gray-opaque" tristable phenotypic switching system in candida albicans: roles of non-genetic diversity in host adaptation. PLoS Biol 12:e1001830

Tong Y, Cao C, Xie J, Ni J, Guan G, Tao L, Zhang L, Huang G (2014) N-acetylglucosamine-induced white-to-opaque switching in Candida albicans is independent of the Wor2 transcription factor. Fungal Genet Biol 62:71–77

Tsong AE, Miller MG, Raisner RM, Johnson AD (2003) Evolution of a combinatorial transcriptional circuit: a case study in yeasts. Cell 115:389–399

Tuch BB, Mitrovich QM, Homann OR, Hernday AD, Monighetti CK, De La Vega FM, Johnson AD (2010) The transcriptomes of two heritable cell types illuminate the circuit governing their differentiation. PLoS Genet 6:e1001070

Whiteway M, Bachewich C (2007) Morphogenesis in Candida albicans. Annu Rev Microbiol 61:529–553

Xie J, Du H, Guan G, Tong Y, Kourkoumpetis TK, Zhang L, Bai FY, Huang G (2012) N-Acetylglucosamine Induces White-to-Opaque Switching and Mating in Candida tropicalis, Providing New Insights into Adaptation and Fungal Sexual Evolution. Eukaryot Cell 11:773–782

Xie J, Tao L, Nobile CJ, Tong Y, Guan G, Sun Y, Cao C, Hernday AD, Johnson AD, Zhang L et al (2013) White-opaque switching in natural MTLa/alpha isolates of Candida albicans: evolutionary implications for roles in host adaptation, pathogenesis, and sex. PLoS Biol 11: e1001525

Yi S, Sahni N, Daniels KJ, Lu KL, Srikantha T, Huang G, Garnaas AM, Soll DR (2011) Alternative mating type configurations (a/alpha versus a/a or alpha/alpha) of Candida albicans result in alternative biofilms regulated by different pathways. PLoS Biol 9:e1001117

Yue H, Hu J, Guan G, Tao L, Du H, Li H, Huang G (2016) Discovery of the gray phenotype and white-gray-opaque tristable phenotypic transitions in *Candida dubliniensis*. Virulence 7:230–242

Zordan RE, Galgoczy DJ, Johnson AD (2006) Epigenetic properties of white-opaque switching in Candida albicans are based on a self-sustaining transcriptional feedback loop. Proc Natl Acad Sci U S A 103:12807–12812

Zordan RE, Miller MG, Galgoczy DJ, Tuch BB, Johnson AD (2007) Interlocking transcriptional feedback loops control white-opaque switching in Candida albicans. PLoS Biol 5:e256

Chapter 9
Modification of the Host Epigenome by Parasitic Protists

Inessa Gendlina, Natalie Silmon de Monerri, and Kami Kim

Abstract Protozoan parasites compose a large group of ubiquitous unicellular eukaryotic organisms that closely interact with, and frequently reside within, a larger host. These parasitic protists rely on their host for nutrients, energy, and biomaterials. The host–parasite interaction is complex, as parasites strive to achieve a delicate balance of survival and replication without inducing host death. The host, in turn, tries to protect itself by various means including activation of death pathways in order to limit parasite spread. Therefore, successful parasites have developed highly evolved tactics in order to avoid host immune recognition and intracellular killing and subvert the host to their needs. To this end, various mechanisms of hijacking of host processes via parasite-derived or secreted effectors have been described. It has recently come to light that parasites also induce alterations to the host epigenomic landscape. Changes in host DNA methylation, histone posttranslational modifications, nucleosome positioning, chromatin assembly, and regulation of transcription have been noted in the parasitized host. To date, only a few parasite-derived effectors have been shown to directly modify host chromatin, and it remains to be elucidated whether parasite-induced alterations to the host epigenomic landscape are brought on specifically by parasites or are due to the host response. Finally, while various parasites target different components of host epigenomic landscape, common themes in subversion of host pathways and process emerge. We aim to review what is known about parasite modulation of host epigenome and touch on some conserved themes in this host–parasite interplay.

Keywords Parasite • Protist • Host • Epigenome • Noncoding RNA • Chromatin remodeling • Histone • DNA methylation • *Toxoplasma gondii* • Theileria • Leishmania • Plasmodium • Cryptosporidium • Microsporidia • Eimeria

---

I. Gendlina • N. Silmon de Monerri • K. Kim (✉)
Departments of Medicine, Pathology and Microbiology & Immunology, Albert Einstein College of Medicine and Montefiore Medical Center, Bronx, NY 10461, USA
e-mail: kami.kim@einstein.yu.edu

W. Doerfler, J. Casadesús (eds.), *Epigenetics of Infectious Diseases*, Epigenetics and Human Health, DOI 10.1007/978-3-319-55021-3_9

9.1 General Comments

9.1.1 Overview of Pathogenic Protists

Protozoan parasites are a vast group of distinct unicellular eukaryotic organisms, capable of infecting humans, animals, and insects, with unique and complex life cycles. Considering the enormous diversity of these organisms, we will limit discussion below to include examples of protozoans able to cause disease in humans and animals. Many of these parasites reside and replicate inside host cells for at least part of their life, inducing significant changes in cellular processes including the epigenetic landscape. Individual parasites have distinct interactions with the host. Some such as *Leishmania* spp. and *Theileria* inhabit the cytoplasm, while others such as *Toxoplasma gondii* form and reside in a parasitophorous vacuole. *Theileria,* in particular, which targets mainly cattle, live within the host cytoplasm of leukocytes and utilize the host cell division apparatus, inducing continuous proliferation and immortalization of the host cells (Spooner et al. 1989), while *Leishmania* species, obligate intracellular parasites targeting macrophages of mammals, reside within host-derived phagolysosomal vacuoles that are adapted to avoid and subvert host immune defenses (Lievin-Le Moal and Loiseau 2015). Some organisms, such as *T. gondii,* possess a highly evolved armament of effectors that are translocated across the parasitophorous vacuole and specifically target host processes (Boothroyd and Dubremetz 2008; Fentress and Sibley 2011). Parasites induce vast changes in host transcription, as has been demonstrated during infection with *T. gondii* (Blader et al. 2001; Chaussabel et al. 2003; Jia et al. 2013), *Plasmodium*-infected hepatocytes (Albuquerque et al. 2009; Chattopadhyay et al. 2011; Kaushansky et al. 2013), and most recently host cells of the avian malaria parasites (Videvall et al. 2015). These infections affect pathways involved in metabolism, cell death, differentiation, and cell cycle (Albuquerque et al. 2009). Despite major life cycle differences, parasites commonly exploit the close association with their mammalian host to achieve defense of self and subversion of the host.

In addition to invasion and replication, protozoan parasites have evolved an array of strategies to evade the host immune system and promote survival. Extensive remodeling of host cell subcellular structure is a feature of many host–parasite interactions. These include incorporating parasite protein into the cell membrane, restructuring the host cytoskeleton, sequestering mitochondria, and altering subcellular localization of organelles, forming transvesicular networks and constructing new organelles (Silmon de Monerri and Kim 2014). *T. gondii*, as an example, reorganizes host ER and mitochondria, relocalizing them to the parasitophorous vacuole (Sinai et al. 1997), and has also evolved to alter host metabolism and subvert energy and metabolic machinery to its cause (Wiley et al. 2010; Menendez et al. 2015). Along with these structural and metabolic changes to the host cell, reprogramming of the host cell transcriptome following infection or exposure is well documented for a significant number of infectious organisms. Distinct transcriptional changes occur in host cells following infection or exposure, in a pathogen-specific manner, and these changes may be long-lasting or transient (Chaussabel et al. 2003). Studies have additionally demonstrated that the host transcriptome may be differentially altered depending on the life cycle stage of the pathogen (Fouts and Boothroyd 2007). Together, these studies suggest that observed transcriptional effects are unique to the specific host–pathogen interaction. Importantly, while many changes in host gene expression are organism specific, overall there appears to be conservation of host pathways targeted by pathogens during infection.

9.1.2 Overview of Host Epigenetic Landscape

Eukaryotic genomes are folded into highly controlled chromatin complexes composed of well-organized hierarchical structures of DNA wound around histone-containing nucleosome complexes. Chromatin composed of nucleosomes then further folds into secondary and tertiary structures to allow efficient and ordered DNA packaging (Luger et al. 2012). The specific confirmation of chromatin renders DNA open and accessible (euchromatin) or tightly compacted and inaccessible (heterochromatin) for transcription factor and RNA polymerase binding. Changes that occur in the structure of chromatin are considered to be "epigenetic," i.e., not encoded in the DNA, and include both short- and long-term alterations to chromatin without change to the underlying DNA sequence. Traditionally, there are considered to be three main types of epigenetic regulation, which include DNA methylation, histone posttranslational modifications, and noncoding RNAs, initially described in the context of cell differentiation (Spivakov and Fisher 2007) (Fig. 9.1).

The best-studied mechanism of epigenetic modification that arises on the DNA itself is methylation. DNA methylation plays a role in regulation of gene expression and is typically associated with transcriptional repression, though recent studies implicate hydroxymethylation in transcriptional activation (Ito et al. 2010). Methylation events occur either de novo or as part of genome maintenance by replicating the methylation pattern of the complementary strand. In eukaryotes, cytosine bases are methylated by DNA methyl transferase (DNMT) enzymes. In mammalian genomes, DNA methylation mainly occurs on CpG dinucleotides, which are often located in CpG-rich regions known as CpG islands (CGI), frequently found at transcriptional start sites. Methylated nucleotides can be further converted from 5-methylcytosine to 5-hydroxymethylcytosine by a family of TET (Ten-eleven translocation) enzymes resulting in hydroxymethylation. Hydroxymethylation was initially thought of as a step towards demethylation, although hydroxymethylation itself has been implicated in stem cell differentiation (Dawlaty et al. 2014). Methylation of promoter regions silences genes by blocking transcription initiation, while methylation in gene bodies may facilitate elongation and block abnormal transcriptional initiation (Jjingo et al. 2012). Additionally, DNA methylation is thought to play a role in splicing, and methylation at centromeres may be

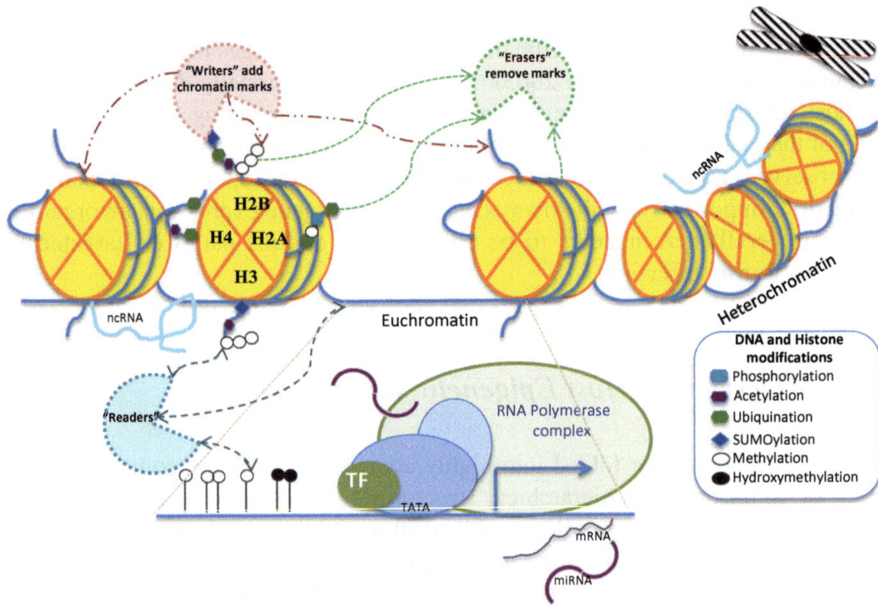

Fig. 9.1 Overview of host epigenetic landscape. Schematic representation of DNA and nucleosome complexes. The host epigenetic landscape is shaped by modifications including histone PTMs, DNA methylation, chromatin remodeling, and ncRNAs, including miRNAs. These modifications are applied by groups of enzymes referred to as "Writers" and "Erasers," where "Writers" add chromatin and histone marks (HAT, HMT, DNMT, kinases), while "Erasers" remove those marks (HDAC, HDM, NuRD, phosphatase). Combinations of specific marks afford precise regulation of DNA accessibility and transcriptional regulation, as they are recognized by "Readers"—specific transcription factors and polymerase machinery. Abbreviations: *PTM* post-translational modifications, *ncRNA* noncoding RNA, *mRNA* messenger RNA, *miRNA* microRNA, *HDAC* histone deacetylase, *DNMT* DNA methyl transferase, *HAT* histone acetyl transferase, *HMT* histone methyl transferase, *HDM* histone demethylase, *NuRD* Nucleosome remodeling deacetylase; Lollipop symbols represent methylation (*white*) and hydroxymethylation (*black*)

involved in overall chromosomal stability. This pattern of genome methylation, while inherited and more stable at CGIs, also undergoes changes during development and aging especially when occurring at the non-CGI regions (Jones 2012). A number of excellent reviews on the function and mechanism of DNA methylation are available (Ndlovu et al. 2011; Pelizzola and Ecker 2011; Jjingo et al. 2012; Jones 2012; Pastor et al. 2013; Dawlaty et al. 2014).

While DNA methylation is generally associated with gene silencing, posttranslational modification (PTM) of histones (or "chromatin marks") can either activate or repress transcription. Nucleosomes form basic structural units of chromatin, and DNA packaging into nucleosomes is essential for controlling DNA accessibility. Each nucleosome is an octamer composed of dimers of H2A, H2B, H3, and H4 histone proteins. Histones, particularly their N-terminal tails, are subject to complex posttranslational modifications, which collectively serve to recruit other proteins to chromatin in order to mediate changes in transcription (Strahl and Allis 2000). Among the histone modifications that have been described are phosphorylation, methylation, acetylation, ubiquitylation, and SUMOylation. These modifications are tightly choreographed as different modifications can target the same residues in a competitive manner. Chromatin marks are applied or removed by specific enzymes commonly referred to as "writers" [histone acetyltransferases (HAT), histone methyltransferases (HMT), kinases] or "erasers" [histone deacetylases (HDAC), histone demethylases (HDM), phosphatases]. Erasers and writers tightly regulate the binding affinity of histones to DNA and further control organization of nucleosome complexes. Specific combinations of PTMs on nucleosomes allow specificity for DNA interactions with and recognition by other protein complexes, and nucleosomes themselves serve as platforms for further regulation or chromatin access by "readers" and chromatin modifiers, including ncRNAs, which together form a multiprotein macromolecular complex (Jenuwein and Allis 2001). In addition, particular PTMs are associated with specific locations within the genome and serve as a foundation for readers by recruiting additional structural and regulatory assemblies. For example, H3K4me3 is found predominantly in promoters of active genes and plays a role in recruitment of the transcriptional machinery being recognized by plant homeodomain (PHD) finger domain-containing proteins (Chi et al. 2010), while H3K27me3 is enriched at promoters of repressed genes and together with CGI methylation marks gene silencing. Misregulation of these specific modifications has been shown to play a role in cancer (Chi et al. 2010). The total combination of histone PTMs alters the affinity of histones for DNA, and modulation of histone–DNA interactions regulates DNA winding and therefore the accessibility of DNA to transcription factors.

In addition to modifications of histones and nucleotides directly, the accessibility of DNA and position of nucleosomes is altered by transcription factors, their associated complexes such as RNA polymerase, as well as the Polycomb complex, which, when bound to the DNA targets, interfere with binding of histones and other transcriptional regulatory proteins. This chromatin remodeling also modulates histone nucleosome positioning and movement along the DNA, causing destabilization, reassembly, and eviction of nucleosomes (Struhl and Segal 2013). It should be noted that the ordering of such events is not clear, as nucleosomes mediate recruitment of other machinery, which in turn may prevent nucleosome binding.

A newly emerging arena in the study of epigenetics is evaluation of functions of noncoding RNAs (ncRNAs). The ncRNAs play a major role in posttranscriptional regulation and genome maintenance and are involved in a wide range of regulatory processes including DNA methylation, histone PTMs, DNA silencing, formation of the molecular scaffolding necessary for chromatin structure and stability, and posttranscriptional regulation of mRNAs (Joh et al. 2014; Fitzgerald and Caffrey 2014; Scaria and Pasha 2012). Additionally, antisense transcripts may play a role in coordinating chromatin and histone marks by recruiting DNMT or histone-modifying enzymes (Faghihi and Wahlestedt 2009). The ncRNAs can be long or small, and small ncRNAs are further divided into microRNA (miRNA), small interfering RNA (siRNA), and PIWI interacting RNA (piRNA). MicroRNAs (miRNAs) are small noncoding ssRNAs ~20–25 nucleotides in length that are best known for regulation of posttranscriptional mRNA processing and play a key role in mRNA and gene expression, while long noncoding RNAs (lncRNA) are

>200 nucleotides in size and interact with mRNAs, miRNAs, and RNA-binding proteins (RBP) to further regulate protein expression. Significant adjustment of protein expression occurs at the level of mRNA by miRNAs, RNA-binding proteins, and lncRNAs, and regulation at the level of mRNA allows for specific and rapid alteration of protein levels. These posttranscriptional modifications mediated by ncRNAs fine-tune regulation of mRNA stability and translation, especially for genes involved in immune and inflammatory responses, such as IFN gamma. RNA-specific regulation of components of immune response and inflammation has been studied in great detail, and readers are referred to excellent reviews on the roles of noncoding RNA in immune regulation (Fitzgerald and Caffrey 2014; Schwerk and Savan 2015).

All of these tightly orchestrated arrangements of DNA methylation, histone PTMs, nucleosome positioning, and other types of chromatin remodeling form a precise signature for regulation of transcription and are targeted by pathogens for their purpose. Reshaping of the host epigenome is an emerging mechanism of host modulation exploited by a variety of pathogens (Silmon de Monerri and Kim 2014; Cheeseman and Weitzman 2015). Unlike viruses, protozoa and bacteria do not insert DNA into the host genome and so have evolved different strategies to influence chromatin and gene regulation. In bacteria, secreted proteins known as nucleomodulins target host chromatin and transcription, altering downstream signaling pathways. Nucleomodulins such as AnkA from *Anaplasma* spp. directly target host DNA and recruit host chromatin and histone-modifying enzymes (HDAC) to globally alter host chromatin (Bierne et al. 2012; Sinclair et al. 2014). Intracellular parasitic protists, similar to intracellular bacteria and viruses, reside within the host cell either within some type of parasitophorous vacuole (*T. gondii*) or free in the host cytoplasm, thus avoiding direct recognition by antibodies and cells of immune system. Parasites then interface with their host via parasite-derived effector proteins, which can be secreted in a targeted manner (e.g., *T. gondii* ROP or GRA proteins) or delivered via exosomes. In turn, parasites utilize host nutrients, metabolites, and energy sources that can be transported or diffuse into the parasite's niche. However, intracellular parasites are subjected to other host defenses including host apoptosis and have devised ways of scavenging or subverting nutrient and energy pathways, while avoiding host defenses and preventing host demise. This review focuses on the epigenetic changes and chromatin remodeling that occur in the host cell following infection or exposure, and various mechanisms used by protozoan parasites to hijack the host transcriptome (Fig. 9.2 and Table 9.1). Additionally, we explore common themes in the host processes that are targeted and postulate biological implications of these alterations.

9.2 Alteration of DNA Methylation

Parasitic protists have evolved mechanisms of modulating host chromatin state by either amending host cytosine methylation or by modifying enzymes that bring about these modifications. Expressly, *Leishmania, Plasmodium*, and *T. gondii*

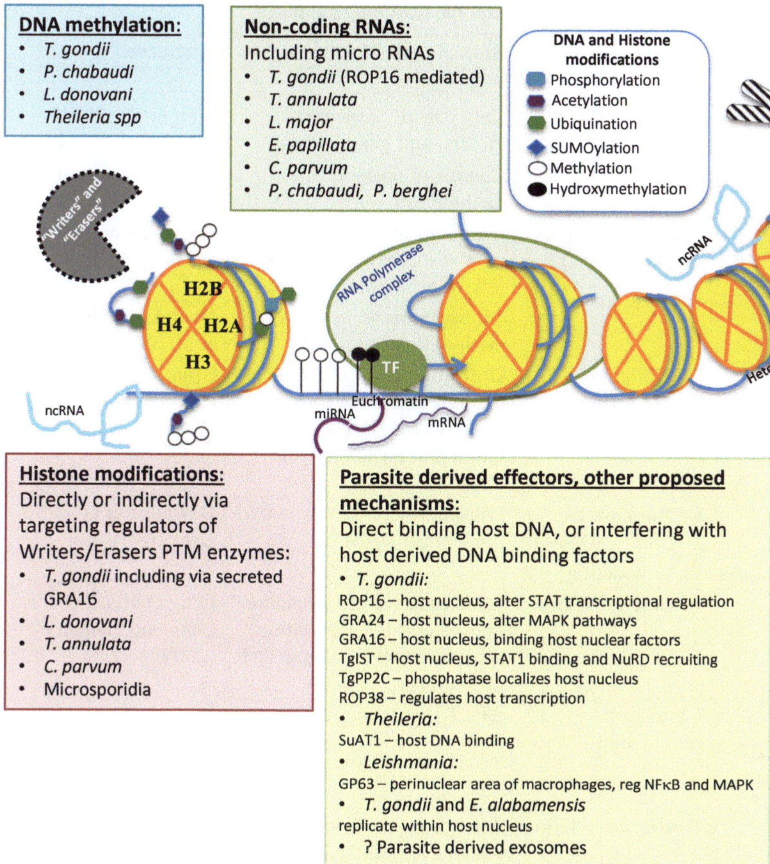

Fig. 9.2 Host epigenetic machinery targeted by parasites. Parasites induce specific alterations to the host epigenome by targeting key mechanisms of landscape design, including regulation of DNA methylation, regulation of enzymes responsible for PTMs, and interfering with chromatin accessibility and nucleosome positioning. Details of mechanisms and references are provided in the text

parasites have been shown to induce explicit changes in host DNA methylation. In a study evaluating methylation changes in macrophages infected with *L. donovani*, Marr et al. found diffuse changes in methylation at CGI in promoters as well as gene bodies (Marr et al. 2014). They noted significant changes in regions involved in regulation of key pathways of host response including NFkB, JAK/STAT, MAPK, mTOR, chemokine signaling, and others (Marr et al. 2014; Arango Duque and Descoteaux 2015). Similarly, genome-wide analysis of methylation of host promoters during infection of mice with *P. chabaudi* notes changes in the methylation of promoters of a number of genes, including toll-like receptor genes (Al-Quraishy et al. 2013). Finally, studies have focused on evaluating link between host behavior and infection, and a recent study implicates *T. gondii*-induced changes in DNA methylation as the cause for change in behavior of infected rats (Hari Dass and Vyas 2014). In response to *T. gondii* infection, hypomethylation of the arginine

Table 9.1 Select parasites and their effect on the host epigenome

Parasite (effector)	Specific host target	Biologic effect	Reference
DNA methylation			
L. donovani	Multiple, HDAC4	CpG islands, gene promoters, and gene bodies	Marr et al. (2014)
P. chabaudi		Change in promoter methylation	Al-Quraishy et al. (2013)
T. gondii	Arginine vasopressin promoter	Increased production arginine vasopressin—may be link to behavioral change	Hari Dass and Vyas (2014)
T. gondii		Global increase in DNA methylation in testes	Dvorakova-Hortova et al. (2014)
T. gondii	E3 Ubiquitin Ligase UHRF1	Downregulation of DNA methyltransferase I leading to increased H3 phosphorylation and host cell cycle arrest	Brunet et al. (2008), Unoki et al. (2009)
Theileria spp.	Casein kinase 2 (CK2) →DNMT	Phosphorylates DNA methyl transferase DNMT3	Dessauge et al. (2005b)
Histone and chromatin remodeling			
T. gondii	Histone H3 PTMs	Decreased H3S10 phosphorylation and H3 acetylation, specifically at IL10 and TNF promoters	Leng et al. (2009), Leng and Denkers (2009)
T. gondii (TgIST)	Chromatin remodeling	STAT regulation regions	Lang et al. (2012), Olias et al. (2016), Gay et al. (2016)
T. gondii	Histone acetylation	Alterations in histone and nuclear protein lysine acetylation	Bouchut et al. (2015)
T. gondii	UHRF1 E3 ubiquitin ligase → Phospho H3	Downregulation of UHRF1, associated with accumulation of phosphorylated H3 associated with mitosis and cyclin expression	Brunet et al. (2008), Unoki et al. (2009)
T. gondii (GRA 16)	Nuclear PP2A phosphatase, HAUSP deubiquitinase	Alter histone ubiquitin PTM of histone	Bougdour et al. (2013)
L. donovani	HDAC4 methylation	Alteration in HDAC4 gene body and upregulation of expression	Marr et al. (2014)
T. annulata	SMYD3 methyltransferase	Induced expression of SMYD3 methylates H3 histone	Cock-Rada et al. (2012)
T. annulata	HDAC9 downregulation		Kinnaird et al. (2013)

(continued)

Table 9.1 (continued)

Parasite (effector)	Specific host target	Biologic effect	Reference
T. annulata	PARP family	DNA binding, modifying DNA-binding proteins, CTCF, and histones, alteration in DNA methylation	Kinnaird et al. (2013)
C. parvum	? HDAC	Alterations in CX3CL1 chemokine controlled by HDAC	Zhou et al. (2013)
Microsporidia Nosema ceranae	Histones H3-like, H4 expression	Expression changes in histones in honeybee midgut epithelium	Aufauvre et al. (2014)
Noncoding RNA			
T. gondii	Alterations in miRNA profile, miR-132	Alteration in host miRNA in brains of infected humans and mice	Thirugnanam et al. (2013), Xu et al. (2013), Xiao et al. (2014), Li et al. (2015)
T. gondii (ROP16)	miR-146a, miR-155	Upregulated in brains of infected mice	Cannella et al. (2014)
T. gondii	miR-17-92, miR-106b-25	Altered expression in human macrophages and fibroblasts	Zeiner et al. (2010), Cai et al. (2013, 2014)
T. annulata	miR-155	Upregulated in transformed leukocytes	Marsolier et al. (2013)
C. parvum	miR-424, miR-503, miR-98, let-7	Changes in host miRNA profile in infected cholangiocytes	Zhou et al. (2009), Chen et al. (2007), Hu et al. (2009, 2010)
E. papillata	miRNA profile	Upregulated in mouse intestinal epithelia	Dkhil et al. (2011)
L. major	miRNA profile	Significant alteration in host miRNAs	Lemaire et al. (2013)
P. chabaudi	miRNA profile	Alteration in host hepatocyte miRNAs	Delic et al. (2011)
P. berghei	miRNA profile	Alteration in mosquito vector miRNAs	Biryukova et al. (2014)

DNMT DNA Methyltransferase, *HDAC* histone deacetylase, *PARP* Poly-ADP ribose polymerase, *PTM* posttranslational modification

vasopressin promoter was observed, which was implicated in increased production of vasopressin in the medial amygdala region of the brain, the region that perceives fear. Specifically, rats infected with *T. gondii* show a reduced aversion to cats, instead demonstrating attraction (Berdoy et al. 2000). Changes in behavior patterns and changes in the brain have been described in rats (Flegr and Markos 2014; Hari Dass and Vyas 2014; Vyas 2015), and this phenomenon may also affect humans as *T. gondii*-infected men perceive the smell of cat urine as being more pleasant as compared to uninfected males (Flegr et al. 2011). Infection with *T. gondii* not only alters behavior but also dramatically reduces reproductive fitness in mice (Dvorakova-Hortova et al. 2014). Specifically, *T. gondii* infection was associated with increased testicular global DNA methylation, as well as increased DNA methylation of genes involved in spermatogenesis (Dvorakova-Hortova et al.

2014). It is interesting to note that in the context of alteration in methylation and behavior, mice infected with *T. gondii* also have decreased levels of serum testosterone (Kankova et al. 2011), especially since regulation of vasopressin expression in medial amygdala by promoter CPG methylation was shown to be regulated by testosterone (Auger et al. 2011). Overall, these studies explore *T. gondii*-induced changes in host DNA methylation and propose specific physiologic consequences to the host. In an alternative approach, *Theileria* parasites appear to utilize an indirect method of altering host DNA methylation by targeting DNMTs. In a study examining how *Theileria* spp. manipulate signaling, it was found that parasites induce constitutive activation of casein kinase 2 (CK2) (Dessauge et al. 2005b). CK2 has numerous roles in transcriptional regulation, including regulating DNA methylation by phosphorylating DNA methyltransferase DNMT3 (Deplus et al. 2014) as well as playing a role in PI3-K activation and the MEK/ERK and Akt/PKB pathways (Dessauge et al. 2005b). By inducing CK2 and regulating DNMT3, *Theileria* induce alteration of the host DNA methylation landscape. It should be noted that apart from *Theileria*-induced CK2, the mechanisms underlying parasite-induced changes in host DNA methylation are largely unknown. Nevertheless, these data suggest that protozoan parasites have evolved to regulate host processes to alter genome methylation patterns that modify function of key cellular processes, including signaling pathways, behavior, and reproduction.

9.3 Histone Modification and Chromatin Remodeling

Parasites have evolved mechanisms to specifically induce epigenetic changes in the host histone code. These include alteration of host histone PTMs either directly or by regulating enzymes that impact these modifications, and alteration of expression of individual histones including variants histones that may differ in DNA-binding affinity (Siggens and Ekwall 2014). A handful of studies have attempted to elucidate host epigenetic changes in response to *T. gondii* infection. In an evaluation of macrophages following infection with *T. gondii*, there was notable impairment of histone 3 (H3) phosphorylation (at Serine 10 residue) and H3 acetylation at the IL10 and TNFα promoters (Leng et al. 2009; Leng and Denkers 2009), as well as impairment of chromatin remodeling at STAT1-regulatory regions (Lang et al. 2012). There was additional interference with chromatin remodeling at the TNF-alpha promoter preventing binding of RNA polymerase transcriptional machinery (Leng et al. 2009). Recent work demonstrated that *T. gondii* secreted factor TgIST (*T. gondii* inhibitor of STAT1 transcriptional activity) translocates to the host cell nucleus where it directly interacts with STAT1 protein promoting its nuclear sequestration, as well as associates with Mi-2/NuRD (nucleosome remodeling deacetylase complex) to facilitate chromatin remodeling and inhibition of transcription (Olias et al. 2016; Gay et al. 2016). In a study focused specifically on evaluating changes in lysine acetylation in cortical astrocytes infected with *T. gondii*, changes were noted in lysine acetylation of nuclear proteins including proteins that function in chromatin biology including histones, as well as proteins involved in

RNA processing and transcription (Bouchut et al. 2015). Specifically, data demonstrated more than twofold increase in acetylation of core histones including histone H3, H4, H2A.Z, among others, while other members of histone cluster and histone-like proteins demonstrated greater then twofold decrease in acetylation (Bouchut et al. 2015). Since histone acetylation is associated with transcriptional regulation, specifically activation (Berger 2007), such substantial alteration in host histone acetylation following parasite infection implies active modulation of the host epigenome, though the exact mechanism is not yet known. Additional work demonstrated that infection by *T. gondii* leads to downregulation of the host UHRF1 E3 ubiquitin ligase gene, accompanied by accumulation of phosphorylated histone H3, a mitotic histone mark, and reduction of host cell cyclin levels (Brunet et al. 2008; Unoki et al. 2009). *T. gondii* also subverts host transcription via GRA16, which travels to the nucleus and forms a complex with host PP2A phosphatase and HAUSP deubiquitinase (Bougdour et al. 2013), which are known to sway ubiquitin PTM balance on nuclear proteins including histones (Khoronenkova et al. 2011; Bougdour et al. 2014). Thus, *T. gondii* specifically targets host nuclear proteins and PTM machinery to promote remodeling of the epigenome.

Other intracellular parasites have also been shown to influence host histones. In *L. donovani*-infected macrophages, analysis of host DNA methylation revealed significant alteration in the methylation of the HDAC4 gene body associated with upregulation of HDAC4 expression (Marr et al. 2014). In an alternative example, *Nosema ceranae*—a member of *Microsporidia,* a diverse group of ~200 genera of obligate intracellular pathogens that infect a wide range of animals, fish, and insects, induce increased expression of histone H3-like and histone H4 in midgut epithelia of honeybees (Aufauvre et al. 2014; Calderon et al. 2015). *Cryptosporidium parvum* is another Apicomplexan parasite that primarily invades mucosal surfaces. Study of host epithelial immune regulation following infection with *C. parvum* revealed alteration in CX3CL1 chemokine that is at least in part directed by HDAC (Zhou et al. 2013). Finally, *Theileria*-transformed leukocytes demonstrate upregulation of matrix metalloproteinase (MMP-9), which is important in cancer cell migration and metastases. This *mmp9* regulation is in part achieved by inducing expression of SMYD3 methyltransferase in infected leukocytes, which methylates histone H3 (H3K4me3) at the *mmp9* promoter leading to transcriptional activation (Cock-Rada et al. 2012). Additionally, an expression microarray of lymphosarcoma cells infected with *T. annulata* revealed significant downregulation in HDAC9 expression (Kinnaird et al. 2013). Together, emerging data assert that various parasites have evolved mechanisms of specifically targeting host chromatin structure and assembly by targeting histone expression and posttranslational modifications.

9.4 Noncoding RNAs

Considering the key role for ncRNAs in regulation of host processes, it is not surprising that pathogens have evolved to target ncRNAs. Viruses widely utilize lncRNA and miRNA for transcriptional regulation to subvert host metabolic pathways (Scaria and Pasha 2012). Specifically, Kaposi's Sarcoma Herpesvirus (KHSV) encodes a miRNA that alters host cell metabolism in part by downregulating EGLN2 and HSPA9, components of the mitochondrial import machinery, which induce a glycolytic shift in host metabolism via stabilization of HIF1a—a master regulator of oxygen sensing and metabolism (Yogev et al. 2014). Evidence is now emerging that other pathogens including parasites modulate similar pathways during host cell infection. Unlike viruses, however, parasites are not known to secrete ncRNA but instead are hypothesized to modulate host-derived miRNAs by regulating their expression (Hakimi and Cannella 2011), as well as RNA PTMs such as methylation, further confounding complexity of host epigenetic regulation (Joh et al. 2014).

Infections with a number of parasites have been shown to alter host ncRNAs. A handful of investigations have evaluated alterations in host ncRNA specifically miRNA in the brain during *T. gondii* infection. In a study of human brain cancers, *T. gondii* infection was shown to alter host miRNA to facilitate carcinogenesis (Thirugnanam et al. 2013), while a microarray analysis of host neuroepithelioma cells infected with different strains of *T. gondii* revealed strain-specific alteration in host transcription (Xiao et al. 2011). Similarly, analysis of mouse brains after infection with *T. gondii* revealed a subset of nine host miRNAs that appear to be explicitly induced by infection (Xu et al. 2013). One of these differentially expressed ncRNAs is miR-132. Mammalian miR-132 is involved in regulation of neuronal synapses and plays a key role in a number of neurologic and psychiatric disorders including schizophrenia, depression, and Parkinson's disease, and it is therefore intriguing that miR-132 is targeted by *T. gondii* (Bicker et al. 2014). Interestingly, change in expression of miR-132 was different depending on the chronicity of infection. During acute infection, there was upregulation of miR-132, thought to contribute to modulation of dopamine signaling in brains of infected mice (Xiao et al. 2014), while there was significant downregulation of miR-132 in brains of chronically infected mice (Li et al. 2015). In addition to miR-132, host miR-146a and miR-155 were also strongly upregulated in brains of mice during chronic infection with *T. gondii,* in ROP16-dependent manner (Cannella et al. 2014). miR-146 is known to dampen the TLR4 response via NFκB-dependent TRAF6 and IRAK1, and miR-155 modules TLR signaling (Schwerk and Savan 2015). Furthermore, miR-155 belongs to the oncomiR group of cancer-associated microRNAs, which have been shown associated with malignant cells, with miR-155 specifically associated with cMyc overexpression (Esquela-Kerscher and Slack 2006). Similar to *T. gondii*, *Theileria* parasites also induce expression of host miR-155 (Marsolier et al. 2013; Cannella et al. 2014). In *Theileria*-transformed leukocytes, there is upregulation of miR-155 regulated by cJun and AP1 transcription factors, which in turn was shown to repress expression of DET1 important in cJun ubiquitination and stabilization (Marsolier et al. 2013).

In addition to changes noted in neuronal cells, *T. gondii* has also been shown to alter the expression of host microRNAs during infection of human fibroblast cells, especially miR-17-92 and miR-106b-25, both oncomiRs important in regulating cell cycle and apoptosis (Zeiner et al. 2010). Similarly, miRNA profiling of

T. gondii-infected human macrophages revealed several host miRNAs important in apoptosis, including miR-17-92, whose expression is altered in a STAT3-regulated manner (Cai et al. 2013, 2014). Likewise, miRNA expression profiling of *C. parvum*-infected cholangiocytes (bile duct epithelial cells) revealed broad alterations in the host miRNA profile (Zhou et al. 2009). Specifically, there was notable suppression of transcription of host miRNAs (miR-424 and miR-503) mediated by hijacking histone deacetylases and NFkB signaling pathways (Zhou et al. 2013). Additionally, infection of human cholangiocytes with *C. parvum* led to alteration in host expression of miR-98 and let-7 miRNA oncomiRs (Chen et al. 2007; Hu et al. 2009, 2010). Other parasites have also induced changes in host epigenomic landscape via modulation of miRNAs. miRNA microarray analysis revealed upregulation of a number of mouse intestinal epithelial cell miRNAs during infection with coccidian *Eimeria papillata* (Dkhil et al. 2011), while analysis of miRNA expression in *L. major*-infected human macrophages revealed downregulation of 64 of 365 miRNAs, especially those involved in BCL, p53, NFκB, TLR, and HIF1α signaling pathways (Lemaire et al. 2013). Further studies are needed to tease out whether these shifts in host miRNA profile favor parasite virulence or host defense.

Several studies have examined the role of ncRNAs during *Plasmodium* infection. *Plasmodium* parasites have complex interplay with their hosts, inducing alterations in the host miRNA profile, as well as themselves being subject to host miRNA regulation (Cohen et al. 2015). In a mouse model of malaria, specific changes in mouse hepatocyte miRNA expression during *P. chabaudi* infection have been elucidated (Delic et al. 2011). Analysis of mouse hepatocytes after infection with *P. chabaudi* induced upregulation of 3 and downregulation of 16 distinct miRNAs, and this pattern was similar both during primary infection and reinfection, suggesting that a distinct set of host miRNAs are involved in the response to infection (Delic et al. 2011). A study evaluating the miRNA profile of the infected mosquito vector likewise found alterations in levels of distinct miRNAs in response to blood meal with *P. berghei* (Biryukova et al. 2014). While the precise mechanism that *Plasmodium* parasites utilize to induce these miRNA changes is unknown, such changes in host ncRNA landscape must afford some advantage to either the parasite or the host.

In addition to modulation of host ncRNAs, there is a suggestion that parasites themselves may encode ncRNAs that target host processes. Some parasites in fact possess small RNA processing machinery and small RNA repertoires (Braun et al. 2010). Sacar et al. conducted a computational analysis of *T. gondii* RNAs and noted mammalian like hairpin structures, which they hypothesized could be delivered to the host to modulate host transcription (Sacar et al. 2014). The actual role of these hairpins in pathogenesis and parasite–host interplay is unknown. Thus, parasites have evolved mechanisms to perturb host ncRNAs, especially microRNA regulatory pathways that control the immune and inflammatory response to infection. These examples demonstrate that parasites target key host pathways including those involved in immune response, by affecting host ncRNA specifically miRNAs.

9.5 Protozoan Effectors Reshape the Host Epigenome

Epigenetic changes in the host during infection may be due to a direct effect by parasite-derived factors or an indirect effect where parasites target host regulators of the epigenetic landscape to induce the observed changes. Some parasite-derived factors influence the host cell by acting on genome and regulatory pathways through cytoplasmic signaling without entering the nucleus, while a number effectors have been shown to participate in reshaping of the host epigenetic landscape by directly interacting with host DNA and transcription (analogous to bacterial nucleomodulins). Additionally, pathogen-derived effectors can closely resemble host factors, a mechanism known as molecular mimicry that has recently been reviewed (Aliberti et al. 2003; Via et al. 2015). Similar to bacterial pathogens, parasites encode proteins that target the host epigenome (Cheeseman and Weitzman 2015). Although some of these proteins target to the host nucleus, only a few are known to directly interact with host chromatin. In a characteristic example, *Theileria* parasites encode AT hook DNA-binding proteins TashA and SuAT1. SuAT1, which contains a nuclear localization motif, is found in the nucleus of infected host cells and participates in control of cell cycle as well as functions to alter host cell morphology (Swan et al. 2001, 2003; Shiels et al. 2004). *T. gondii* parasites secrete a large number of dense granule (GRA) and rhoptry (ROP) proteins into the host cell. These target host cell processes in the cytoplasm, nucleus, and other subcellular compartments and induce dramatic changes in subcellular morphology, signaling, and transcriptional remodeling (Boothroyd and Dubremetz 2008; English et al. 2015; Hakimi and Bougdour 2015). Specifically, *T. gondii* GRA24 localizes to the host nucleus, where it augments host MAPK signaling by inducing autophosphorylation of p38a MAPK, inducing alteration in Erg and cFos transcription (Braun et al. 2013; Bougdour et al. 2014). It should be noted that the GRA24 kinase interacting motif closely mimics those of host p38, ERK, and JNK factors. Similarly, GRA16 mediates host transcriptional dysregulation by directly binding host nuclear factors and altering the activity of PP2A and HAUSP to induce HAUSP-dependent degradation of p53, an important transcriptional regulator of cell cycle (Bougdour et al. 2013). Another parasite-secreted factor, GRA15, participates in activation of the host NFκB pathway (Rosowski et al. 2011; Hakimi and Bougdour 2015). Secreted kinase ROP16 localizes to the host nucleus where it activates STAT3 and STAT6 transcription leading to restriction of host cell growth (Saeij et al. 2007; Butcher et al. 2011). In addition, ROP16 is responsible for a large number of transcriptional changes and inhibition of cytokine signaling. Recently identified TgIST protein also localizes to host nucleus where it interacts with both STAT1 and NuRD complex, mediating transcriptional repression (Olias et al. 2016; Gay et al. 2016). Another rhoptry protein, *T. gondii* TgPP2C, is a protein phosphatase that is targeted to the host nucleus, and while its exact function is not yet known, parasites knocked out for this gene exhibit mild growth defect (Gilbert et al. 2007). Finally, *T. gondii* ROP38 downregulates host transcription, especially MAPK, STAT, and Fos signaling pathways (Peixoto et al. 2010). Similarly,

Leishmania GP63 metalloprotease, which regulates host NFκB/AP1 and MAPK signaling, localizes to a perinuclear area of host macrophages (Isnard et al. 2015), where it may contribute to changes in host transcription (Arango Duque and Descoteaux 2015; Isnard et al. 2015).

Although intracellular parasites possess a number of effectors that modulate the host epigenome, the molecular mechanisms for parasite-induced changes in the host epigenetic landscape remain unknown. An emerging area of great interest in host–pathogen interactions is centered on discovery of parasite-derived extracellular vesicles that appear to be similar to eukaryotic exosomes used for cell–cell communication. Parasite-derived exosomes that may target the host have been described for *Leishmania*, *Trichomonas*, *Trypanosomes*, and *Plasmodium* parasites (Mantel and Marti 2014; Coakley et al. 2015; Schorey et al. 2015). For example, *Leishmania*-derived GP63 and EF1α are found in parasite-derived exosomes (Silverman et al. 2010; Silverman and Reiner 2011), and *Trichomonas vaginalis* extracellular parasites secrete exosome-like vesicles containing proteins and RNA that modulate host response and adhesion (Twu et al. 2013). Furthermore, RNA transfer has been shown to be mediated via exosomes during cell–cell communication, and recent studies of nematodes have demonstrated transfer of small RNAs in the parasite-derived exosomes (Coakley et al. 2015). It would, therefore, be intriguing to ponder whether parasites utilize exosomal ncRNA transfer to modulate their host, especially since *Leishmania*-derived exosomes have been shown to harbor conserved ncRNAs (Lambertz et al. 2015).

In addition to secretion of specific host-targeted effectors, several parasites have been observed to replicate inside host nuclei. Actively dividing *T. gondii* were observed in the nucleus of various cell types, where they appear to develop in the absence of a vacuolar membrane (Azab et al. 1973; Barbosa et al. 2005). *Eimeria alabamensis*, a related Apicomplexan, have also been observed inside nuclei of intestinal villi (Nishida et al. 2009). Similarly, some microsporidia undergo intranuclear replication (Palenzuela et al. 2014). The biological significance of these observations is unknown, but may represent alternative pathways for parasites to develop and potentially influence the host nucleus.

9.6 Commonly Targeted Pathways

While various parasites employ distinct mechanisms for reshaping epigenomes, targeting of key canonical pathways has emerged as a common theme in the host–parasite interaction. As one would predict, these pathways are highly conserved and are involved in immune modulation, cell cycle progression, metabolism, and overall cell signaling, specifically including regulation by Jak/STAT, NFκB, MAPK pathways, IFN-gamma signaling, and HIF1α. Some of these mechanisms have recently been reviewed (Luder et al. 2009; Melo et al. 2011; Cheeseman and Weitzman 2015; Hakimi and Bougdour 2015; Luder et al. 2015). While a number of alterations to pathways occur in the host cell cytoplasm via protein modification, we will focus on specific epigenetic mechanisms, involving targeted alteration to chromatin structure, including subversion of transcription. While different pathogens target many of the same host pathways, typically each has a unique mechanism. Some parasites induce upregulation of a target protein, while others sequester inhibitors or target stabilization mechanisms. HIF1a and NFkB pathways are often perturbed as detailed below. There is significant cross talk between signaling pathways that together tightly orchestrate control of the cell. We present a brief overview linking examples of parasite alterations to the host epigenome with manipulation of major host pathways (Fig. 9.3).

Mitogen-activated protein kinases (MAPK) are a large family of serine/threonine kinases that transmit extracellular signals via a cytoplasmic signal transduction pathway to modulate essential cellular processes including apoptosis, stress response, and survival. One of the final steps in the pathway involves phosphorylation and activation of Erk kinase, JNK kinase, or p38, which, as dimers, translocate into the nucleus to regulate transcription of genes involved in stress response, apoptosis, and inflammation. MAPK also phosphorylate and regulate other transcription factors including cFos, cMyc, STAT3, and p53 to regulate apoptosis (Yang et al. 2013; Dhillon et al. 2007). Downstream factors of the MAPK cascade participate in shaping of the host epigenome, as specifically cJun interacts with the nucleosome remodeling complex (Aguilera et al. 2011). Considering the key role of MAPK signaling in cellular processes and responses to various stimuli, it is not surprising that parasites have evolved strategies to modulate and subvert this signal transduction pathway. At least one of the mechanisms by which parasites achieve these regulatory changes is alteration in host DNA methylation as shown to occur specifically at MAPK pathway targets for *L. donovani* (Marr et al. 2014), as well as targeted by CK2 kinase activated by *Theilera* (Dessauge et al. 2005b). *T. gondii* GRA24 specifically binds and promotes activation of MAPK/p38 causing nuclear translocation and phosphorylation of its targets including cytokines involved in inflammatory response, overall creating a proinflammatory state (Braun et al. 2013), while ROP38 kinase causes downregulation of transcription of the MAPK pathways (Peixoto et al. 2010). These molecules illustrate how the balance of *T. gondii* parasite factors can modulate the host MAPK pathway. *Theileria* parasites also modulate MAPK signaling, and cells infected and transformed by *Theileria* demonstrate constitutive activation of JNK and AP1 transcription factor (Chaussepied et al. 1998; Lizundia et al. 2007; Hayashida et al. 2010). cFos and cJun are ubiquitous transcription factors downstream of MAPK/MEK pathways, involved in regulation of a wide range of essential cellular processes. Both families contain several proteins, and cFos and cJun transcription factors combine to form an Activator Protein 1 (AP1) transcription factor, which binds DNA. Recent work has specifically shown that *Theileria* parasites secrete a peptidyl-prolyl isomerase (PIN1) homologue into the host cell that causes cJun stabilization via degradation of host ubiquitin ligase FBW7, leading to oncogenic transformation of the host cell (Marsolier et al. 2015). *Leishmania* parasites induce cleavage of the cJun component of the AP1 transcription factor via parasite-derived GP63 protein (Contreras et al. 2010). Finally, *L. Mexicana* alters dendritic cell signaling leading to

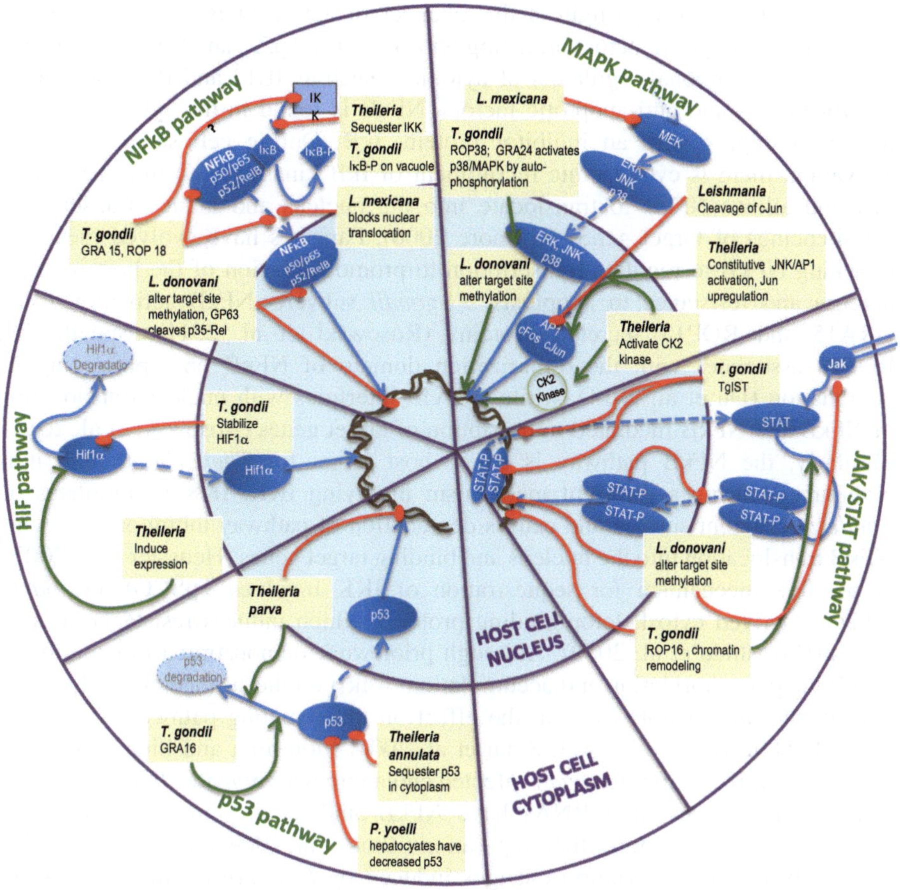

Fig. 9.3 Host regulatory pathways targeted by parasitic infection. Parasites alter the host epigenome by manipulating transcriptional regulators of key host processes involved in immune response, cell cycle, death pathways, metabolism, and other signal transduction events. Shown is a cartoon schematic of select pathways and how they are targeted by parasites. Parasites target host NFkB, p53, and MAPK transcriptional regulation in part to subvert host cell death and cytokine response pathways. Additionally, parasites subvert host metabolism as regulated by the HIF1 pathway. Details and references are described in the text. Red connectors denote inhibition; green arrows represent activation

inactivation of host MAPK and inhibition of phosphorylation of downstream p38 and ERK (Contreras et al. 2014). These signal transduction changes eventually lead to epigenomic remodeling, as direct interactions of downstream MAPK proteins with histone modifications and chromatin remodeling have been described, providing a mechanistic link between signal transduction cascade and chromatin remodeling (Aguilera et al.) Thus, while utilizing very different mechanisms, distinct parasites target the MAPK pathway to control host processes and the epigenetic landscape.

Nuclear Factor Kappa-light-chain-enhancer of activated B cells (NFκB) is a family of transcription factors including NFκB (p50 and p52) and Rel members that are key transcriptional regulators of cytokines such as IL12 and IFNγ as well as growth factors and anti-apoptotic factors. NFκB is found in the cytoplasm in an inactive complex with an inhibitor protein, IκB. During cell surface receptor activation, there is cytoplasmic recruitment of IκB kinase (IKK) that inactivates IκB and allows NFκB to translocate into the nucleus and induce transcription (or silencing) of target genes (Gilmore 2006). Parasites have evolved means of hijacking NFκB to regulate transcription to promote evasion of the host immune defense and resistance to apoptosis. *T. gondii* subverts NFκB activation using GRA15 and ROP18 secreted proteins (Rosowski et al. 2011). Specifically, ROP18 associates with the dimerization domain of NFκB p65, promoting its degradation (Du et al. 2014), while GRA15 interferes with nuclear translocation of NFκB and NFκB-mediated transcription of target genes (Rosowski et al. 2011). Similarly, the NFκB pathway is a key host target for *Theileria*, though it is manipulated using a different mechanism involving IKK. IKK accumulates on *Theileria* schizont surface and causes degradation of pathway inhibitors, allowing NFκB translocation into the nucleus and binding target genes (Heussler et al. 2002). A possible mechanism for sequestration of IKK involves TpSCOP (*T. parva* schizont-derived cytoskeleton-binding protein), which induces resistance to apoptosis (Hayashida et al. 2010). Although prior work demonstrated modulation of NFκB via phosphorylation and accumulation of IkB on the surface of the *T. gondii* parasitophorous vacuole as well, the effect on the signaling pathway is not well defined (Molestina et al. 2003; Sinai et al. 2004; Molestina and Sinai 2005a, b). Analysis of miRNA in *C. parvum*-infected cholangiocytes revealed that some of the differentially expressed miRNAs have NFkB-binding sites in their promoters, suggesting a mechanism for their regulation (Zhou et al. 2009, 2013). Additionally, as already noted, methylation changes induced by *L. donovani* in macrophages occur in the NFkB pathway (Marr et al. 2014), while GP63 cleaves the p35-RelA subunit of NFkB (Gregory et al. 2008), presumably to alter host NFκB regulated transcription. In an alternate mechanism, *L. mexicana* prevents nuclear translocation of AP1 and NFκB components in infected dendritic cells (Contreras et al. 2014). Finally, microarray analysis of lymphosarcoma cells infected with *T. annulata* revealed significant alteration in a number of key transcriptional regulators, including AP1 subunits FOS and JUN, as well as NFκB, all of which were activated during infection (Kinnaird et al. 2013). Alterations in key host processes implicated in cell growth, cytokine signaling, cell division, motility, and death were also observed. Studies in cancer cells established that JNK and NFκB signaling play opposite roles and together impose a tightly regulated balance in transcription. By altering either NFκB or MAPK/JNK signaling, parasites shift host cell fate to promote their survival.

p53 is an important tumor suppressor protein and transcription factor involved in regulation of apoptosis, cell cycle, and DNA repair. It is also subject to extensive posttranslational modifications including phosphorylation, ubiquitination, acetylation, and methylation (Kruse and Gu 2009). Considering its cornerstone role, p53 is another major parasite target. As mentioned, *T. gondii* modulates host p53 by targeted degradation via GRA16 with PP2A phosphatase and HAUSP deubiquitinase (Bougdour et al. 2013). In a remarkable example of hijacking of host apoptosis, *T. annulata* causes immortalization of host leukocytes, in part by targeting p53 (Haller et al. 2010). Contrary to *T. gondii*, *Theileria* inactivates p53 protein by sequestering it in the cytoplasm on the schizont membrane, preventing its translocation into the nucleus. Curing cells of *Theileria* infection by drug treatment results in p53 translocation to the nucleus (Haller et al. 2010). Additionally, *Theileria parva*-transformed leukocytes upregulate MDM2, a major regulator of p53. MDM2 binding blocks p53 transcriptional activity and promotes p53 ubiquitination and degradation, such that there is an overall decrease in p53 (Hayashida et al. 2013). Similarly, hepatocyte-infected *Plasmodium yoelii* parasites have decreased levels of p53 (Kaushansky et al. 2013). Akin to p53, parasites target the transcription factor cMyc, a central controller of a large number of genes involved in cell cycle, apoptosis, differentiation, and metabolism. Significant host upregulation and stabilization of cMyc occurs during infection with *T. gondii*, possibly in a JNK-mediated manner (Franco et al. 2014). Similarly, cMyc is stabilized by phosphorylation by CK2 in *Theileria*-transformed leukocytes, promoting anti-apoptotic signaling (Dessauge et al. 2005a). Overall, data suggest that parasites modulate the host epigenome by secreting specialized effectors or sequestering and hijacking key transcriptional regulators of cell cycle and apoptosis, effectively disabling transcription of pro-apoptotic factors to ensure their own survival.

Janus kinase–signal transducer and activator of transcription (JAK-STAT) pathway transmits extracellular messaging from cytokines and IFNγ bound to cell receptors directly into transcriptional regulation by binding to promoters of target genes involved in growth and immune response. STAT proteins are located in the cytoplasm and are inactive until they are recruited to an activated receptor, become phosphorylated by an associated JAK kinase, dimerize (either homo- or heterodimers of different STAT proteins), and translocate to the nucleus where they bind to specific IFNγ activation sequences, thus causing transcription (or repression) of the target gene (Aaronson and Horvath 2002). Similar to p53, STAT proteins can undergo phosphorylation by other regulatory proteins, including MAPK kinases, which can alter the efficiency of STAT–DNA interactions. Host IFN gamma (IFNg) signaling is one of the main mechanisms utilized in resistance and elimination of invading parasites, and protozoan parasites have evolved mechanisms to avoid IFNg-directed death (Suzuki et al. 1988; Yarovinsky 2014; Luder et al. 2015). *T. gondii* hijack host signaling cascades to make host cells unresponsive to IFNγ, by interfering with STAT signaling and its DNA binding to IFNγ response elements. Specifically, *T. gondii* induces alteration in the host epigenetic landscape leading to impairment of histone acetylation at IFNγ regulated promoters and improper assembly of chromatin regulatory machinery at the IFN-γ-targeted STAT1 response elements (Kim et al. 2007; Lang et al. 2012; Rosowski and Saeij 2012). Additionally, early studies of genetic crosses demonstrated that *T. gondii* specifically targets STAT pathways in the infected host cell (Saeij et al.

2007). Secreted *T. gondii* protein Rop16 localizes to the host nucleus and subverts host signaling machinery by directly phosphorylating STAT3 and STAT6 proteins, leading to restriction of host cell growth and a number of other transcriptional changes (Saeij et al. 2007; Ong et al. 2010; Butcher et al. 2011; Denkers et al. 2012). In addition, Rop16 phosphorylates STAT1, rendering it inactive and therefore subverting the host cell IFNγ response (Rosowski and Saeij 2012). In eukaryotic cells, transcriptionally active regions are marked by histone H3 and H4 lysine acetylation at the N-terminal tails, important for assembly of transcriptional apparatus. However, macrophages infected with *T. gondii* do not exhibit acetylation of lysine residues in histones of IFNγ-responsive promoters while use of HDAC (deacetylase) inhibitor restored the IFNγ response (Lang et al. 2012). Recent studies aiming to identify the mechanism utilized by *T. gondii* to inhibit STAT1/IFNg signaling noted another regulatory mechanism involving removal of STAT1 from the nuclear-cytoplasmic cycling pool by maintaining it as chromatin bound and preventing disassociation of STAT1 from DNA (Rosowski et al. 2014). Recently identified TgIST appears to be involved in this STAT1 chromatin binding as well as *T. gondii* related transcriptional repression (Olias et al. 2016; Gay et al. 2016). In a different tactic, evaluation of mouse dendritic cells infected with *T. gondii* revealed phosphorylation and nuclear translocation of STAT1 without binding to IFN response elements. This STAT1 rearrangement was induced by parasite invasion but not dependent on parasite replication (Schneider et al. 2013). All together, these mechanisms provide a clue as to how *T. gondii* subvert host IFNγ response via epigenetic and transcriptional dysregulation of STAT signaling, thus promoting parasite survival and growth. Targeting of STAT signaling is utilized by other parasites as well; e.g., *Leishmania* alter DNA methylation of infected macrophages at CpG islands, specifically disrupting JAK/STAT and MAPK signaling (Marr et al. 2014). Furthermore, *L. donovani* infection of macrophages also induced inhibition of JAK/STAT signaling, in part by SHP-1 phosphatase-induced blockade of Jak phosphorylation, as well as induced reduction of Interferon Regulatory factor 1 (IRF1), suggesting parasite-induced impairment in STATa nuclear translocation (Olivier et al. 2005; Matte and Descoteaux 2010).

Intracellular parasites rely on the host cell for their nutritional needs and therefore cause shifts in overall host metabolism. Hypoxia-inducible factor-1 (HIF1) is a master regulator of transcription in response to changes in host oxygen, iron, and glucose availability. The stability of the HIF1-alpha (HIF1α) subunit is tightly regulated in the cytoplasm, such that alteration in overall host state attenuates HIF1a degradation and allows HIF1 (α + β heterodimer) to translocate to the nucleus where it binds to HREs (hypoxia response elements). HREs regulate transcription of genes involved in metabolism and glucose utilization. HIF1α itself is further regulated by posttranslational modifications. Early microarray analyses revealed significant alterations in the expression of genes involved in host metabolism in response to *T. gondii* infection, and it was subsequently shown that parasites alter HIF function by modulating its expression and stability, presumably to subvert host metabolic processes and key metabolite targeting (Blader et al. 2001; Wiley et al. 2010; Singh et al. 2012; Medjkane et al. 2014; Menendez et al.

2015; Metheni et al. 2015). In *Theileria*-infected and transformed leukocytes, there is a notable shift in host metabolism towards glycolysis, known as the Warburg effect or aerobic glycolysis. This is largely controlled by HIF1, which in turn is regulated by NFκB and AP-1 that are also altered during infection (Metheni et al. 2015). Along with the STAT and JUN pathways, HIF1α is specifically targeted and stabilized during *T. gondii* infection (Spear et al. 2006; Wiley et al. 2010). In a similar manner, *Leishmania* parasites alter HIF1 function by upregulating HIF1 expression and stabilization of HIF1 against degradation (Singh et al. 2012). MicroRNA-210 (hypoxamir) is a major hypoxia-inducible miRNA, whose expression is regulated by HIF1α, which plays a role in modulating mitochondrial respiration and alteration in cell proliferation (Chan et al. 2012). Expression of miR-210 was significantly increased during human macrophage infection with *L. major* parasites, and its upregulation in macrophages was dependent on HIF1α (Lemaire et al. 2013). Whether these alterations in host metabolism are a result of direct parasite targeting or whether they are fundamental to the host response to infection remains to be elucidated. The dysregulation of host metabolic state allows redirecting of nutrients, energy, and metabolic intermediates to promote parasite growth. It is also worth noting that alpha ketoglutarate (αKG) is a key intermediate of the mitochondrial TCA cycle and is a cofactor for a large number of enzymes involved in essential host processes, including TET enzymes that regulate DNA CpG methylation, JMDM1 JmjC histone demethylases (Tsukada et al. 2006), and PHD2 enzyme (prolyl hydroxylase domain 2) which is directly involved in HIF1α stability (Semenza 2007). Other work in cancer biology also links metabolic intermediates directly to changes in epigenome (Moussaieff et al. 2015). Therefore, by altering host cell metabolism and inducing a shift away from mitochondrial respiration, parasites cause reduction in available αKG, potentially altering host cell processes that directly regulate DNA methylation and the epigenetic landscape.

Parasites regulate a number of other key host processes including cell death pathways such as apoptosis and cell cycle progression. Gene expression and signaling are at the heart of cell cycle progression. Perturbation of cell cycle checkpoints and regulation of apoptosis are used by intracellular parasites to ensure their survival. Intracellular parasites are protected from immune recognition, and the infected cell may undergo apoptosis to curb infection. Intracellular parasitic protists including *Toxoplasma*, *Leishmania*, *Theileria*, and *Cryptosporidia* inhibit apoptosis of the infected host cell (Heussler et al. 2001). Cells infected with *T. gondii* are resistant to extracellular induction of apoptosis (Nash et al. 1998). Additionally, *T. gondii* parasites appear to actively interfere with host death pathways. During infection, *T. gondii* modulates genes involved in apoptotic pathways, primarily NFκB signaling. Degradation of pro-apoptotic BCL2 proteins altered miRNA and STAT (STAT3, miR17-92, and Bim) signaling, and degradation of pro-apoptotic p53 also contributes to inhibition of apoptosis, promoting host cell and parasite survival (Carmen and Sinai 2011; Cai et al. 2014). *Theileria* and *Cryptosporidium* similarly stabilize the host NFκB pathway to abrogate apoptosis

(Heussler et al. 2001). In an analogous manner, there is a notable decrease in pro-apoptotic p53 in *P. yoelii*-infected hepatocytes, supporting cell survival (Kaushansky et al. 2013). While utilizing similar mechanisms in inhibition of apoptosis, these parasites induce distinct effects upon the host cell cycle. *T. gondii* induces arrest in the host cell cycle by downregulating expression of UHRF1 E3 ubiquitin ligase, as well as by manipulating ERK kinase, leading to induction of G1/S phase progression and blockage in G2/M transition (Brunet et al. 2008; Molestina et al. 2008; Unoki et al. 2009). *Leishmania* parasites induce host cell cycle arrest at an earlier stage, during G0 to S transition, via downregulation of cyclin-dependent kinases and upregulation of cyclin kinase inhibitors p21 and p27 (Kuzmenok et al. 2005). *Theileria* parasites reside directly in the host cytoplasm of leukocytes and coopt the host cell division apparatus to induce continuous uncontrolled proliferation and oncogenic transformation of the host cells, coupling host cell division to parasite division (Spooner et al. 1989). *Theileria* subvert host cell cycling mainly by activation of the NFκB pathway, but microarray analysis of infected cells revealed changes in mRNA levels of a significant proportion of host genes, underscoring the complexity of host–parasite interactions (Shiels et al. 2006; Durrani et al. 2012; Kinnaird et al. 2013). Intracellular parasites commonly actively target and modulate host cell cycle and apoptosis pathways to facilitate parasite survival and replication. Overall, intracellular parasites have developed sophisticated mechanisms for targeting and subverting key host processes involved in chromatin assembly and structure, significantly altering the host epigenomic landscape.

9.7 Concluding Remarks

Protozoan parasites have a complex relationship with their hosts, relying on them for nutrients and metabolic products, while avoiding host immune defenses and preventing their demise. To that effect, various pathogens including viruses, bacteria, and eukaryotic parasite utilize similar means of subverting their host. The host–pathogen interaction has been studied extensively, and through recent work it has become apparent that significant alterations occur in the host epigenetic landscape during infection. Some of these changes are achieved by specific secretion of protozoan proteins into the host, which may alter host epigenetics directly by modulating chromatin packaging and transcription of specific genes, or indirectly by modifying activity of vital host proteins or host miRNA. Furthermore, while there are global changes to the overall epigenomic landscape of the host, a number of regulatory alterations occur in regions encoding conserved elements of essential cell processes such as cell signaling, death pathways, metabolism, and growth. Additionally, alteration in availability and function of key host transcription factors further tempers the host epigenetic landscape. In this context, it is apparent that by disturbing regulation of the signaling cascades, parasites induce perturbations in the host epigenome.

The term "host" has been used here to describe any cell infected by a parasite, but in fact, notable and distinct changes to the epigenome occur in a wide range of infected organisms from mammals to insect vectors and in a wide range of cell types including cells of the immune system, intestinal epithelia, and neurons. It is intriguing that epigenetic changes have been described in cells that are not directly infected, suggesting a significant role for cell–cell communication in disease pathogenesis. Subversion of immune cells may promote parasite dissemination and survival. Moreover, changes to the host epigenome can be transient or long term, as seen in *Theileria*-cured leukocytes (Kinnaird et al. 2013). It is intriguing to speculate exactly what role long-term alteration, especially those occurring in the immune cells, may play in cellular memory and cell-mediated immunity, and whether these changes are further inherited and become part of "epigenetic memory." Finally, while some of the host changes are due to a direct parasite effect, others should be attributed to the host response to the infection. This concept has been described for *T. gondii* infection wherein three general groups of genes or processes are modulated—those necessary for parasite survival or "pro-parasite," those necessary for host defense or "pro-host," and "bystander" genes that do not appear to be directly necessary for either (Blader et al. 2001; Blader and Saeij 2009).

Our understanding of the mechanisms for how pathogens reshape the host epigenetic landscape is still rudimentary. It will be interesting to see whether pathogens induce unique or universal epigenetic signatures of infection on the host that can be used in clinical diagnoses and treatment. There clearly is an intricately laced regulatory web modulating host epigenetic landscape, and further work is needed to dissect the exact mechanisms and causal interrelationships.

Acknowledgements Supported by NIH grants R01AI087625 (KK), R21AI101801(KK), and T32AI070117 (IG). We apologize to authors whose work we did not include due to space limitations.

References

Aaronson DS, Horvath CM (2002) A road map for those who don't know JAK-STAT. Science 296:1653–1655. doi:10.1126/science.1071545

Aguilera C, Nakagawa K, Sancho R, Chakraborty A, Hendrich B, Behrens A (2011) c-Jun N-terminal phosphorylation antagonises recruitment of the Mbd3/NuRD repressor complex. Nature 469:231–235. doi:10.1038/nature09607

Albuquerque SS, Carret C, Grosso AR et al (2009) Host cell transcriptional profiling during malaria liver stage infection reveals a coordinated and sequential set of biological events. BMC Genomics 10:270. doi:10.1186/1471-2164-10-270

Aliberti J, Valenzuela JG, Carruthers VB et al (2003) Molecular mimicry of a CCR5 binding-domain in the microbial activation of dendritic cells. Nat Immunol 4:485–490. doi:10.1038/ni915

Al-Quraishy S, Dkhil MA, Abdel-Baki AA, Delic D, Santourlidis S, Wunderlich F (2013) Genome-wide screening identifies Plasmodium chabaudi-induced modifications of DNA methylation status of Tlr1 and Tlr6 gene promoters in liver, but not spleen, of female C57BL/6 mice. Parasitol Res 112:3757–3770. doi:10.1007/s00436-013-3565-2

Arango Duque G, Descoteaux A (2015) Leishmania survival in the macrophage: where the ends justify the means. Curr Opin Microbiol 26:32–40. doi:10.1016/j.mib.2015.04.007

Aufauvre J, Misme-Aucouturier B, Vigues B, Texier C, Delbac F, Blot N (2014) Transcriptome analyses of the honeybee response to Nosema ceranae and insecticides. PLoS One 9:e91686. doi:10.1371/journal.pone.0091686

Auger CJ, Coss D, Auger AP, Forbes-Lorman RM (2011) Epigenetic control of vasopressin expression is maintained by steroid hormones in the adult male rat brain. Proc Natl Acad Sci U S A 108:4242–4247. doi:10.1073/pnas.1100314108

Azab ME, Rifaat MA, Salem SA, Zaghloul I, Morsy TA (1973) The intranuclear development of Toxoplasma. Z Parasitenkd 42:39–42

Barbosa HS, Ferreira-Silva MF, Guimaraes EV, Carvalho L, Rodrigues RM (2005) Absence of vacuolar membrane involving Toxoplasma gondii during its intranuclear localization. J Parasitol 91:182–184. doi:10.1645/GE-276R

Berdoy M, Webster JP, Macdonald DW (2000) Fatal attraction in rats infected with Toxoplasma gondii. Proc Biol Sci 267:1591–1594. doi:10.1098/rspb.2000.1182

Berger SL (2007) The complex language of chromatin regulation during transcription. Nature 447:407–412. doi:10.1038/nature05915

Bicker S, Lackinger M, Weiss K, Schratt G (2014) MicroRNA-132, -134, and -138: a microRNA troika rules in neuronal dendrites. Cell Mol Life Sci 71:3987–4005. doi:10.1007/s00018-014-1671-7

Bierne H, Hamon M, Cossart P (2012) Epigenetics and bacterial infections. Cold Spring Harb Perspect Med 2:a010272. doi:10.1101/cshperspect.a010272

Biryukova I, Ye T, Levashina E (2014) Transcriptome-wide analysis of microRNA expression in the malaria mosquito Anopheles gambiae. BMC Genomics 15:557. doi:10.1186/1471-2164-15-557

Blader IJ, Manger ID, Boothroyd JC (2001) Microarray analysis reveals previously unknown changes in Toxoplasma gondii-infected human cells. J Biol Chem 276:24223–24231. doi:10.1074/jbc.M100951200

Blader IJ, Saeij JP (2009) Communication between Toxoplasma gondii and its host: impact on parasite growth, development, immune evasion, and virulence. APMIS 117:458–476. doi:10.1111/j.1600-0463.2009.02453.x

Boothroyd JC, Dubremetz JF (2008) Kiss and spit: the dual roles of Toxoplasma rhoptries. Nat Rev Microbiol 6:79–88. doi:10.1038/nrmicro1800

Bouchut A, Chawla AR, Jeffers V, Hudmon A, Sullivan WJ Jr (2015) Proteome-wide lysine acetylation in cortical astrocytes and alterations that occur during infection with brain parasite Toxoplasma gondii. PLoS One 10:e0117966. doi:10.1371/journal.pone.0117966

Bougdour A, Durandau E, Brenier-Pinchart MP et al (2013) Host cell subversion by Toxoplasma GRA16, an exported dense granule protein that targets the host cell nucleus and alters gene expression. Cell Host Microbe 13:489–500. doi:10.1016/j.chom.2013.03.002

Bougdour A, Tardieux I, Hakimi MA (2014) Toxoplasma exports dense granule proteins beyond the vacuole to the host cell nucleus and rewires the host genome expression. Cell Microbiol 16:334–343. doi:10.1111/cmi.12255

Braun L, Brenier-Pinchart MP, Yogavel M et al (2013) A Toxoplasma dense granule protein, GRA24, modulates the early immune response to infection by promoting a direct and sustained host p38 MAPK activation. J Exp Med 210:2071–2086. doi:10.1084/jem.20130103

Braun L, Cannella D, Ortet P et al (2010) A complex small RNA repertoire is generated by a plant/fungal-like machinery and effected by a metazoan-like Argonaute in the single-cell human parasite Toxoplasma gondii. PLoS Pathog 6:e1000920. doi:10.1371/journal.ppat.1000920

Brunet J, Pfaff AW, Abidi A et al (2008) Toxoplasma gondii exploits UHRF1 and induces host cell cycle arrest at G2 to enable its proliferation. Cell Microbiol 10:908–920. doi:10.1111/j.1462-5822.2007.01093.x

Butcher BA, Fox BA, Rommereim LM et al (2011) Toxoplasma gondii rhoptry kinase ROP16 activates STAT3 and STAT6 resulting in cytokine inhibition and arginase-1-dependent growth control. PLoS Pathog 7:e1002236. doi:10.1371/journal.ppat.1002236

Cai Y, Chen H, Jin L, You Y, Shen J (2013) STAT3-dependent transactivation of miRNA genes following Toxoplasma gondii infection in macrophage. Parasit Vectors 6:356. doi:10.1186/1756-3305-6-356

Cai Y, Chen H, Mo X et al (2014) Toxoplasma gondii inhibits apoptosis via a novel STAT3-miR-17-92-Bim pathway in macrophages. Cell Signal 26:1204–1212. doi:10.1016/j.cellsig.2014.02.013

Calderon EJ, Cushion MT, Xiao L, Lorenzo-Morales J, Matos O, Kaneshiro ES, Weiss LM (2015) The 13th International Workshops on Opportunistic Protists (IWOP13). J Eukaryot Microbiol 62:701–709. doi:10.1111/jeu.12221

Cannella D, Brenier-Pinchart MP, Braun L et al (2014) miR-146a and miR-155 delineate a MicroRNA fingerprint associated with Toxoplasma persistence in the host brain. Cell Rep 6:928–937. doi:10.1016/j.celrep.2014.02.002

Carmen JC, Sinai AP (2011) The differential effect of toxoplasma gondii infection on the stability of BCL2-family members involves multiple activities. Front Microbiol 2:1. doi:10.3389/fmicb.2011.00001

Chan YC, Banerjee J, Choi SY, Sen CK (2012) miR-210: the master hypoxamir. Microcirculation 19:215–223. doi:10.1111/j.1549-8719.2011.00154.x

Chattopadhyay R, de la Vega P, Paik SH, Murata Y, Ferguson EW, Richie TL, Ooi GT (2011) Early transcriptional responses of HepG2-A16 liver cells to infection by Plasmodium falciparum sporozoites. J Biol Chem 286:26396–26405. doi:10.1074/jbc.M111.240879

Chaussabel D, Semnani RT, McDowell MA, Sacks D, Sher A, Nutman TB (2003) Unique gene expression profiles of human macrophages and dendritic cells to phylogenetically distinct parasites. Blood 102:672–681. doi:10.1182/blood-2002-10-3232

Chaussepied M, Lallemand D, Moreau MF, Adamson R, Hall R, Langsley G (1998) Upregulation of Jun and Fos family members and permanent JNK activity lead to constitutive AP-1 activation in Theileria-transformed leukocytes. Mol Biochem Parasitol 94:215–226

Cheeseman K, Weitzman JB (2015) Host-parasite interactions: an intimate epigenetic relationship. Cell Microbiol 17:1121–1132. doi:10.1111/cmi.12471

Chen XM, Splinter PL, O'Hara SP, LaRusso NF (2007) A cellular micro-RNA, let-7i, regulates Toll-like receptor 4 expression and contributes to cholangiocyte immune responses against Cryptosporidium parvum infection. J Biol Chem 282:28929–28938. doi:10.1074/jbc.M702633200

Chi P, Allis CD, Wang GG (2010) Covalent histone modifications—miswritten, misinterpreted and mis-erased in human cancers. Nat Rev Cancer 10:457–469. doi:10.1038/nrc2876

Coakley G, Maizels RM, Buck AH (2015) Exosomes and other extracellular vesicles: the new communicators in parasite infections. Trends Parasitol 31:477–489. doi:10.1016/j.pt.2015.06.009

Cock-Rada AM, Medjkane S, Janski N et al (2012) SMYD3 promotes cancer invasion by epigenetic upregulation of the metalloproteinase MMP-9. Cancer Res 72:810–820. doi:10.1158/0008-5472.CAN-11-1052

Cohen A, Combes V, Grau GE (2015) MicroRNAs and Malaria—a dynamic interaction still incompletely understood. J Neuroinfect Dis 6(1). pii:165

Contreras I, Estrada JA, Guak H et al (2014) Impact of Leishmania mexicana infection on dendritic cell signaling and functions. PLoS Negl Trop Dis 8:e3202. doi:10.1371/journal.pntd.0003202

Contreras I, Gomez MA, Nguyen O, Shio MT, McMaster RW, Olivier M (2010) Leishmania-induced inactivation of the macrophage transcription factor AP-1 is mediated by the parasite metalloprotease GP63. PLoS Pathog 6:e1001148. doi:10.1371/journal.ppat.1001148

Dawlaty MM, Breiling A, Le T et al (2014) Loss of Tet enzymes compromises proper differentiation of embryonic stem cells. Dev Cell 29:102–111. doi:10.1016/j.devcel.2014.03.003

Delic D, Dkhil M, Al-Quraishy S, Wunderlich F (2011) Hepatic miRNA expression reprogrammed by Plasmodium chabaudi malaria. Parasitol Res 108:1111–1121. doi:10.1007/s00436-010-2152-z

Denkers EY, Bzik DJ, Fox BA, Butcher BA (2012) An inside job: hacking into Janus kinase/signal transducer and activator of transcription signaling cascades by the intracellular protozoan Toxoplasma gondii. Infect Immun 80:476–482. doi:10.1128/IAI.05974-11

Deplus R, Blanchon L, Rajavelu A et al (2014) Regulation of DNA methylation patterns by CK2-mediated phosphorylation of Dnmt3a. Cell Rep 8:743–753. doi:10.1016/j.celrep.2014.06.048

Dessauge F, Lizundia R, Baumgartner M, Chaussepied M, Langsley G (2005a) Taking the Myc is bad for Theileria. Trends Parasitol 21:377–385. doi:10.1016/j.pt.2005.06.003

Dessauge F, Lizundia R, Langsley G (2005b) Constitutively activated CK2 potentially plays a pivotal role in Theileria-induced lymphocyte transformation. Parasitology 130(Suppl):S37–S44. doi:10.1017/S0031182005008140

Dhillon AS, Hagan S, Rath O, Kolch W (2007) MAP kinase signalling pathways in cancer. Oncogene 26:3279–3290. doi:10.1038/sj.onc.1210421

Dkhil M, Abdel-Baki AA, Delic D, Wunderlich F, Sies H, Al-Quraishy S (2011) Eimeria papillata: upregulation of specific miRNA-species in the mouse jejunum. Exp Parasitol 127:581–586. doi:10.1016/j.exppara.2010.11.002

Du J, An R, Chen L et al (2014) Toxoplasma gondii virulence factor ROP18 inhibits the host NF-kappaB pathway by promoting p65 degradation. J Biol Chem 289:12578–12592. doi:10.1074/jbc.M113.544718

Durrani Z, Weir W, Pillai S, Kinnaird J, Shiels B (2012) Modulation of activation-associated host cell gene expression by the apicomplexan parasite Theileria annulata. Cell Microbiol 14:1434–1454. doi:10.1111/j.1462-5822.2012.01809.x

Dvorakova-Hortova K, Sidlova A, Ded L et al (2014) Toxoplasma gondii decreases the reproductive fitness in mice. PLoS One 9:e96770. doi:10.1371/journal.pone.0096770

English ED, Adomako-Ankomah Y, Boyle JP (2015) Secreted effectors in Toxoplasma gondii and related species: determinants of host range and pathogenesis? Parasite Immunol 37:127–140. doi:10.1111/pim.12166

Esquela-Kerscher A, Slack FJ (2006) Oncomirs—microRNAs with a role in cancer. Nat Rev Cancer 6:259–269. doi:10.1038/nrc1840

Faghihi MA, Wahlestedt C (2009) Regulatory roles of natural antisense transcripts. Nat Rev Mol Cell Biol 10:637–643. doi:10.1038/nrm2738

Fentress SJ, Sibley LD (2011) The secreted kinase ROP18 defends Toxoplasma's border. Bioessays 33:693–700. doi:10.1002/bies.201100054

Fitzgerald KA, Caffrey DR (2014) Long noncoding RNAs in innate and adaptive immunity. Curr Opin Immunol 26:140–146. doi:10.1016/j.coi.2013.12.001

Flegr J, Lenochova P, Hodny Z, Vondrova M (2011) Fatal attraction phenomenon in humans: cat odour attractiveness increased for toxoplasma-infected men while decreased for infected women. PLoS Negl Trop Dis 5:e1389. doi:10.1371/journal.pntd.0001389

Flegr J, Markos A (2014) Masterpiece of epigenetic engineering—how Toxoplasma gondii reprogrammes host brains to change fear to sexual attraction. Mol Ecol 23:5934–5936. doi:10.1111/mec.13006

Fouts AE, Boothroyd JC (2007) Infection with Toxoplasma gondii bradyzoites has a diminished impact on host transcript levels relative to tachyzoite infection. Infect Immun 75:634–642. doi:10.1128/IAI.01228-06

Franco M, Shastri AJ, Boothroyd JC (2014) Infection by Toxoplasma gondii specifically induces host c-Myc and the genes this pivotal transcription factor regulates. Eukaryot Cell 13:483–493. doi:10.1128/EC.00316-13

Gay G, Braun L, Brenier-Pinchart MP, Vollaire J, Josserand V, Bertini RL, Varesano A, Touquet B, De Bock PJ, Coute Y, Tardieux I, Bougdour A, Hakimi MA (2016) Toxoplasma gondii

TgIST co-opts host chromatin repressors dampening STAT1-dependent gene regulation and IFN-gammamediated host defenses. J Exp Med 213:1779–1798. doi:10.1084/jem.20160340
Gilbert LA, Ravindran S, Turetzky JM, Boothroyd JC, Bradley PJ (2007) Toxoplasma gondii targets a protein phosphatase 2C to the nuclei of infected host cells. Eukaryot Cell 6:73–83. doi:10.1128/EC.00309-06
Gilmore TD (2006) Introduction to NF-kappaB: players, pathways, perspectives. Oncogene 25:6680–6684. doi:10.1038/sj.onc.1209954
Gregory DJ, Godbout M, Contreras I, Forget G, Olivier M (2008) A novel form of NF-kappaB is induced by Leishmania infection: involvement in macrophage gene expression. Eur J Immunol 38:1071–1081. doi:10.1002/eji.200737586
Hakimi MA, Bougdour A (2015) Toxoplasma's ways of manipulating the host transcriptome via secreted effectors. Curr Opin Microbiol 26:24–31. doi:10.1016/j.mib.2015.04.003
Hakimi MA, Cannella D (2011) Apicomplexan parasites and subversion of the host cell microRNA pathway. Trends Parasitol 27:481–486. doi:10.1016/j.pt.2011.07.001
Haller D, Mackiewicz M, Gerber S et al (2010) Cytoplasmic sequestration of p53 promotes survival in leukocytes transformed by Theileria. Oncogene 29:3079–3086. doi:10.1038/onc.2010.61
Hari Dass SA, Vyas A (2014) Toxoplasma gondii infection reduces predator aversion in rats through epigenetic modulation in the host medial amygdala. Mol Ecol 23:6114–6122. doi:10.1111/mec.12888
Hayashida K, Hattori M, Nakao R et al (2010) A schizont-derived protein, TpSCOP, is involved in the activation of NF-kappaB in Theileria parva-infected lymphocytes. Mol Biochem Parasitol 174:8–17. doi:10.1016/j.molbiopara.2010.06.005
Hayashida K, Kajino K, Hattori M, Wallace M, Morrison I, Greene MI, Sugimoto C (2013) MDM2 regulates a novel form of incomplete neoplastic transformation of Theileria parva infected lymphocytes. Exp Mol Pathol 94:228–238. doi:10.1016/j.yexmp.2012.08.008
Heussler VT, Kuenzi P, Rottenberg S (2001) Inhibition of apoptosis by intracellular protozoan parasites. Int J Parasitol 31:1166–1176
Heussler VT, Rottenberg S, Schwab R et al (2002) Hijacking of host cell IKK signalosomes by the transforming parasite Theileria. Science 298:1033–1036. doi:10.1126/science.1075462
Hu G, Zhou R, Liu J, Gong AY, Chen XM (2010) MicroRNA-98 and let-7 regulate expression of suppressor of cytokine signaling 4 in biliary epithelial cells in response to Cryptosporidium parvum infection. J Infect Dis 202:125–135. doi:10.1086/653212
Hu G, Zhou R, Liu J, Gong AY, Eischeid AN, Dittman JW, Chen XM (2009) MicroRNA-98 and let-7 confer cholangiocyte expression of cytokine-inducible Src homology 2-containing protein in response to microbial challenge. J Immunol 183:1617–1624. doi:10.4049/jimmunol.0804362
Isnard A, Christian JG, Kodiha M, Stochaj U, McMaster WR, Olivier M (2015) Impact of Leishmania infection on host macrophage nuclear physiology and nucleopore complex integrity. PLoS Pathog 11:e1004776. doi:10.1371/journal.ppat.1004776
Ito S, D'Alessio AC, Taranova OV, Hong K, Sowers LC, Zhang Y (2010) Role of Tet proteins in 5mC to 5hmC conversion, ES-cell self-renewal and inner cell mass specification. Nature 466:1129–1133. doi:10.1038/nature09303
Jenuwein T, Allis CD (2001) Translating the histone code. Science 293:1074–1080. doi:10.1126/science.1063127
Jia B, Lu H, Liu Q, Yin J, Jiang N, Chen Q (2013) Genome-wide comparative analysis revealed significant transcriptome changes in mice after Toxoplasma gondii infection. Parasit Vectors 6:161. doi:10.1186/1756-3305-6-161
Jjingo D, Conley AB, Yi SV, Lunyak VV, Jordan IK (2012) On the presence and role of human gene-body DNA methylation. Oncotarget 3:462–474
Joh RI, Palmieri CM, Hill IT, Motamedi M (2014) Regulation of histone methylation by noncoding RNAs. Biochim Biophys Acta 1839:1385–1394. doi:10.1016/j.bbagrm.2014.06.006

Jones PA (2012) Functions of DNA methylation: islands, start sites, gene bodies and beyond. Nat Rev Genet 13:484–492. doi:10.1038/nrg3230

Kankova S, Kodym P, Flegr J (2011) Direct evidence of Toxoplasma-induced changes in serum testosterone in mice. Exp Parasitol 128:181–183. doi:10.1016/j.exppara.2011.03.014

Kaushansky A, Ye AS, Austin LS et al (2013) Suppression of host p53 is critical for Plasmodium liver-stage infection. Cell Rep 3:630–637. doi:10.1016/j.celrep.2013.02.010

Khoronenkova SV, Dianova II, Parsons JL, Dianov GL (2011) USP7/HAUSP stimulates repair of oxidative DNA lesions. Nucleic Acids Res 39:2604–2609. doi:10.1093/nar/gkq1210

Kim SK, Fouts AE, Boothroyd JC (2007) Toxoplasma gondii dysregulates IFN-gamma-inducible gene expression in human fibroblasts: insights from a genome-wide transcriptional profiling. J Immunol 178:5154–5165

Kinnaird JH, Weir W, Durrani Z, Pillai SS, Baird M, Shiels BR (2013) A Bovine lymphosarcoma cell line infected with Theileria annulata exhibits an irreversible reconfiguration of host cell gene expression. PLoS One 8:e66833. doi:10.1371/journal.pone.0066833

Kruse JP, Gu W (2009) Modes of p53 regulation. Cell 137:609–622. doi:10.1016/j.cell.2009.04.050

Kuzmenok OI, Chiang SC, Lin YC, Lee ST (2005) Retardation of cell cycle progression of macrophages from G1 to S phase by ICAM-L from Leishmania. Int J Parasitol 35:1547–1555. doi:10.1016/j.ijpara.2005.08.006

Lambertz U, Oviedo Ovando ME, Vasconcelos EJ, Unrau PJ, Myler PJ, Reiner NE (2015) Small RNAs derived from tRNAs and rRNAs are highly enriched in exosomes from both old and new world Leishmania providing evidence for conserved exosomal RNA Packaging. BMC Genomics 16:151. doi:10.1186/s12864-015-1260-7

Lang C, Hildebrandt A, Brand F, Opitz L, Dihazi H, Luder CG (2012) Impaired chromatin remodelling at STAT1-regulated promoters leads to global unresponsiveness of Toxoplasma gondii-infected macrophages to IFN-gamma. PLoS Pathog 8:e1002483. doi:10.1371/journal.ppat.1002483

Lemaire J, Mkannez G, Guerfali FZ et al (2013) MicroRNA expression profile in human macrophages in response to Leishmania major infection. PLoS Negl Trop Dis 7:e2478. doi:10.1371/journal.pntd.0002478

Leng J, Butcher BA, Egan CE, Abi Abdallah DS, Denkers EY (2009) Toxoplasma gondii prevents chromatin remodeling initiated by TLR-triggered macrophage activation. J Immunol 182:489–497

Leng J, Denkers EY (2009) Toxoplasma gondii inhibits covalent modification of histone H3 at the IL-10 promoter in infected macrophages. PLoS One 4:e7589. doi:10.1371/journal.pone.0007589

Li YE, Kannan G, Pletnikov MV, Yolken RH, Xiao J (2015) Chronic infection of Toxoplasma gondii downregulates miR-132 expression in multiple brain regions in a sex-dependent manner. Parasitology 142:623–632. doi:10.1017/S003118201400167X

Lievin-Le Moal V, Loiseau PM (2015) Leishmania hijacking of the macrophage intracellular compartments. FEBS J. doi:10.1111/febs.13601

Lizundia R, Chaussepied M, Naissant B et al (2007) The JNK/AP-1 pathway upregulates expression of the recycling endosome rab11a gene in B cells transformed by Theileria. Cell Microbiol 9:1936–1945. doi:10.1111/j.1462-5822.2007.00925.x

Luder CG, Stanway RR, Chaussepied M, Langsley G, Heussler VT (2009) Intracellular survival of apicomplexan parasites and host cell modification. Int J Parasitol 39:163–173. doi:10.1016/j.ijpara.2008.09.013

Luder CG, Sumpf K, Nast R (2015) Releasing the Brake on IFN-gamma Signaling on Infection. Trends Parasitol 31:456–459. doi:10.1016/j.pt.2015.08.006

Luger K, Dechassa ML, Tremethick DJ (2012) New insights into nucleosome and chromatin structure: an ordered state or a disordered affair? Nat Rev Mol Cell Biol 13:436–447. doi:10.1038/nrm3382

Mantel PY, Marti M (2014) The role of extracellular vesicles in Plasmodium and other protozoan parasites. Cell Microbiol 16:344–354. doi:10.1111/cmi.12259
Marr AK, MacIsaac JL, Jiang R, Airo AM, Kobor MS, McMaster WR (2014) Leishmania donovani infection causes distinct epigenetic DNA methylation changes in host macrophages. PLoS Pathog 10:e1004419. doi:10.1371/journal.ppat.1004419
Marsolier J, Perichon M, DeBarry JD et al (2015) Theileria parasites secrete a prolyl isomerase to maintain host leukocyte transformation. Nature 520:378–382. doi:10.1038/nature14044
Marsolier J, Pineau S, Medjkane S et al (2013) OncomiR addiction is generated by a miR-155 feedback loop in Theileria-transformed leukocytes. PLoS Pathog 9:e1003222. doi:10.1371/journal.ppat.1003222
Matte C, Descoteaux A (2010) Leishmania donovani amastigotes impair gamma interferon-induced STAT1alpha nuclear translocation by blocking the interaction between STAT1alpha and importin-alpha5. Infect Immun 78:3736–3743. doi:10.1128/IAI.00046-10
Medjkane S, Perichon M, Marsolier J, Dairou J, Weitzman JB (2014) Theileria induces oxidative stress and HIF1alpha activation that are essential for host leukocyte transformation. Oncogene 33:1809–1817. doi:10.1038/onc.2013.134
Melo MB, Jensen KD, Saeij JP (2011) Toxoplasma gondii effectors are master regulators of the inflammatory response. Trends Parasitol 27:487–495. doi:10.1016/j.pt.2011.08.001
Menendez MT, Teygong C, Wade K, Florimond C, Blader IJ (2015) siRNA Screening identifies the Host Hexokinase 2 (HK2) gene as an important Hypoxia-Inducible Transcription Factor 1 (HIF-1) target gene in Toxoplasma gondii-infected cells. MBio 6. doi:10.1128/mBio.00462-15
Metheni M, Lombes A, Bouillaud F, Batteux F, Langsley G (2015) HIF-1alpha induction, proliferation and glycolysis of Theileria-infected leukocytes. Cell Microbiol. doi:10.1111/cmi.12421
Molestina RE, El-Guendy N, Sinai AP (2008) Infection with Toxoplasma gondii results in dysregulation of the host cell cycle. Cell Microbiol 10:1153–1165. doi:10.1111/j.1462-5822.2008.01117.x
Molestina RE, Payne TM, Coppens I, Sinai AP (2003) Activation of NF-kappaB by Toxoplasma gondii correlates with increased expression of antiapoptotic genes and localization of phosphorylated IkappaB to the parasitophorous vacuole membrane. J Cell Sci 116:4359–4371. doi:10.1242/jcs.00683
Molestina RE, Sinai AP (2005a) Detection of a novel parasite kinase activity at the Toxoplasma gondii parasitophorous vacuole membrane capable of phosphorylating host IkappaBalpha. Cell Microbiol 7:351–362. doi:10.1111/j.1462-5822.2004.00463.x
Molestina RE, Sinai AP (2005b) Host and parasite-derived IKK activities direct distinct temporal phases of NF-kappaB activation and target gene expression following Toxoplasma gondii infection. J Cell Sci 118:5785–5796. doi:10.1242/jcs.02709
Moussaieff A, Kogan NM, Aberdam D (2015) Concise review: energy metabolites: key mediators of the epigenetic state of pluripotency. Stem Cells 33:2374–2380. doi:10.1002/stem.2041
Nash PB, Purner MB, Leon RP, Clarke P, Duke RC, Curiel TJ (1998) Toxoplasma gondii-infected cells are resistant to multiple inducers of apoptosis. J Immunol 160:1824–1830
Ndlovu MN, Denis H, Fuks F (2011) Exposing the DNA methylome iceberg. Trends Biochem Sci 36:381–387. doi:10.1016/j.tibs.2011.03.002
Nishida T, Hatama S, Ishikawa Y, Kadota K (2009) Intranuclear coccidiosis in a calf. J Vet Med Sci 71:1109–1113
Olias P, Etheridge RD, Zhang Y, Holtzman MJ, Sibley LD (2016) Toxoplasma effector recruits the Mi-2/NuRD complex to repress STAT1 transcription and block IFN-γ-dependent gene expression. Cell Host Microbe 20:72–82. doi:10.1016/j.chom.2016.06.006
Olivier M, Gregory DJ, Forget G (2005) Subversion mechanisms by which Leishmania parasites can escape the host immune response: a signaling point of view. Clin Microbiol Rev 18:293–305. doi:10.1128/CMR.18.2.293-305.2005

Ong YC, Reese ML, Boothroyd JC (2010) Toxoplasma rhoptry protein 16 (ROP16) subverts host function by direct tyrosine phosphorylation of STAT6. J Biol Chem 285:28731–28740. doi:10.1074/jbc.M110.112359

Palenzuela O, Redondo MJ, Cali A, Takvorian PM, Alonso-Naveiro M, Alvarez-Pellitero P, Sitja-Bobadilla A (2014) A new intranuclear microsporidium, Enterospora nucleophila n. sp., causing an emaciative syndrome in a piscine host (Sparus aurata), prompts the redescription of the family Enterocytozoonidae. Int J Parasitol 44:189–203. doi:10.1016/j.ijpara.2013.10.005

Pastor WA, Aravind L, Rao A (2013) TETonic shift: biological roles of TET proteins in DNA demethylation and transcription. Nat Rev Mol Cell Biol 14:341–356. doi:10.1038/nrm3589

Peixoto L, Chen F, Harb OS et al (2010) Integrative genomic approaches highlight a family of parasite-specific kinases that regulate host responses. Cell Host Microbe 8:208–218. doi:10.1016/j.chom.2010.07.004

Pelizzola M, Ecker JR (2011) The DNA methylome. FEBS Lett 585:1994–2000. doi:10.1016/j.febslet.2010.10.061

Rosowski EE, Lu D, Julien L, Rodda L, Gaiser RA, Jensen KD, Saeij JP (2011) Strain-specific activation of the NF-kappaB pathway by GRA15, a novel Toxoplasma gondii dense granule protein. J Exp Med 208:195–212. doi:10.1084/jem.20100717

Rosowski EE, Nguyen QP, Camejo A, Spooner E, Saeij JP (2014) Toxoplasma gondii Inhibits gamma interferon (IFN-gamma)- and IFN-beta-induced host cell STAT1 transcriptional activity by increasing the association of STAT1 with DNA. Infect Immun 82:706–719. doi:10.1128/IAI.01291-13

Rosowski EE, Saeij JP (2012) Toxoplasma gondii clonal strains all inhibit STAT1 transcriptional activity but polymorphic effectors differentially modulate IFNgamma induced gene expression and STAT1 phosphorylation. PLoS One 7:e51448. doi:10.1371/journal.pone.0051448

Sacar MD, Bagci C, Allmer J (2014) Computational prediction of microRNAs from Toxoplasma gondii potentially regulating the hosts' gene expression. Genomics Proteomics Bioinformatics 12:228–238. doi:10.1016/j.gpb.2014.09.002

Saeij JP, Coller S, Boyle JP, Jerome ME, White MW, Boothroyd JC (2007) Toxoplasma co-opts host gene expression by injection of a polymorphic kinase homologue. Nature 445:324–327. doi:10.1038/nature05395

Scaria V, Pasha A (2012) Long non-coding RNAs in infection biology. Front Genet 3:308. doi:10.3389/fgene.2012.00308

Schneider AG, Abi Abdallah DS, Butcher BA, Denkers EY (2013) Toxoplasma gondii triggers phosphorylation and nuclear translocation of dendritic cell STAT1 while simultaneously blocking IFNgamma-induced STAT1 transcriptional activity. PLoS One 8:e60215. doi:10.1371/journal.pone.0060215

Schorey JS, Cheng Y, Singh PP, Smith VL (2015) Exosomes and other extracellular vesicles in host-pathogen interactions. EMBO Rep 16:24–43. doi:10.15252/embr.201439363

Schwerk J, Savan R (2015) Translating the Untranslated Region. J Immunol 195:2963–2971. doi:10.4049/jimmunol.1500756

Semenza GL (2007) Hypoxia-inducible factor 1 (HIF-1) pathway. Sci STKE 2007:cm8. doi:10.1126/stke.4072007cm8

Shiels B, Langsley G, Weir W, Pain A, McKellar S, Dobbelaere D (2006) Alteration of host cell phenotype by Theileria annulata and Theileria parva: mining for manipulators in the parasite genomes. Int J Parasitol 36:9–21. doi:10.1016/j.ijpara.2005.09.002

Shiels BR, McKellar S, Katzer F et al (2004) A Theileria annulata DNA binding protein localized to the host cell nucleus alters the phenotype of a bovine macrophage cell line. Eukaryot Cell 3:495–505

Siggens L, Ekwall K (2014) Epigenetics, chromatin and genome organization: recent advances from the ENCODE project. J Intern Med 276:201–214. doi:10.1111/joim.12231

Silmon de Monerri NC, Kim K (2014) Pathogens hijack the epigenome: a new twist on host-pathogen interactions. Am J Pathol 184:897–911. doi:10.1016/j.ajpath.2013.12.022

Silverman JM, Clos J, de' Oliveira CC et al (2010) An exosome-based secretion pathway is responsible for protein export from Leishmania and communication with macrophages. J Cell Sci 123:842–852. doi:10.1242/jcs.056465

Silverman JM, Reiner NE (2011) Leishmania exosomes deliver preemptive strikes to create an environment permissive for early infection. Front Cell Infect Microbiol 1:26. doi:10.3389/fcimb.2011.00026

Sinai AP, Payne TM, Carmen JC, Hardi L, Watson SJ, Molestina RE (2004) Mechanisms underlying the manipulation of host apoptotic pathways by Toxoplasma gondii. Int J Parasitol 34:381–391. doi:10.1016/j.ijpara.2003.11.009

Sinai AP, Webster P, Joiner KA (1997) Association of host cell endoplasmic reticulum and mitochondria with the Toxoplasma gondii parasitophorous vacuole membrane: a high affinity interaction. J Cell Sci 110(Pt 17):2117–2128

Sinclair SH, Rennoll-Bankert KE, Dumler JS (2014) Effector bottleneck: microbial reprogramming of parasitized host cell transcription by epigenetic remodeling of chromatin structure. Front Genet 5:274. doi:10.3389/fgene.2014.00274

Singh AK, Mukhopadhyay C, Biswas S, Singh VK, Mukhopadhyay CK (2012) Intracellular pathogen Leishmania donovani activates hypoxia inducible factor-1 by dual mechanism for survival advantage within macrophage. PLoS One 7:e38489. doi:10.1371/journal.pone.0038489

Spear W, Chan D, Coppens I, Johnson RS, Giaccia A, Blader IJ (2006) The host cell transcription factor hypoxia-inducible factor 1 is required for Toxoplasma gondii growth and survival at physiological oxygen levels. Cell Microbiol 8:339–352. doi:10.1111/j.1462-5822.2005.00628.x

Spivakov M, Fisher AG (2007) Epigenetic signatures of stem-cell identity. Nat Rev Genet 8:263–271. doi:10.1038/nrg2046

Spooner RL, Innes EA, Glass EJ, Brown CG (1989) Theileria annulata and T. parva infect and transform different bovine mononuclear cells. Immunology 66:284–288

Strahl BD, Allis CD (2000) The language of covalent histone modifications. Nature 403:41–45. doi:10.1038/47412

Struhl K, Segal E (2013) Determinants of nucleosome positioning. Nat Struct Mol Biol 20:267–273. doi:10.1038/nsmb.2506

Suzuki Y, Orellana MA, Schreiber RD, Remington JS (1988) Interferon-gamma: the major mediator of resistance against Toxoplasma gondii. Science 240:516–518

Swan DG, Stadler L, Okan E et al (2003) TashHN, a Theileria annulata encoded protein transported to the host nucleus displays an association with attenuation of parasite differentiation. Cell Microbiol 5:947–956

Swan DG, Stern R, McKellar S et al (2001) Characterisation of a cluster of genes encoding Theileria annulata AT hook DNA-binding proteins and evidence for localisation to the host cell nucleus. J Cell Sci 114:2747–2754

Thirugnanam S, Rout N, Gnanasekar M (2013) Possible role of Toxoplasma gondii in brain cancer through modulation of host microRNAs. Infect Agent Cancer 8:8. doi:10.1186/1750-9378-8-8

Tsukada Y, Fang J, Erdjument-Bromage H, Warren ME, Borchers CH, Tempst P, Zhang Y (2006) Histone demethylation by a family of JmjC domain-containing proteins. Nature 439:811–816. doi:10.1038/nature04433

Twu O, de Miguel N, Lustig G, Stevens GC, Vashisht AA, Wohlschlegel JA, Johnson PJ (2013) Trichomonas vaginalis exosomes deliver cargo to host cells and mediate hostratioparasite interactions. PLoS Pathog 9:e1003482. doi:10.1371/journal.ppat.1003482

Unoki M, Brunet J, Mousli M (2009) Drug discovery targeting epigenetic codes: the great potential of UHRF1, which links DNA methylation and histone modifications, as a drug target in cancers and toxoplasmosis. Biochem Pharmacol 78:1279–1288. doi:10.1016/j.bcp.2009.05.035

Via A, Uyar B, Brun C, Zanzoni A (2015) How pathogens use linear motifs to perturb host cell networks. Trends Biochem Sci 40:36–48. doi:10.1016/j.tibs.2014.11.001

Videvall E, Cornwallis CK, Palinauskas V, Valkiunas G, Hellgren O (2015) The avian transcriptome response to malaria infection. Mol Biol Evol 32:1255–1267. doi:10.1093/molbev/msv016

Vyas A (2015) Mechanisms of Host Behavioral Change in Toxoplasma gondii Rodent Association. PLoS Pathog 11:e1004935. doi:10.1371/journal.ppat.1004935

Wiley M, Sweeney KR, Chan DA et al (2010) Toxoplasma gondii activates hypoxia-inducible factor (HIF) by stabilizing the HIF-1alpha subunit via type I activin-like receptor kinase receptor signaling. J Biol Chem 285:26852–26860. doi:10.1074/jbc.M110.147041

Xiao J, Jones-Brando L, Talbot CC Jr, Yolken RH (2011) Differential effects of three canonical Toxoplasma strains on gene expression in human neuroepithelial cells. Infect Immun 79:1363–1373. doi:10.1128/IAI.00947-10

Xiao J, Li Y, Prandovszky E et al (2014) MicroRNA-132 dysregulation in Toxoplasma gondii infection has implications for dopamine signaling pathway. Neuroscience 268:128–138. doi:10.1016/j.neuroscience.2014.03.015

Xu MJ, Zhou DH, Nisbet AJ, Huang SY, Fan YF, Zhu XQ (2013) Characterization of mouse brain microRNAs after infection with cyst-forming Toxoplasma gondii. Parasit Vectors 6:154. doi:10.1186/1756-3305-6-154

Yang SH, Sharrocks AD, Whitmarsh AJ (2013) MAP kinase signalling cascades and transcriptional regulation. Gene 513:1–13. doi:10.1016/j.gene.2012.10.033

Yarovinsky F (2014) Innate immunity to Toxoplasma gondii infection. Nat Rev Immunol 14:109–121. doi:10.1038/nri3598

Yogev O, Lagos D, Enver T, Boshoff C (2014) Kaposi's sarcoma herpesvirus microRNAs induce metabolic transformation of infected cells. PLoS Pathog 10:e1004400. doi:10.1371/journal.ppat.1004400

Zeiner GM, Norman KL, Thomson JM, Hammond SM, Boothroyd JC (2010) Toxoplasma gondii infection specifically increases the levels of key host microRNAs. PLoS One 5:e8742. doi:10.1371/journal.pone.0008742

Zhou R, Gong AY, Chen D, Miller RE, Eischeid AN, Chen XM (2013) Histone deacetylases and NF-kB signaling coordinate expression of CX3CL1 in epithelial cells in response to microbial challenge by suppressing miR-424 and miR-503. PLoS One 8:e65153. doi:10.1371/journal.pone.0065153

Zhou R, Hu G, Liu J, Gong AY, Drescher KM, Chen XM (2009) NF-kappaB p65-dependent transactivation of miRNA genes following Cryptosporidium parvum infection stimulates epithelial cell immune responses. PLoS Pathog 5:e1000681. doi:10.1371/journal.ppat.1000681

Chapter 10
Epigenetic Regulation in *T. brucei*: Changing Coats Is a Chance to Survive

Ana C. Pena, Francisco Aresta-Branco, and Luísa M. Figueiredo

Abstract *Trypanosoma brucei* is a unicellular protozoan transmitted by *Glossina* spp. ("tsetse") flies and the causative agent of Human African trypanosomiasis. This parasite is famous for undergoing antigenic variation, one of the most sophisticated strategies to escape the immune response of its mammalian hosts. Antigenic variation depends on the tight control of the expression of variant surface glycoproteins (VSGs), which form a dense coat that covers the parasite. To perform antigenic variation, *T. brucei* needs to meet two essential requirements: (1) to express a single VSG gene, among a genetic repertoire of ~2000 members, and (2) to periodically switch the expressed VSG gene. In recent years, several chromatin-associated factors have been found to be important to control VSG gene expression. This chapter focuses on the epigenetic regulation of gene expression in *T. brucei*, particularly in antigenic variation.

Keywords Trypanosomes • Antigenic variation • VSG • Chromatin • Epigenetics • Monoallelic expression • Pol I

Abbreviations

Tb	*Trypanosoma brucei*
ASF1A, ASF1B	Anti-silencing factor 1A, 1B
BES	Bloodstream expression site
BSF	Bloodstream form
BDF3	Bromodomain factor 3
CAF-1b	Chromatin assembly factor 1b
CenH3	Centromere-specific histone H3 variant
CITFA	Class I transcription factor A
ChIP	Chromatin immunoprecipitation
DAC1-3	Histone deacetylase 1-3
DOT1A, DOT1B	Disruptor of telomeric silencing 1A, 1B

A.C. Pena • F. Aresta-Branco • L.M. Figueiredo (✉)
Instituto de Medicina Molecular (IMM), Faculdade de Medicina da Universidade de Lisboa, Av. Professor Egas Moniz, 1649-028 Lisbon, Portugal
e-mail: lmf@medicina.ulisboa.pt

W. Doerfler, J. Casadesús (eds.), *Epigenetics of Infectious Diseases*, Epigenetics and Human Health, DOI 10.1007/978-3-319-55021-3_10

ELP3b	Elongator protein 3b
ESAG	Expression site-associated gene
ESB	Expression site body
FACT	Facilitates chromatin transcription complex
FYRP	Phenylalanine/tyrosine rich protein
GPI	Glycosylphosphatidylinositol
HAT	Human African trypanosomiasis
HAT1-3	Histone acetyltransferase 1-3
ISWI	Imitation switch chromatin remodeler
J	β-D-glucosyl-hydroxymethyluracil
JBP1-2	J binding protein 1-2
JGT	Base J-associated glucosyltransferase
Ku70/Ku80	Ku heterodimer
MCM-BP	Mini-chromosome maintenance-binding protein
NLP	Nucleoplasmin-like protein
NUP-1	Nuclear lamin-like protein
ORC1	Origin replication complex 1/cell division cycle 6-like protein
PIP5K	Phosphatidylinositol, 5-kinase
PIP5Pase	Phosphatidylinositol 5-phosphatase
Pol I, Pol II	RNA polymerase I, II
PTM	Posttranslational modification
PTU	Polycistronic transcription unit
RAP1	Repressor/activator of protein 1
RCCP	Regulator of chromosome condensation 1-like protein
rDNA	Ribosomal DNA
RRM1	RNA recognition motif-containing protein
SIR2rp1	Silent information regulator 2-related protein 1
SIZ1	SUMO E3 ligase
TDP-1	Trypanosome DNA-binding protein
TERT	Telomerase reverse transcriptase
TIF2	TRF-interacting factor 2
TRF	Telomeric repeat-binding factor
TSS	Transcription start site
TTS	Transcription termination site
VSG	Variant surface glycoprotein

10.1 Introduction

Trypanosoma brucei is a unicellular, flagellate protozoan parasite (supergroup Excavata, phylum Euglenozoa) that is the causative agent of Human African trypanosomiasis (HAT), one of the world's most important neglected diseases. *T. brucei* is exclusively extracellular, colonizing the lymph, blood, and interstitial spaces of several mammalian hosts and being transmitted through the bite of the blood-feeding flies of the genus *Glossina*, also called "tsetse" (it means "flies") by African natives. The parasite is confined to sub-Saharan Africa, where two subspecies, *T. b. gambiense* and *T. b. rhodesiense*, can infect humans and cause HAT.

Historically HAT has been called "sleeping sickness" due to the characteristic shift of the normal sleep/wake cycle that develops in infected individuals at the late stage of the disease. The late stage of HAT occurs when *T. brucei* parasites cross the blood–brain barrier and invade the central nervous system, which typically coincides with the onset of meningoencephalitis and several neurological disorders including mental, motor system, and sensory system disturbances (Fig. 10.1). HAT is clinically complex and mostly lethal if left untreated. Diagnosis is difficult and the available treatments are often ineffective and toxic (Kennedy 2013; Brun et al. 2010). HAT represents a serious health burden in Africa with ~ 70 million people currently estimated to be at risk of the disease (WHO 2015). Another *T. brucei* subspecies, *T. b. brucei*, is noninfective to humans but infects wild and domestic mammals, causing a disease called "nagana." In African rural

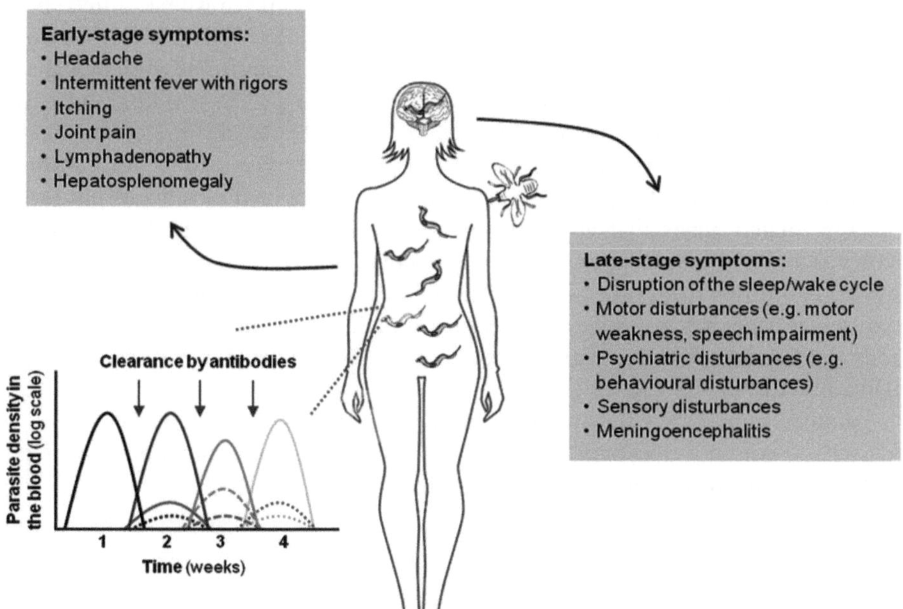

Fig. 10.1 Clinical features of Human African trypanosomiasis (HAT). Most common clinical signs of early (or hemolymphatic) and late (or encephalitic) stage of HAT are indicated. At early stage, *T. brucei* colonizes the interstitial spaces, lymph, and bloodstream of its mammalian host. Due to antigenic variation and succeeding rounds of parasite clearance by host antibodies, the number of parasites in the blood cyclically rises and drops with each peak of parasite density corresponding to a different VSG variant expressed by the majority of the trypanosome population (different shades of *gray* and *dotted lines*). After several weeks or months, the parasite invades the central nervous system leading to the late stage of HAT that gave the disease its well-known name of "sleeping sickness"

populations, nagana is a serious socioeconomic problem, leading to ~ 3 million losses in livestock per year (www.fao.org/ag/againfo/programmes/en/paat/home.html).

During *T. brucei* life cycle, parasites quickly adapt to the different environments they encounter in the mammal and the fly, in order to balance survival, proliferation, and transmissibility (Matthews et al. 2015; Dyer et al. 2013). In the course of its life cycle, *T. brucei* alternates between the mammal host and the fly vector, passing through several developmental stages characterized by differences in ultrastructure, gene expression, metabolism, and surface protein composition.

10.2 The Unusual *T. brucei* Genome and Transcription

African trypanosomes branched early from the main eukaryotic lineage, highly diverging from the animal, fungi, and plant groups (Cavalier-Smith 2010). Such evolutionary history is certainly one of the reasons why this parasite acquired some unusual biological features. Despite its small genome (35 Mb/haploid DNA content), *T. brucei* karyotype consists in more than 120 chromosomes: 11 megabase chromosomes (1–6 Mb), 1–5 intermediate chromosomes (200–900 kb), and ~100 mini-chromosomes (30–150 kb) (Melville et al. 2000; Wickstead et al. 2004). Remarkably, the vast majority of *T. brucei* genome is organized in polycistronic transcription units (PTUs), which consist of large directional clusters of not functionally related genes that are transcribed polycistronically (Siegel et al. 2009; Kolev et al. 2010). Associated with this unusual transcription mechanism, trypanosomes also differ from other eukaryotes regarding mRNA maturation. The polycistronic precursor transcripts are processed at their 5′ end via *trans*-splicing, which adds a m^7G-capped 39-nucleotide spliced leader RNA to the 5′UTR of each mRNA (Ullu and Tschudi 1991).

Another remarkable feature of *T. brucei* is that RNA polymerase I (Pol I) transcribes not only ribosomal DNA (rDNA) but also genes that encode for major surface proteins of the parasite (Gunzl et al. 2003).

10.3 Antigenic Variation Is Crucial for Immune Evasion

As an exclusive extracellular parasite, *T. brucei* cell surface represents the first line of defense against the host immune system. In the mammalian infective bloodstream form (BSF), the parasite cell surface is composed of about 10^7 identical copies of variant surface glycoprotein (VSG) (Cross 1975). VSGs form homodimers and are anchored to the cell membrane by a glycosylphosphatidylinositol (GPI) anchor (Ferguson et al. 1988). VSGs are encoded by a multigene family of ~2000 members. For a *VSG* gene to be transcribed, it needs to be present or recombined into a bloodstream expression site (BES), a specialized subtelomeric locus transcribed by Pol I. There are ~15 BESs in the genome residing at the subtelomeres of megabase and intermediate chromosomes (Hertz-Fowler et al. 2008). In each parasite, only one BES is transcriptionally active, while all the others are silent. Hundreds of inactive *VSG* copies are organized in tandem arrays at non-BES loci in proximal subtelomeric positions in megabase chromosomes or as single genes in mini-chromosomes (Williams et al. 1982; Berriman et al. 2005; Cross et al. 2014).

All known *T. brucei* subspecies undergo antigenic variation. Even though antigenic variation in trypanosomes is not the sole mechanism by which these parasites defend from the mammalian immune system (Pays et al. 2014), it is determinant for the establishment of persistent infections and stands as a model for the study of antigenic variation. Antigenic variation is frequently exploited by viruses, bacteria, and other eukaryotic pathogens such as the human malaria parasite *Plasmodium falciparum* to circumvent the immune response of its mammalian hosts (Deitsch et al. 2009).

In *T. brucei*, antigenic variation relies on two main features: (1) a single VSG is expressed at the surface and (2) the expressed VSG switches periodically. During *T. brucei* infection, the mammalian host primarily mounts an adaptive humoral response against the VSG. The surface VSG epitopes are recognized by B-cells and activate a VSG-specific antibody response that drives rapid parasite clearance (Black et al. 2010). Meanwhile, before immune clearance takes place, a few parasites in the original population switch to a new VSG variant, which is not recognized by the produced antibodies. As a result, parasites expressing the new VSG variant will not be eliminated by the first immune clearance response, and they will substitute the previous parasite population. As successive parasite populations are removed and newly switched variants emerge, periodic waves of parasite density appear in the blood (Mugnier et al. 2015). Such episodes of parasite clearance and relapse are a hallmark of *T. brucei* infections and were observed for the first time in humans more than a century ago (Ross and Thomson 1910) (Fig. 10.1).

VSG switching can occur via different molecular mechanisms. The most frequent mechanism is the activation of a silent *VSG* gene by homologous recombination (HR) into the active BES (Robinson et al. 1999). Recombination is the only way to access the *VSG* repertoire of ~2000 genes or pseudogenes that lie mainly in subtelomeric non-BES loci. The most frequent HR switching mechanism is the so-called duplicative gene conversion. This involves duplication and insertion of an inactive BES-linked or array *VSG* into the active BES, replacing the previously active *VSG* gene. A new *VSG* might also be assembled by segmental gene conversion in which segments of *VSG* genes and pseudogenes are copied and recombine to generate novel functional "mosaic" *VSGs* that will replace the previously active *VSG*. In addition, HR can also mediate a telomere exchange in which reciprocal crossover between two telomeres (no DNA sequence is lost) and their associated *VSGs* takes place. VSG switching can also be recombination independent. This less common mechanism consists in the transcriptional activation of a different BES and silencing of the previous one with no involvement of DNA rearrangements by recombination. This is termed transcriptional or in situ switching (McCulloch et al. 2015).

It is important to note that *T. brucei* switches VSGs ex vivo in axenic cultures, indicating that VSG switching does not depend on the immune response to occur (Doyle et al. 1980). Moreover, it was observed that antigenic variation still exists in immune-deficient mice (Myler et al. 1985). This supports the view that VSG switching is mostly a stochastic process in which antibodies act chiefly as a selective force. However, the fact that switching rates are ~1000-fold higher in fly-transmitted infections argues that there might exist an environmental cue that prompts switching or that adaptation to culture conditions leads to changes in gene expression that eventually result in a lower switching rate (Turner 1997).

10.4 Epigenetics in *T. brucei*

10.4.1 Nuclear and Chromatin Organization

The main nuclear architectural features typical of eukaryotes are observed in *T. brucei* such as a double bilayered nuclear envelope and conserved nuclear pore complexes (DeGrasse et al. 2009). Also, like all other eukaryotes, rDNA is transcribed and ribosomes are assembled in the nucleolus, a specialized nuclear subcompartment. *T. brucei* has another nuclear subcompartment apparently exclusive of this organism, the Expression Site Body (ESB). This is an extranucleolar Pol I territory, functionally distinct from the nucleolus, where active BES is transcribed in BSF parasites (Navarro and Gull 2001; Chaves et al. 1998).

Unlike most eukaryotes, trypanosome chromosomes do not visibly condense during metaphase and a 30-nm fiber has never been observed. In fact, overall, *T. brucei* DNA is less compacted within the nucleus when compared with higher eukaryotes (Hecker and Gander 1985; Hecker et al. 1994). Differences in the degree of chromatin compaction also occur between parasite life cycle stages, with BSFs displaying a more compacted chromatin than insect-stage procyclic forms (Schlimme et al. 1993). These differences may accommodate the rapid changes in gene expression required for parasite development throughout its life cycle and may be due to the presence of divergent histones (Hecker et al. 1994).

10.4.2 Histones and Chromatin Regulators

Although histones (H2A, H2B, H3, H4, and H1) are highly conserved proteins across evolution, the histones of *T. brucei* are very divergent, particularly within its N-terminal tails (Thatcher and Gorovsky 1994; Kasinsky et al. 2001). As a result,

T. brucei lacks many of the well-conserved histone posttranslational modifications (PTMs), while containing some apparently exclusive of this parasite. Examples of the latter are a complex pattern of acetylations on multiple lysines of H2A C-terminus and abundant levels of methylated alanines at H2A, H2B, and H4 N-termini. Conserved histone PTMs present in *T. brucei* are the trimethylation of H3 lysine 4 (H3K4me3) and mono-, di-, or trimethylation of H3K76 (H3K76me/m2/m3) (K79 in other eukaryotes); in addition, acetylations of H4 at K4, K10, and K14 might be the homologues of conserved lysine acetylations at nearby positions in other eukaryotes (K5, K12, and K16) (Janzen et al. 2006a, b; Mandava et al. 2007).

Histone variants are common in several eukaryotes, and they may differ from canonical histones by only a few amino acids. *T. brucei* possesses a homologue of H2A.Z variant and three specific histone variants, the H3.V, H4.V, and H2B.V (Lowell and Cross 2004; Lowell et al. 2005). Besides, it contains several H1 sequence variants, which can be subgrouped into three classes according to their N-terminal sequences (Gruter and Betschart 2001). Each H1 class displays distinct mRNA levels in BSFs (Pena et al. 2014), but it is unknown whether they are functionally distinct. Curiously, the parasite also appears to lack a centromere-specific histone H3 variant (CenH3) (Lowell and Cross 2004; Berriman et al. 2005), a hallmark of eukaryotic kinetochores, possessing a set of rather unconventional kinetochore proteins (Akiyoshi and Gull 2014).

A milestone in trypanosome research was the finding that several histone variants and histone PTMs associate with putative Pol II transcription start sites (TSSs) and transcription termination sites (TTSs). Chromatin immunoprecipitation (ChIP) sequencing revealed that H2A.Z, H2B.V, acetylated H4K10, trimethylated H3K4, and a putative acetyl-binding protein, the bromodomain factor 3 (TbBDF3), are enriched at TSSs, whereas H3.V and H4.V preferentially locate at potential TTSs (Siegel et al. 2009; Wright et al. 2010).

DNA methylation (5-methylcytosine) levels are very low in *T. brucei*, when compared to most eukaryotes (Militello et al. 2008). On the other hand, *T. brucei* and evolutionarily close relatives (including the *Euglena* algae) contain an unusual DNA modification: β-D-glucosyl-hydroxymethyluracil, also termed base J (Gommers-Ampt et al. 1993; van Leeuwen et al. 1998; Dooijes et al. 2000). Base J is crucial for transcription initiation and termination in the related protozoans *Trypanosoma cruzi* and *Leishmania* spp. (Ekanayake and Sabatini 2011; van Luenen et al. 2012), but there is only evidence for a minor role in transcriptional control in *T. brucei* (Reynolds et al. 2014).

A number of chromatin-associated factors have been identified in *T. brucei* such as histone-modifying enzymes, chromatin-remodeling enzymes, and chromatin architectural proteins. Epigenetic regulation mediated by these factors impacts several biological processes of the parasite such as the cell cycle, parasite differentiation, antigenic variation, DNA repair, and telomere integrity (Table 10.1). The importance of epigenetic regulation on the general biology of *T. brucei* has been reviewed elsewhere (Figueiredo et al. 2009; Alsford et al. 2012). This chapter

Table 10.1 Histones and chromatin-associated proteins involved in gene expression regulation in *Trypanosoma brucei*

Class	Name	Functions	Reference
Canonical histones	H1[a], H3[b]	Maintaining BESs silent[a,b], maintaining other Pol I loci silent[a], VSG switching[a], cell cycle[b]	Povelones et al. (2012), Pena et al. (2014), Alsford and Horn (2012)
Histone variants	H2A.Z, H3.V, H4.V	Mark the TSS and TTS at Pol II PTUs	Lowell and Cross (2004), Lowell et al. (2005), Siegel et al. (2009)
Histone deacetylases	DAC1[a], DAC3[b], SIR2rp1[c]	Telomeric silencing[a,c], maintaining BESs silent[b], DNA repair[c]	Ingram and Horn (2002), Wang et al. (2010), Garcia-Salcedo et al. (2003), Alsford et al. (2007)
Histone acetyltransferases	HAT1*[a], HAT2[b], HAT3[c] ELP3b*[d]	Telomeric silencing[a], VSG switching[c], rDNA transcription[d], cell cycle[a,b], DNA repair[c], H4K10ac[b], H4K4ac[c]	Siegel et al. (2008), Kawahara et al. (2008), Alsford and Horn (2011), Glover and Horn (2014)
Histone methyltransferases	DOT1A[a], DOT1B[b]	VSG switching[b], cell cycle [a,b] differentiation to PFs[b]; H3K76 me/me2[a] H3K76me3[b]	Janzen et al. (2006b), Figueiredo et al. (2008), Gassen et al. 2012
SUMOylation enzymes	SIZ1	Maintaining BES active, recruitment of Pol I	Lopez-Farfan et al. (2014)
Chromatin remodelers	ISWI[a], NLP [b], RCCP[c], FYRP[d]	Maintaining BESs silent[a–d]/ active[b], maintaining other Pol I loci silent[a]	Hughes et al. (2007), Stanne et al. (2011, 2015)
Histone chaperones	ASF1A[a], ASF1B[b], CAF-1b[c], FACT[d], NLP[e]	Telomeric silencing[b,c], maintaining BESs silent[b–e]/ active[d,e], maintaining other Pol I loci silent[e], cell cycle [a–d]	Li et al. (2007), Narayanan et al. (2011), Denninger et al. (2010), Alsford and Horn (2012), Denninger and Rudenko (2014), Pascoalino et al. (2014)
Chromatin architectural proteins	TDP-1	Maintaining BES active, maintaining other Pol I loci active	Narayanan and Rudenko (2013), (Aresta-Branco et al. 2016)
Histone-binding proteins	BDF3	Mark the TSS at Pol II PTUs	Siegel et al. (2009)
DNA-modifying enzymes	JBP1[a], JBP2 [b], JGT[c]	Base J synthesis[a,b,c], transcription termination[a,b]	Cross et al. (2002), DiPaolo et al. (2005), Bullard et al. (2014), Reynolds et al. (2014)
Telomeric proteins	RAP1[a], TIF2[b], TRF[c], TERT[d]	Maintaining BESs silent[a], VSG switching[b–d], cell cycle [a,c], telomere and subtelomere integrity[b–d]	Conway et al. (2002), Janzen et al. (2004), Li et al. (2005), Dreesen and Cross (2006), Yang et al. (2009), Hovel-Miner et al. (2012), Pandya et al. (2013); Jehi et al. (2014a, b)

(continued)

Table 10.1 (continued)

Class	Name	Functions	Reference
Nuclear architecture	NUP-1[a] PIP5Pase[b], PIP5K[c]	Maintaining BESs silent[a–c], maintaining other Pol I loci silent[a,b], VSG switching[a,c], Pol I localization[b,c], nucleus integrity[a], cell cycle[a], chromosome positioning[a–c]	DuBois et al. (2012), Cestari and Stuart (2015)
Replication proteins	Cohesin complex[a], MCM-BP[b] ORC1[c]	Inheritance of active BES[a], maintaining BESs silent[b,c], maintaining other Pol I loci silent[b,c], VSG switching[a,c], widespread mRNA abundance[c], cell cycle[a–c]	Landeira et al. (2009), Kim et al. (2013), Benmerzouga et al. (2013), Tiengwe et al. (2012)
RNA-binding proteins	RRM1	Widespread mRNA abundance	Naguleswaran et al. (2015)

*No experimental evidence for the respective enzymatic/chromatin binding activity
[a, b, c, d, e]indicate the roles identified for the respective factor

focuses on the impact of epigenetic regulation in antigenic variation, namely in controlling monoallelic VSG expression and VSG switching.

10.5 Epigenetic Control of *T. brucei* Antigenic Variation

Most of what is known so far about chromatin organization and epigenetics in *T. brucei* is related to *VSG* genes. The focus on these genes is not only due to the importance of antigenic variation for *T. brucei* but also because BESs can be regulated at the transcription level, while Pol II PTUs are thought to be transcribed constitutively. A BES can range in size between 45 and 60 kb, and it contains a Pol I promoter that drives polycistronic transcription of around 10 expression site-associated genes (*ESAGs*) and a single *VSG* gene. *VSG* is always the last gene of a BES (~1 kb upstream from the telomere), and it is flanked upstream by an array of 70 bp repeats (Hertz-Fowler et al. 2008). The function of most ESAGs is unknown even though some of them are known to be important for the survival of BSF parasites inside the mammal host such as ESAG6/ESAG7 dimer (transferrin receptor) and ESAG4 (receptor-like adenylate cyclase that inhibits host innate immunity) (Steverding et al. 1994; Salmon et al. 2012).

There is evidence that Pol I is recruited to the BES promoter by Class I Transcription Factor A (CITFA), a complex of eight subunits consisting of the dynein light chain LC8 and seven other polypeptides (Nguyen et al. 2012; Kirkham et al. 2015). CITFA binds preferentially to the promoter of the active BES compared to silent BES promoters, and this correlates with higher levels of Pol I occupancy and promoter-proximal primary transcripts (Nguyen et al. 2014), suggesting the existence of higher rates of transcription initiation at the active BES promoter (Nguyen et al. 2014; Kirkham et al. 2015). However, repression of transcription initiation at silent BES is not the sole reason why gene expression is silenced at these loci. In fact, there is evidence that low levels of transcription at silent BES are likely a result of inefficient RNA processing and export from the nucleus (Vanhamme et al. 2000; Kassem et al. 2014).

Monoallelic expression of a BES has been the subject of thorough investigation. The parasite needs to maintain active transcription of a single BES, while silencing expression of all the others. Besides, the active/silent transcriptional state of BES needs to be rapidly reversible to ensure successful in situ switching in a short period of time in order to avoid prolonged display of multiple VSGs at the surface. The sequencing of the *T. brucei* genome including the BES loci (Berriman et al. 2005; Hertz-Fowler et al. 2008), together with the development of genetic tools to mark active or silent BESs and perform gene knockdown/knockout in *T. brucei,* has led to the discovery in recent years of several chromatin-related factors important for antigenic variation.

10.5.1 How Are BESs Maintained Silent?

Silent BESs possess a very compact and nucleosome-enriched chromatin (Figueiredo and Cross 2010; Stanne and Rudenko 2010), which most likely ensures a layer of control regarding monoallelic expression. In recent years, several studies identified and characterized epigenetic factors that, once depleted, result in partial transcriptional derepression of silent BESs. Interestingly, for the majority of such factors, transcriptional derepression is limited to the BES promoter regions, without detectable changes in the transcript levels of the terminally located *VSG* genes. Such factors are summarized in Table 10.1 and include the core histone H3 and linker histone H1, the subunits TbSpt16 and TbPob3 of the histone chaperone TbFACT, the histone chaperones TbASF1A, TbCAF-1b, and TbNLP, the histone deacetylase TbDAC3, and the interacting partners of the chromatin remodeler TbISWI, RCCP, and FYRP (as well as TbNLP, already mentioned). The role of some of these factors in BES silencing is also tightly associated with cell cycle progression since derepression of silent BES promoter regions occurs only in S-phase (CAF-1b depleted mutant) or G2/M (histone H3, TbSpt16, and TbCAF-1b depleted mutants) (Alsford and Horn 2012; Denninger et al. 2010).

Other epigenetic factors are important to silence BESs all the way until the telomere, repressing not only the genes nearby the BES promoter but also the silent *VSG* genes. These factors include the histone H3K76 methyltransferase TbDOT1B, the chromatin remodeler TbISWI, the telomeric protein TbRAP1, the lamin-like TbNUP-1, and two proteins of the replication machinery, TbORC1 and TbMCM-BP (Table 10.1). Moreover, deficiency in TbDOT1B, TbRAP1, TbORC1, or TbMCM-BP results in a significant percentage of trypanosomes expressing more than one VSG at the cell surface (Figueiredo et al. 2008; Yang et al. 2009; Benmerzouga et al. 2013; Kim et al. 2013). In contrast, some factors such as the histone deacetylase TbSir2rp1 and the putative histone acetlytransferase TbHAT1 appear to be important exclusively for maintaining telomere-proximal silencing but not for silencing the BESs (Alsford et al. 2007; Kawahara et al. 2008), indicating the existence of multiple mechanisms to control transcription of BESs.

Differential nuclear localization of silent BESs may also play an important role in BES silencing. FISH analysis suggests that most telomeres co-localize with TbNUP-1, a lamin-like protein localized at the nuclear periphery that is likely a nuclear lamina component (DuBois et al. 2012). Depletion of TbNUP-1 disrupts nuclear morphology and leads to re-localization of telomeres from megabase chromosomes to nuclear blebs, with a concomitant derepression of silent BESs, indicating that nuclear structure is important for silencing gene expression at BESs. Two enzymes of the inositol phosphate pathway (TbPIP5Pase and TbPIP5K) were recently shown to be required for BES silencing, for the correct positioning of telomeres and Pol I in the nucleus, supporting the impact of nuclear architecture in BES silencing. It is, however, unknown how these enzymes or their associated metabolites epigenetically affect the telomeres and BESs (Cestari and Stuart 2015).

10.5.2 How Is a BES Maintained Active?

The structure and dynamics of the active BES is poorly understood and very few epigenetic regulators have been shown to control it.

Unlike silent BESs, the active BES possesses an open and nucleosome-depleted chromatin (Figueiredo et al. 2008; Stanne and Rudenko 2010). Three proteins have been identified as being important for maintaining the transcriptional status of the active BES: TbTDP1, TbNLP, a TbFACT subunit (TbSpt16), and a cohesin component (TbSCC1). TbTDP1 is a high mobility group box protein that binds along the entire BES, and that is necessary to keep chromatin open when transcription is halted (Aresta-Branco et al. 2016). Its depletion drives chromatin condensation and a decrease in active *VSG* transcript levels (Narayanan and Rudenko 2013). Depletion of TbNLP also reduces gene expression at the promoter region of active BES, although it is not clear if this results from an indirect consequence of silent BES derepression also resulting from TbNLP depletion or from a double role of this putative chaperone in active and silent BESs. On the other hand, the histone deacetylase DAC1 antagonizes telomeric silencing, but has no attributed role in gene expression at the active BES (Wang et al. 2010), again indicating that BES transcriptional status is epigenetically controlled by mechanisms independent from telomere silencing.

Interestingly, depletion of TbSSC1, a component of the cohesin complex, revealed that cohesion is necessary for the epigenetic inheritance of the active transcriptional status of BES probably by mediating a delayed separation of the sister chromatids from the active BES locus during early mitosis (Landeira et al. 2009).

Posttranslational modifications also seem important to maintain the status of the active BES. Accumulation of SUMOylated chromatin-associated proteins occurs throughout the BES including in the RPA1 subunit of Pol I. Reduced levels of this modification by depletion of a sumoylation enzyme, the E3 ligase TbSIZ1, reduce the recruitment of Pol I to the active BES and a consequent transcriptional downregulation (Lopez-Farfan et al. 2014).

Given that the active BES localizes in the ESB, it is possible that it retains factors that are essential for the chromatin and transcriptional status necessary for *VSG* expression, thus contributing to monoallelic expression of the unique BES that locates in the ESB. Even though no specific factor has been associated with the ESB, TbTDP-1 and SUMOylated proteins are enriched in this subnuclear body, in agreement with their role as positive regulators of the active BES (Narayanan and Rudenko 2013; Lopez-Farfan et al. 2014).

10.5.3 How Do BESs Undergo Transcriptional Switching?

BES transcriptional switching (or in situ switching) consists in silencing the active BES with a concomitant activation of a silent BES. Because chromatin of the active and silent BESs is dramatically different, BES switching involves important modifications in chromatin structure, which are likely well coordinated with changes in the transcriptional machinery. Besides, switching is probably triggered by an early event, whose nature remains unknown.

While some chromatin-associated factors are mostly involved in recombination-mediated switching (H1, TbHAT3, TbTRF, and TbTIF2), we will focus on those important for the epigenetic regulation of transcriptional switching.

All factors important for this type of switching also play a role in VSG monoallelic expression consistent with an interdependent control of both processes: cohesin represses in situ switching without affecting recombination-mediated switching (Landeira et al. 2009); TbORC1 represses preferentially BES transcriptional switching (Benmerzouga et al. 2013); TbNUP1 was proposed to keep silent BESs away from the ESB, therefore preventing switching events (DuBois et al. 2012); and TbDOT1B is considered a central factor in promoting fast in situ switching events since its depletion causes cells to express more than one VSG at the cell surface up to several weeks until VSG coat replacement is completed (Figueiredo et al. 2008). More recently, inositol phosphate pathway was also found to regulate switching as TbPIP5K represses both recombination-dependent and in situ switching (Cestari and Stuart 2015) (Fig. 10.2).

Transcriptional switching needs to be a fast but well-coordinated process to minimize the time in which more than one VSG is exposed at the surface. In fact, when two different BESs marked with drug resistance genes were forced to be simultaneously expressed, rapid and permanent back and forth switches between the two BES were apparently prompted to guarantee drug resistance (Chaves et al. 1999). The authors postulated the existence of natural switching intermediates in

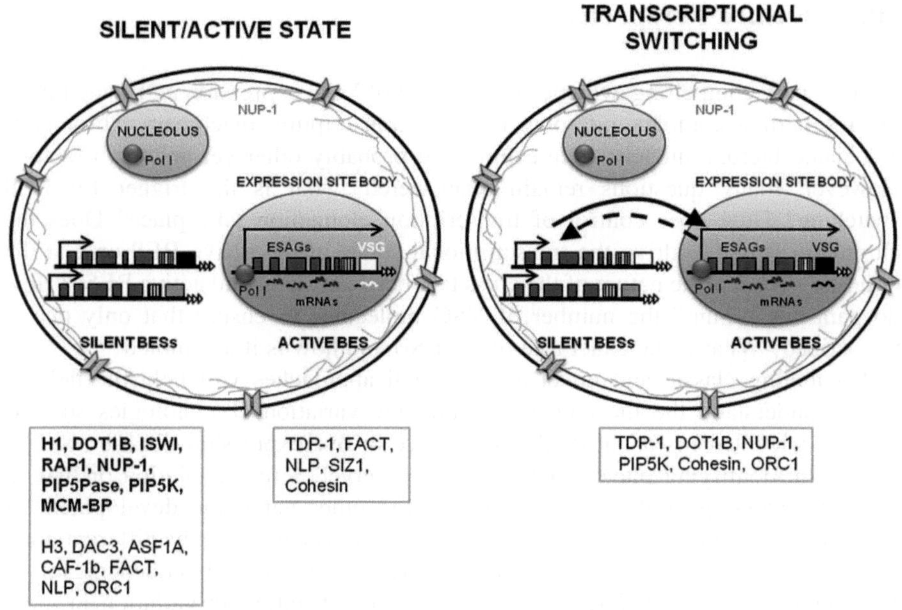

Fig. 10.2 Chromatin-associated factors involved in the maintenance of BES transcriptional state and BES switching. Two illustrations represent the transcriptional status of silent and active BESs inside the nucleus of *T. brucei* BSFs in steady state (*left*) or during an in situ switch (*right*). *Straight arrows* indicate direction of Pol I transcription at BES promoters. Low and high transcriptional activity is indicated by *short* or *long arrows*, respectively. *Curved arrows* indicate the two BES undergoing a transcriptional switch. Factors involved in each process are listed in a box below each respective illustration. Factors necessary for repression of the entire BES (*left*) are highlighted in bold

which silent BESs are in a pre-active state that facilitates a transcriptional interchange.

Recently, two studies presented more details about this intermediate state, although with different approaches. A first study showed that the ectopic expression of a second *VSG* gene from a non-BES locus promotes attenuation of transcription from the whole active BES (Batram et al. 2014). This *VSG* silencing/attenuation is reversible and mediated by TbDOT1B. This study proposes that activation of a silent *VSG* may act as a trigger for transcriptional switching. A second study (from our lab) employed transcriptional silencing of the active BES to provoke transcriptional switching (Aresta-Branco et al. 2016). This trigger resulted in the maintenance of an open, transcription-independent chromatin structure of the active BES, which is dependent on TbTDP-1. Both studies suggest that transcriptional attenuation/open chromatin maintenance is an intermediate state that allows probing for the integrity and functionality of a new BES before full commitment for switching. Furthermore, in the second study, probing before commitment for switching or reversal for initial active BES was verified.

10.6 Future Perspectives

There is now compelling evidence that control of VSG expression and switching is the result of a complex interplay between transcription machinery, chromatin-associated factors, nuclear architecture, and probably other yet unknown factors. However, many questions remain unanswered: what is the trigger for BES switching? How does control of transcription elongation take place? Does the chromatin structure drive the transcriptional status of the active BES or is it the opposite? What is the nature of the cross talk between silent and active BESs? How do parasites "count" the number of VSG molecules to ensure that only one is transcribed? What is the exact role of the ESB and how is it assembled?

Combining classic and novel experimental approaches will help the field to further understand the mechanism of antigenic variation. Technologies such as genome-wide screens using RNA interference or overexpression libraries will be helpful in identifying players that are not conserved across evolution or that an educated guessing could not propose. On the other hand, the development of transfection for pleomorphic strains of *T. brucei* which are fully differentiation competent (such as AnTat1.1E clone from EATRO1125) and sequencing of its VSGnome (Cross et al. 2014) will allow studies of antigenic variation in more natural conditions, in which switching frequencies are typically higher and thus experimentally more tractable. The use of such strains will also allow testing phenotypes in mouse models because infections last longer than with laboratory-adapted monomorphic *T. brucei* strains (Mugnier et al. 2015).

In the short term, perhaps the field will answer a few "hanging" questions. First, whether base J plays a role in antigenic variation. Base J is more abundant in silent BESs than active BES (van Leeuwen et al. 1997). However, a J-null cell line apparently did not show any phenotype in terms of regulation of BES transcriptional silencing (Cliffe et al. 2010). A second exciting question in the field is whether noncoding RNAs are important, especially in mediating gene silencing. Such mechanism has been proposed for *P. falciparum*, in which distinct noncoding transcripts emanating from the *var* gene intron are associated with *var* gene silencing (Epp et al. 2009) and activation (Amit-Avraham et al. 2015; Epp et al. 2009).

Epigenetic regulation of virulence genes is not exclusive of trypanosomes. Other protozoan parasites such as *Plasmodium* spp. and *Toxoplasma gondii* (toxoplasmosis parasite) also use epigenetic regulation as a common means for controlling gene expression (Croken et al. 2012). Identifying the players involved in gene regulation of important virulence factors will not only broaden our knowledge about the biology of these fascinating pathogens, but it may reveal novel targets for therapy.

Since the final submission of this book chapter, new studies were published on epigenetic regulation in trypanosomes. New factors were discovered to be necessary for VSG monoallelic expression—bromodomain proteins TbBDF2 and TbBDF3 (Schulz et al. 2015), VSG exclusion-1 (TbVEX1) (Glover et al. 2016) and the noncoding telomeric repeat-containing RNA (TERRA) (Nanavaty et al. 2017).

The Rudenko group showed that the ESB appears to be a Pol I nucleated structure, similar to the nucleolus (Kerry et al. 2017). New evidence clarified the role of base J and H3.V as bona-fide Pol II transcription termination marks (Reynolds et al. 2016; Schulz et al. 2016) and characterized the position of nucleosomes throughout the chromosomes of these ancient organisms (Maree et al. 2017).

References

Akiyoshi B, Gull K (2014) Discovery of unconventional kinetochores in kinetoplastids. Cell 156 (6):1247–1258. doi:10.1016/j.cell.2014.01.049

Alsford S, duBois K, Horn D, Field MC (2012) Epigenetic mechanisms, nuclear architecture and the control of gene expression in trypanosomes. Expert Rev Mol Med 14:e13. doi:10.1017/erm.2012.7

Alsford S, Horn D (2011) Elongator protein 3b negatively regulates ribosomal DNA transcription in african trypanosomes. Mol Cell Biol 31(9):1822–1832. doi:10.1128/MCB.01026-10

Alsford S, Horn D (2012) Cell-cycle-regulated control of VSG expression site silencing by histones and histone chaperones ASF1A and CAF-1b in *Trypanosoma brucei*. Nucleic Acids Res. doi:10.1093/nar/gks813

Alsford S, Kawahara T, Isamah C, Horn D (2007) A sirtuin in the African trypanosome is involved in both DNA repair and telomeric gene silencing but is not required for antigenic variation. Mol Microbiol 63(3):724–736. doi:10.1111/j.1365-2958.2006.05553.x

Amit-Avraham I, Pozner G, Eshar S, Fastman Y, Kolevzon N, Yavin E, Dzikowski R (2015) Antisense long noncoding RNAs regulate var gene activation in the malaria parasite *Plasmodium falciparum*. Proc Natl Acad Sci U S A 112(9):E982–E991. doi:10.1073/pnas.1420855112

Aresta-Branco F, Pimenta S, Figueiredo LMA (2016) A transcription-independent epigenetic mechanism is associated with antigenic switching in *Trypanosoma brucei*. Nucleic Acids Res 44(7):3131–3146

Batram C, Jones NG, Janzen CJ, Markert SM, Engstler M (2014) Expression site attenuation mechanistically links antigenic variation and development in *Trypanosoma brucei*. elife 3: e02324

Benmerzouga I, Concepcion-Acevedo J, Kim HS, Vandoros AV, Cross GA, Klingbeil MM, Li B (2013) *Trypanosoma brucei* Orc1 is essential for nuclear DNA replication and affects both VSG silencing and VSG switching. Mol Microbiol 87(1):196–210. doi:10.1111/mmi.12093

Berriman M, Ghedin E, Hertz-Fowler C, Blandin G, Renauld H, Bartholomeu DC, Lennard NJ, Caler E, Hamlin NE, Haas B, Bohme U, Hannick L et al (2005) The genome of the African trypanosome *Trypanosoma brucei*. Science 309(5733):416–422. doi:10.1126/science.1112642

Black SJ, Guirnalda P, Frenkel D, Haynes C, Bockstal V (2010) Induction and regulation of *Trypanosoma brucei* VSG-specific antibody responses. Parasitology 137(14):2041–2049. doi:10.1017/S003118200999165X

Brun R, Blum J, Chappuis F, Burri C (2010) Human African trypanosomiasis. Lancet 375 (9709):148–159. doi:10.1016/S0140-6736(09)60829-1

Bullard W, Lopes da Rosa-Spiegler J, Liu S, Wang Y, Sabatini R (2014) Identification of the glucosyltransferase that converts hydroxymethyluracil to base J in the trypanosomatid genome. J Biol Chem 289(29):20273–20282. doi:10.1074/jbc.M114.579821

Cavalier-Smith T (2010) Kingdoms Protozoa and Chromista and the eozoan root of the eukaryotic tree. Biol Lett 6(3):342–345. doi:10.1098/rsbl.2009.0948

Cestari I, Stuart K (2015) Inositol phosphate pathway controls transcription of telomeric expression sites in trypanosomes. Proc Natl Acad Sci U S A 112(21):E2803–E2812. doi:10.1073/pnas.1501206112

Chaves I, Rudenko G, Dirks-Mulder A, Cross M, Borst P (1999) Control of variant surface glycoprotein gene-expression sites in *Trypanosoma brucei*. EMBO J 18(17):4846–4855. doi:10.1093/emboj/18.17.4846

Chaves I, Zomerdijk J, Dirks-Mulder A, Dirks RW, Raap AK, Borst P (1998) Subnuclear localization of the active variant surface glycoprotein gene expression site in *Trypanosoma brucei*. Proc Natl Acad Sci U S A 95(21):12328–12333

Cliffe LJ, Siegel TN, Marshall M, Cross GA, Sabatini R (2010) Two thymidine hydroxylases differentially regulate the formation of glucosylated DNA at regions flanking polymerase II polycistronic transcription units throughout the genome of *Trypanosoma brucei*. Nucleic Acids Res 38(12):3923–3935. doi:10.1093/nar/gkq146

Conway C, McCulloch R, Ginger ML, Robinson NP, Browitt A, Barry JD (2002) Ku is important for telomere maintenance, but not for differential expression of telomeric VSG genes, in African trypanosomes. J Biol Chem 277(24):21269–21277. doi:10.1074/jbc.M200550200

Croken MM, Nardelli SC, Kim K (2012) Chromatin modifications, epigenetics, and how protozoan parasites regulate their lives. Trends Parasitol 28(5):202–213. doi:10.1016/j.pt.2012.02.009

Cross GA (1975) Identification, purification and properties of clone-specific glycoprotein antigens constituting the surface coat of *Trypanosoma brucei*. Parasitology 71(3):393–417

Cross M, Kieft R, Sabatini R, Dirks-Mulder A, Chaves I, Borst P (2002) J-binding protein increases the level and retention of the unusual base J in trypanosome DNA. Mol Microbiol 46(1):37–47

Cross GA, Kim HS, Wickstead B (2014) Capturing the variant surface glycoprotein repertoire (the VSGnome) of *Trypanosoma brucei* Lister 427. Mol Biochem Parasitol 195(1):59–73. doi:10.1016/j.molbiopara.2014.06.004

DeGrasse JA, DuBois KN, Devos D, Siegel TN, Sali A, Field MC, Rout MP, Chait BT (2009) Evidence for a shared nuclear pore complex architecture that is conserved from the last common eukaryotic ancestor. Mol Cell Proteomics 8(9):2119–2130. doi:10.1074/mcp.M900038-MCP200

Deitsch KW, Lukehart SA, Stringer JR (2009) Common strategies for antigenic variation by bacterial, fungal and protozoan pathogens. Nat Rev Microbiol 7(7):493–503. doi:10.1038/nrmicro2145

Denninger V, Fullbrook A, Bessat M, Ersfeld K, Rudenko G (2010) The FACT subunit TbSpt16 is involved in cell cycle specific control of VSG expression sites in *Trypanosoma brucei*. Mol Microbiol 78(2):459–474. doi:10.1111/j.1365-2958.2010.07350.x

Denninger V, Rudenko G (2014) FACT plays a major role in histone dynamics affecting VSG expression site control in *Trypanosoma brucei*. Mol Microbiol 94(4):945–962. doi:10.1111/mmi.12812

DiPaolo C, Kieft R, Cross M, Sabatini R (2005) Regulation of trypanosome DNA glycosylation by a SWI2/SNF2-like protein. Mol Cell 17(3):441–451. doi:10.1016/j.molcel.2004.12.022

Dooijes D, Chaves I, Kieft R, Dirks-Mulder A, Martin W, Borst P (2000) Base J originally found in kinetoplastida is also a minor constituent of nuclear DNA of *Euglena gracilis*. Nucleic Acids Res 28(16):3017–3021

Doyle JJ, Hirumi H, Hirumi K, Lupton EN, Cross GA (1980) Antigenic variation in clones of animal-infective *Trypanosoma brucei* derived and maintained in vitro. Parasitology 80(2):359–369

Dreesen O, Cross GA (2006) Consequences of telomere shortening at an active VSG expression site in telomerase-deficient *Trypanosoma brucei*. Eukaryot Cell 5(12):2114–2119. doi:10.1128/EC.00059-06

DuBois KN, Alsford S, Holden JM, Buisson J, Swiderski M, Bart JM, Ratushny AV, Wan Y, Bastin P, Barry JD, Navarro M, Horn D et al (2012) NUP-1 Is a large coiled-coil nucleoskeletal protein in trypanosomes with lamin-like functions. PLoS Biol 10(3):e1001287. doi:10.1371/journal.pbio.1001287

Dyer NA, Rose C, Ejeh NO, Acosta-Serrano A (2013) Flying tryps: survival and maturation of trypanosomes in tsetse flies. Trends Parasitol 29(4):188–196. doi:10.1016/j.pt.2013.02.003

Ekanayake D, Sabatini R (2011) Epigenetic regulation of polymerase II transcription initiation in *Trypanosoma cruzi*: modulation of nucleosome abundance, histone modification, and polymerase occupancy by O-linked thymine DNA glucosylation. Eukaryot Cell 10(11):1465–1472. doi:10.1128/EC.05185-11

Epp C, Li F, Howitt CA, Chookajorn T, Deitsch KW (2009) Chromatin associated sense and antisense noncoding RNAs are transcribed from the var gene family of virulence genes of the malaria parasite *Plasmodium falciparum*. RNA 15(1):116–127. doi:10.1261/rna.1080109

Ferguson MA, Homans SW, Dwek RA, Rademacher TW (1988) Glycosyl-phosphatidylinositol moiety that anchors *Trypanosoma brucei* variant surface glycoprotein to the membrane. Science 239(4841 Pt 1):753–759

Figueiredo LM, Cross GA (2010) Nucleosomes are depleted at the VSG expression site transcribed by RNA polymerase I in African trypanosomes. Eukaryot Cell 9(1):148–154. doi:10.1128/EC.00282-09

Figueiredo LM, Cross GA, Janzen CJ (2009) Epigenetic regulation in African trypanosomes: a new kid on the block. Nat Rev Microbiol 7(7):504–513. doi:10.1038/nrmicro2149

Figueiredo LM, Janzen CJ, Cross GA (2008) A histone methyltransferase modulates antigenic variation in African trypanosomes. PLoS Biol 6(7):e161. doi:10.1371/journal.pbio.0060161

Garcia-Salcedo JA, Gijon P, Nolan DP, Tebabi P, Pays E (2003) A chromosomal SIR2 homologue with both histone NAD-dependent ADP-ribosyltransferase and deacetylase activities is involved in DNA repair in *Trypanosoma brucei*. EMBO J 22(21):5851–5862. doi:10.1093/emboj/cdg553

Gassen A, Brechtefeld D, Schandry N, Arteaga-Salas JM, Israel L, Imhof A, Janzen CJ (2012) DOT1A-dependent H3K76 methylation is required for replication regulation in *Trypanosoma brucei*. Nucleic Acids Res 40(20):10302–10311. doi:10.1093/nar/gks801

Glover L, Horn D (2014) Locus-specific control of DNA resection and suppression of subtelomeric VSG recombination by HAT3 in the African trypanosome. Nucleic Acids Res 42(20):12600–12613. doi:10.1093/nar/gku900

Glover L, Hutchinson S, Alsford S, Horn D (2016) VEX1 controls the allelic exclusion required for antigenic variation in trypanosomes. Proc Natl Acad Sci USA 113(26):7225–7230. doi:10.1073/pnas.1600344113

Gommers-Ampt JH, Van Leeuwen F, de Beer AL, Vliegenthart JF, Dizdaroglu M, Kowalak JA, Crain PF, Borst P (1993) Beta-D-glucosyl-hydroxymethyluracil: a novel modified base present in the DNA of the parasitic protozoan *T. brucei*. Cell 75(6):1129–1136. doi:0092-8674(93)90322-H

Gruter E, Betschart B (2001) Isolation, characterisation and organisation of histone H1 genes in African trypanosomes. Parasitol Res 87(12):977–984

Gunzl A, Bruderer T, Laufer G, Schimanski B, Tu LC, Chung HM, Lee PT, Lee MG (2003) RNA polymerase I transcribes procyclin genes and variant surface glycoprotein gene expression sites in *Trypanosoma brucei*. Eukaryot Cell 2(3):542–551

Hecker H, Betschart B, Bender K, Burri M, Schlimme W (1994) The chromatin of trypanosomes. Int J Parasitol 24(6):809–819. doi:0020-7519(94)90007-8

Hecker H, Gander ES (1985) The compaction pattern of the chromatin of trypanosomes. Biol Cell 53(3):199–208

Hertz-Fowler C, Figueiredo LM, Quail MA, Becker M, Jackson A, Bason N, Brooks K, Churcher C, Fahkro S, Goodhead I, Heath P, Kartvelishvili M et al (2008) Telomeric expression sites are highly conserved in *Trypanosoma brucei*. PloS ONE 3(10):e3527. doi:10.1371/journal.pone.0003527

Hovel-Miner GA, Boothroyd CE, Mugnier M, Dreesen O, Cross GA, Papavasiliou FN (2012) Telomere length affects the frequency and mechanism of antigenic variation in *Trypanosoma brucei*. PLoS Pathog 8(8):e1002900. doi:10.1371/journal.ppat.1002900

Hughes K, Wand M, Foulston L, Young R, Harley K, Terry S, Ersfeld K, Rudenko G (2007) A novel ISWI is involved in VSG expression site downregulation in African trypanosomes. EMBO J 26(9):2400–2410. doi:10.1038/sj.emboj.7601678

Ingram AK, Horn D (2002) Histone deacetylases in *Trypanosoma brucei*: two are essential and another is required for normal cell cycle progression. Mol Microbiol 45(1):89–97. doi:3018

Janzen CJ, Fernandez JP, Deng H, Diaz R, Hake SB, Cross GA (2006a) Unusual histone modifications in *Trypanosoma brucei*. FEBS Lett 580(9):2306–2310. doi:10.1016/j.febslet.2006.03.044

Janzen CJ, Hake SB, Lowell JE, Cross GA (2006b) Selective di- or trimethylation of histone H3 lysine 76 by two DOT1 homologs is important for cell cycle regulation in *Trypanosoma brucei*. Mol Cell 23(4):497–507. doi:10.1016/j.molcel.2006.06.027

Janzen CJ, Lander F, Dreesen O, Cross GA (2004) Telomere length regulation and transcriptional silencing in KU80-deficient *Trypanosoma brucei*. Nucleic Acids Res 32(22):6575–6584. doi:10.1093/nar/gkh991

Jehi SE, Li X, Sandhu R, Ye F, Benmerzouga I, Zhang M, Zhao Y, Li B (2014a) Suppression of subtelomeric VSG switching by *Trypanosoma brucei* TRF requires its TTAGGG repeat-binding activity. Nucleic Acids Res 42(20):12899–12911. doi:10.1093/nar/gku942

Jehi SE, Wu F, Li B (2014b) *Trypanosoma brucei* TIF2 suppresses VSG switching by maintaining subtelomere integrity. Cell Res. doi:10.1038/cr.2014.60

Kasinsky HE, Lewis JD, Dacks JB, Ausio J (2001) Origin of H1 linker histones. FASEB J 15 (1):34–42. doi:10.1096/fj.00-0237rev

Kassem A, Pays E, Vanhamme L (2014) Transcription is initiated on silent variant surface glycoprotein expression sites despite monoallelic expression in *Trypanosoma brucei*. Proc Natl Acad Sci U S A 111(24):8943–8948. doi:10.1073/pnas.1404873111

Kawahara T, Siegel TN, Ingram AK, Alsford S, Cross GA, Horn D (2008) Two essential MYST-family proteins display distinct roles in histone H4K10 acetylation and telomeric silencing in trypanosomes. Mol Microbiol 69(4):1054–1068. doi:10.1111/j.1365-2958.2008.06346.x

Kennedy PG (2013) Clinical features, diagnosis, and treatment of human African trypanosomiasis (sleeping sickness). Lancet Neurol 12(2):186–194. doi:10.1016/S1474-4422(12)70296-X

Kerry LE, Pegg EE, Cameron DP, Budzak J, Poortinga G, Hannan KM, Hannan RD, Rudenko G (2017) Selective inhibition of RNA polymerase I transcription as a potential approach to treat African trypanosomiasis. PLoS Negl Trop Dis 11(3):e0005432. doi:10.1371/journal.pntd.0005432

Kim HS, Park SH, Gunzl A, Cross GA (2013) MCM-BP is required for repression of life-cycle specific genes transcribed by RNA polymerase I in the mammalian infectious form of *Trypanosoma brucei*. PloS one 8(2):e57001. doi:10.1371/journal.pone.0057001

Kirkham JK, Park SH, Nguyen TN, Lee JH, Gunzl A (2015) The dynein light chain LC8 is required for RNA polymerase I-mediated transcription in *Trypanosoma brucei*, facilitating assembly and promoter binding of class I transcription factor A. Mol Cell Biol. doi:10.1128/MCB.00705-15

Kolev NG, Franklin JB, Carmi S, Shi H, Michaeli S, Tschudi C (2010) The transcriptome of the human pathogen *Trypanosoma brucei* at single-nucleotide resolution. PLoS Pathog 6(9): e1001090. doi:10.1371/journal.ppat.1001090

Landeira D, Bart JM, Van Tyne D, Navarro M (2009) Cohesin regulates VSG monoallelic expression in trypanosomes. J Cell Biol 186(2):243–254. doi:10.1083/jcb.200902119

Li B, Espinal A, Cross GA (2005) Trypanosome telomeres are protected by a homologue of mammalian TRF2. Mol Cell Biol 25(12):5011–5021. doi:10.1128/MCB.25.12.5011-5021.2005

Li Z, Gourguechon S, Wang CC (2007) Tousled-like kinase in a microbial eukaryote regulates spindle assembly and S-phase progression by interacting with Aurora kinase and chromatin assembly factors. J Cell Sci 120(Pt 21):3883–3894. doi:10.1242/jcs.007955

Lopez-Farfan D, Bart JM, Rojas-Barros DI, Navarro M (2014) SUMOylation by the E3 ligase TbSIZ1/PIAS1 positively regulates VSG expression in *Trypanosoma brucei*. PLoS Pathog 10 (12):e1004545. doi:10.1371/journal.ppat.1004545

Lowell JE, Cross GA (2004) A variant histone H3 is enriched at telomeres in *Trypanosoma brucei*. J Cell Sci 117(Pt 24):5937–5947. doi:10.1242/jcs.01515

Lowell JE, Kaiser F, Janzen CJ, Cross GA (2005) Histone H2AZ dimerizes with a novel variant H2B and is enriched at repetitive DNA in *Trypanosoma brucei*. J Cell Sci 118 (Pt 24):5721–5730. doi:10.1242/jcs.02688

Mandava V, Fernandez JP, Deng H, Janzen CJ, Hake SB, Cross GA (2007) Histone modifications in *Trypanosoma brucei*. Mol Biochem Parasitol 156(1):41–50. doi:10.1016/j.molbiopara.2007.07.005

Maree JP, Povelones ML, Clark DJ, Rudenko G, Patterton HG (2017) Well-positioned nucleosomes punctuate polycistronic pol II transcription units and flank silent VSG gene arrays in *Trypanosoma brucei*. Epigenetics Chromatin 10:14. doi:10.1186/s13072-017-0121-9

Matthews KR, McCulloch R, Morrison LJ (2015) The within-host dynamics of African trypanosome infections. Philos Trans R Soc Lond Ser B Biol Sci 370(1675). doi:10.1098/rstb.2014.0288

McCulloch R, Morrison LJ, Hall JP (2015) DNA recombination strategies during antigenic variation in the African Trypanosome. Microbiol Spectr 3(2). doi:10.1128/microbiolspec

Melville SE, Leech V, Navarro M, Cross GA (2000) The molecular karyotype of the megabase chromosomes of *Trypanosoma brucei* stock 427. Molecular and biochemical parasitology 111 (2):261–273. doi:S0166685100003169

Militello KT, Wang P, Jayakar SK, Pietrasik RL, Dupont CD, Dodd K, King AM, Valenti PR (2008) African trypanosomes contain 5-methylcytosine in nuclear DNA. Eukaryot Cell 7 (11):2012–2016. doi:10.1128/EC.00198-08

Mugnier MR, Cross GA, Papavasiliou FN (2015) The *in vivo* dynamics of antigenic variation in *Trypanosoma brucei*. Science 347(6229):1470–1473. doi:10.1126/science.aaa4502

Myler PJ, Allen AL, Agabian N, Stuart K (1985) Antigenic variation in clones of *Trypanosoma brucei* grown in immune-deficient mice. Infect Immun 47(3):684–690

Naguleswaran A, Gunasekera K, Schimanski B, Heller M, Hemphill A, Ochsenreiter T, Roditi I (2015) *Trypanosoma brucei* RRM1 is a nuclear RNA-binding protein and modulator of chromatin structure. MBio 6(2):e00114. doi:10.1128/mBio.00114-15

Nanavaty V, Sandhu R, Jehi SE, Pandya UM, Li B (2017) *Trypanosoma brucei* RAP1 maintains telomere and subtelomere integrity by suppressing TERRA and telomeric RNA:DNA hybrids. Nucleic Acids Res, March 16. doi: 10.1093/nar/gkx184. [Epub ahead of print]

Narayanan MS, Kushwaha M, Ersfeld K, Fullbrook A, Stanne TM, Rudenko G (2011) NLP is a novel transcription regulator involved in VSG expression site control in *Trypanosoma brucei*. Nucleic Acids Res 39(6):2018–2031. doi:10.1093/nar/gkq950

Narayanan MS, Rudenko G (2013) TDP1 is an HMG chromatin protein facilitating RNA polymerase I transcription in African trypanosomes. Nucleic Acids Res 41(5):2981–2992. doi:10.1093/nar/gks1469

Navarro M, Gull K (2001) A pol I transcriptional body associated with VSG mono-allelic expression in *Trypanosoma brucei*. Nature 414(6865):759–763. doi:10.1038/414759a

Nguyen TN, Muller LS, Park SH, Siegel TN, Gunzl A (2014) Promoter occupancy of the basal class I transcription factor A differs strongly between active and silent VSG expression sites in *Trypanosoma brucei*. Nucleic Acids Res 42(5):3164–3176. doi:10.1093/nar/gkt1301

Nguyen TN, Nguyen BN, Lee JH, Panigrahi AK, Gunzl A (2012) Characterization of a novel class I transcription factor A (CITFA) subunit that is indispensable for transcription by the multifunctional RNA polymerase I of *Trypanosoma brucei*. Eukaryot Cell 11 (12):1573–1581. doi:10.1128/EC.00250-12

Pandya UM, Sandhu R, Li B (2013) Silencing subtelomeric VSGs by *Trypanosoma brucei* RAP1 at the insect stage involves chromatin structure changes. Nucleic Acids Res 41 (16):7673–7682. doi:10.1093/nar/gkt562

Pascoalino B, Dindar G, Vieira-da-Rocha JP, Machado CR, Janzen CJ, Schenkman S (2014) Characterization of two different Asf1 histone chaperones with distinct cellular localizations and functions in *Trypanosoma brucei*. Nucleic Acids Res 42(5):2906–2918. doi:10.1093/nar/gkt1267

Pays E, Vanhollebeke B, Uzureau P, Lecordier L, Perez-Morga D (2014) The molecular arms race between African trypanosomes and humans. Nat Rev Microbiol 12(8):575–584. doi:10.1038/nrmicro3298

Pena AC, Pimentel MR, Manso H, Vaz-Drago R, Pinto-Neves D, Aresta-Branco F, Rijo-Ferreira F, Guegan F, Pedro Coelho L, Carmo-Fonseca M, Barbosa-Morais NL, Figueiredo LM (2014) *Trypanosoma brucei* histone H1 inhibits RNA polymerase I transcription and is important for parasite fitness in vivo. Mol Microbiol 93(4):645–663. doi:10.1111/mmi.12677

Povelones ML, Gluenz E, Dembek M, Gull K, Rudenko G (2012) Histone H1 plays a role in heterochromatin formation and VSG expression site silencing in *Trypanosoma brucei*. PLoS Pathog 8(11):e1003010. doi:10.1371/journal.ppat.1003010

Reynolds D, Cliffe L, Forstner KU, Hon CC, Siegel TN, Sabatini R (2014) Regulation of transcription termination by glucosylated hydroxymethyluracil, base J, in *Leishmania major* and *Trypanosoma brucei*. Nucleic Acids Res 42(15):9717–9729. doi:10.1093/nar/gku714

Reynolds D, Hofmeister BT, Cliffe L, Alabady M, Siegel TN, Schmitz RJ, Sabatini R (2016) Histone H3 variant regulates RNA polymerase II transcription termination and dual strand transcription of siRNA loci in *Trypanosoma brucei*. PLoS Genet 12(1):e1005758. doi:10.1371/journal.pgen.1005758

Robinson NP, Burman N, Melville SE, Barry JD (1999) Predominance of duplicative VSG gene conversion in antigenic variation in African trypanosomes. Mol Cell Biol 19(9):5839–5846

Ross R, Thomson D (1910) A case of sleeping sickness showing regular periodical increase of the parasites disclosed. Br Med J 1(2582):1544–1545

Salmon D, Vanwalleghem G, Morias Y, Denoeud J, Krumbholz C, Lhomme F, Bachmaier S, Kador M, Gossmann J, Dias FB, De Muylder G, Uzureau P et al (2012) Adenylate cyclases of *Trypanosoma brucei* inhibit the innate immune response of the host. Science 337 (6093):463–466. doi:10.1126/science.1222753

Schlimme W, Burri M, Bender K, Betschart B, Hecker H (1993) *Trypanosoma brucei brucei*: differences in the nuclear chromatin of bloodstream forms and procyclic culture forms. Parasitology 107(Pt 3):237–247

Schulz D, Mugnier MR, Paulsen EM, Kim HS, Chung CW, Tough DF, Rioja I, Prinjha RK, Papavasiliou FN, Debler EW (2015) Bromodomain proteins contribute to maintenance of bloodstream form stage identity in the African trypanosome. PLoS Biol 13(12):e1002316. doi:10.1371/journal.pbio.1002316

Schulz D, Zaringhalam M, Papavasiliou FN, Kim HS (2016) Base J and H3.V regulate transcriptional termination in Trypanosoma brucei. PLoS Genet 12(1):e1005762. doi:10.1371/journal.pgen.1005762

Siegel TN, Hekstra DR, Kemp LE, Figueiredo LM, Lowell JE, Fenyo D, Wang X, Dewell S, Cross GA (2009) Four histone variants mark the boundaries of polycistronic transcription units in *Trypanosoma brucei*. Genes Dev 23(9):1063–1076. doi:10.1101/gad.1790409

Siegel TN, Kawahara T, Degrasse JA, Janzen CJ, Horn D, Cross GA (2008) Acetylation of histone H4K4 is cell cycle regulated and mediated by HAT3 in *Trypanosoma brucei*. Mol Microbiol 67 (4):762–771. doi:10.1111/j.1365-2958.2007.06079.x

Stanne TM, Kushwaha M, Wand M, Taylor JE, Rudenko G (2011) TbISWI regulates multiple polymerase I (Pol I)-transcribed loci and is present at Pol II transcription boundaries in *Trypanosoma brucei*. Eukaryot Cell 10(7):964–976. doi:10.1128/EC.05048-11

Stanne T, Narayanan MS, Ridewood S, Ling A, Witmer K, Kushwaha M, Wiesler S, Wickstead B, Wood J, Rudenko G (2015) Identification of the ISWI chromatin remodeling complex of the early branching eukaryote *Trypanosoma brucei*. J Biol Chem. doi:10.1074/jbc.M115.679019

Stanne TM, Rudenko G (2010) Active VSG expression sites in *Trypanosoma brucei* are depleted of nucleosomes. Eukaryot Cell 9(1):136–147. doi:10.1128/EC.00281-09

Steverding D, Stierhof YD, Chaudhri M, Ligtenberg M, Schell D, Beck-Sickinger AG, Overath P (1994) ESAG 6 and 7 products of *Trypanosoma brucei* form a transferrin binding protein complex. Eur J Cell Biol 64(1):78–87

Thatcher TH, Gorovsky MA (1994) Phylogenetic analysis of the core histones H2A, H2B, H3, and H4. Nucleic Acids Res 22(2):174–179

Tiengwe C, Marcello L, Farr H, Dickens N, Kelly S, Swiderski M, Vaughan D, Gull K, Barry JD, Bell SD, McCulloch R (2012) Genome-wide analysis reveals extensive functional interaction between DNA replication initiation and transcription in the genome of *Trypanosoma brucei*. Cell Rep 2(1):185–197. doi:10.1016/j.celrep.2012.06.007

Turner CM (1997) The rate of antigenic variation in fly-transmitted and syringe-passaged infections of *Trypanosoma brucei*. FEMS Microbiol Lett 153(1):227–231. doi:S0378-1097(97)00266-8

Ullu E, Tschudi C (1991) Trans splicing in trypanosomes requires methylation of the 5′ end of the spliced leader RNA. Proc Natl Acad Sci U S A 88(22):10074–10078

van Leeuwen F, Taylor MC, Mondragon A, Moreau H, Gibson W, Kieft R, Borst P (1998) Beta-D-glucosyl-hydroxymethyluracil is a conserved DNA modification in kinetoplastid protozoans and is abundant in their telomeres. Proc Natl Acad Sci U S A 95(5):2366–2371

van Leeuwen F, Wijsman ER, Kieft R, van der Marel GA, van Boom JH, Borst P (1997) Localization of the modified base J in telomeric VSG gene expression sites of *Trypanosoma brucei*. Genes Dev 11(23):3232–3241

van Luenen HG, Farris C, Jan S, Genest PA, Tripathi P, Velds A, Kerkhoven RM, Nieuwland M, Haydock A, Ramasamy G, Vainio S, Heidebrecht T et al (2012) Glucosylated hydroxymethyluracil, DNA base J, prevents transcriptional readthrough in Leishmania. Cell 150(5):909–921. doi:10.1016/j.cell.2012.07.030

Vanhamme L, Poelvoorde P, Pays A, Tebabi P, Van Xong H, Pays E (2000) Differential RNA elongation controls the variant surface glycoprotein gene expression sites of *Trypanosoma brucei*. Mol Microbiol 36(2):328–340

Wang QP, Kawahara T, Horn D (2010) Histone deacetylases play distinct roles in telomeric VSG expression site silencing in African trypanosomes. Mol Microbiol 77(5):1237–1245. doi:10.1111/j.1365-2958.2010.07284.x

WHO (2015) Investing to overcome the global impact of neglected tropical diseases: third WHO report on neglected tropical diseases. Switzerland, Geneva

Wickstead B, Ersfeld K, Gull K (2004) The small chromosomes of *Trypanosoma brucei* involved in antigenic variation are constructed around repetitive palindromes. Genome Res 14(6):1014–1024. doi:10.1101/gr.2227704

Williams RO, Young JR, Majiwa PA (1982) Genomic environment of *T. brucei* VSG genes: presence of a minichromosome. Nature 299(5882):417–421

Wright JR, Siegel TN, Cross GA (2010) Histone H3 trimethylated at lysine 4 is enriched at probable transcription start sites in *Trypanosoma brucei*. Mol Biochem Parasitol 172(2):141–144. doi:10.1016/j.molbiopara.2010.03.013

Yang X, Figueiredo LM, Espinal A, Okubo E, Li B (2009) RAP1 is essential for silencing telomeric variant surface glycoprotein genes in *Trypanosoma brucei*. Cell 137(1):99–109. doi:10.1016/j.cell.2009.01.037

Chapter 11
Epigenetics of Malaria Parasites

Evelien M. Bunnik and Karine G. Le Roch

Abstract Malaria parasites of the *Plasmodium* spp. have complex life cycles that include multiple stages in a vertebrate host and a mosquito vector, requiring strong regulation of their transcriptional programs for life cycle progression and survival. Epigenetic mechanisms controlling gene expression have been shown to be important for parasite proliferation during the intra-erythrocytic developmental cycle, and evidence is emerging for a role of such mechanisms in other life cycle stages. In particular, parasite-specific genes involved in pathogenesis, evasion of host immune responses, invasion of host cells, and life cycle progression are regulated at the epigenetic level. In this chapter, the various processes involved in the regulation of these parasite-specific genes and gene families will be discussed. In addition, the changes observed at the level of local chromatin structure and global nuclear organization will be described. Finally, the promise of epigenetic regulators as targets of novel antimalarial drugs will be addressed. Collectively, these topics provide insight into the unique biology of *Plasmodium* spp. and highlight the areas where additional research is necessary for an increased understanding of gene regulation in this deadly parasite.

Keywords Malaria • Chromosomes • Chromatin • Epigenetics • Genome organization • Gene expression regulation

E.M. Bunnik
Institute for Integrative Genome Biology, Center for Disease Vector Research, Department of Cell Biology and Neuroscience, University of California Riverside, 900 University Avenue, Riverside, CA 92521, USA

Department of Microbiology, Immunology and Molecular Genetics, The University of Texas Health Science Center at San Antonio, 7703 Floyd Curl Drive, San Antonio, TX 78229, USA

K.G. Le Roch (✉)
Institute for Integrative Genome Biology, Center for Disease Vector Research, Department of Cell Biology and Neuroscience, University of California Riverside, 900 University Avenue, Riverside, CA 92521, USA
e-mail: karine.leroch@ucr.edu

W. Doerfler, J. Casadesús (eds.), *Epigenetics of Infectious Diseases*, Epigenetics and Human Health, DOI 10.1007/978-3-319-55021-3_11

11.1 The Malaria Parasite Life Cycle

Malaria is still one of the most deadly infectious diseases in the world, responsible for an estimated 438,000 deaths and 214 million infections in 2014 (WHO 2015). Malaria is most prevalent in sub-Saharan Africa, although large regions of South-East Asia and South America are also affected. Children under the age of five are most likely to experience severe malaria and make up about 70% of the total number of malaria deaths.

In humans, five different parasite species of the *Plasmodium* genus can cause malaria, of which *P. falciparum* is the most prevalent and the most deadly variant. *P. vivax* is less virulent than *P. falciparum*, but can remain in the body in a dormant form that can cause relapses weeks or months after the initial infection. All *Plasmodium* species have similar, complex life cycles that involve alternating infections of the human host and the *Anopheles* mosquito vector (Fig. 11.1). A human infection starts when an infected female mosquito takes a blood meal and injects sporozoites present in her salivary glands into the skin. Within 20 min, these sporozoites travel to the liver and enter hepatocytes. Over a period of 7–10 days, each parasite replicates extensively and releases thousands of merozoites into the bloodstream. These merozoites then infect red blood cells to start the first of multiple subsequent replication cycles.

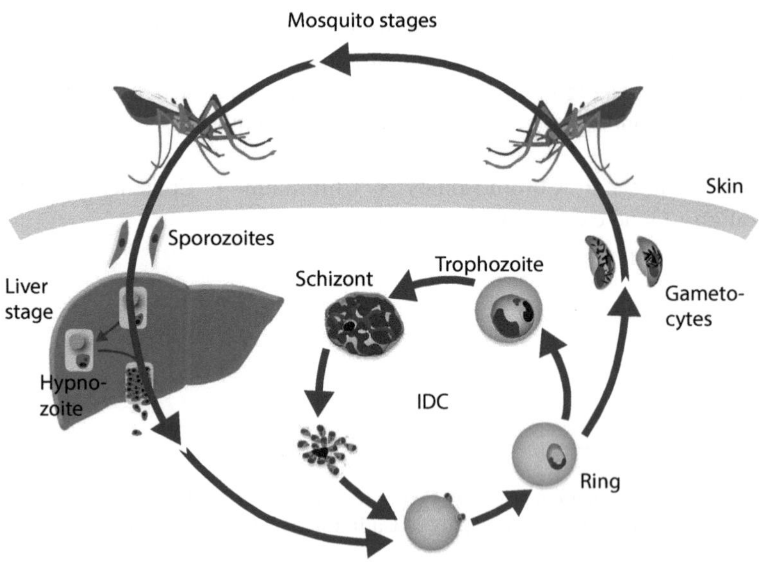

Fig. 11.1 Simplified overview of the *Plasmodium* life cycle. Sporozoites enter humans through the bite of an infected mosquito and subsequently establish liver stage infection. Certain *Plasmodium* species, such as *P. vivax*, can give rise to hypnozoites in the liver, a dormant form of the parasite that can be reactivated months to years later. Upon egress from the liver, the parasite replicates asexually in erythrocytes (intra-erythrocytic developmental cycle, IDC). During the IDC, parasites can commit to gametocyte differentiation, giving rise to a male and female form that can be taken up by a mosquito and reproduce sexually in the mosquito midgut. The parasite then develops through several additional stages to form sporozoites that can be injected into a new human host

During this intra-erythrocytic developmental cycle (IDC), the parasite progresses through three distinct stages, termed ring, trophozoite, and schizont. The ring stage is characterized by remodeling of the host cell to establish the supply of hemoglobin as an energy source, as well as to ensure evasion of the human immune system. During the trophozoite stage, the parasite becomes highly transcriptionally and metabolically active, in preparation for cell division. Finally, the parasite multiplies into 16–32 daughter parasites using a process of asexual replication called schizogony. During schizogony, the nucleus undergoes multiple rounds of division, which is followed by cytokinesis to subdivide the multinucleated parasite into identical daughter cells. Approximately 48 h after invasion of the red blood cell, these daughter parasites burst out of the host cell, ready to invade new red blood cells. During the IDC, environmental stress, such as low nutrient levels, can induce sexual differentiation of parasites into male and female gametocytes. When a mosquito ingests these mature gametocytes, they undergo sexual replication in the mosquito midgut and further develop into the salivary gland sporozoites that can be transmitted to a new human host.

The IDC is the stage responsible for disease in humans and can be maintained in long-term in vitro culture for *P. falciparum*. For these reasons, this parasite species and this stage of the parasite life cycle have been most extensively studied. Consequently, most of our knowledge concerning epigenetic regulation of gene expression in the parasite is restricted to the *P. falciparum* IDC, while little is known about epigenetic profiles in sporozoites, the liver stage, the mosquito stages, or in other *Plasmodium* species. The topics discussed below will therefore predominantly address the IDC of *P. falciparum*, unless mentioned otherwise.

11.2 Transcriptional Regulation in Plasmodium

To understand the role of epigenetics mechanisms of gene regulation, it is important to first explore the transcriptional landscape in *Plasmodium*. In all eukaryotes, transcription is initiated by the assembly of RNA polymerase II and the general transcription factors into the pre-initiation complex (PIC). Recruitment and activation of the PIC is regulated by specific transcription factors (TFs). Only one major family of ~27 specific TFs has been identified in *Plasmodium*, which contain a modified version of the AP2 domain found in plant TFs and are called ApiAP2 (Balaji et al. 2005). Some of these ApiAP2 TFs are thought to control the transitions between the different parasite life stages, such as AP2-G for gametocyte development (Kafsack et al. 2014; Sinha et al. 2014), AP2-O for ookinete development in the mosquito (Yuda et al. 2009), AP2-Sp for sporozoite formation (Yuda et al. 2010), and AP2-L for liver-stage development (Iwanaga et al. 2012). For others, their role in transcription is less clear. Relative to the size of the *Plasmodium* genome, the number of TFs that has been identified is considerably smaller than in most other eukaryotes. Considering the many different forms that the parasite adopts in widely different host environments, it is thought that this small number of TFs is not sufficient to initiate the corresponding differential transcriptional programs.

It has now been established that the majority of the *P. falciparum* genome is constitutively maintained in a euchromatic state, while only the subtelomeric regions and several internal loci are marked as heterochromatin, characterized by the presence of H3K9me3 and Heterochromatin Protein 1 (PfHP1). These heterochromatin regions contain gene families with clonally variant expression patterns that are uniquely dependent on epigenetic mechanisms of transcriptional regulation and are the only genes that are subject to "classical" epigenetic control (Rovira-Graells et al. 2012). Under nonselective growth conditions, a population of genetically identical parasites shows variation in expression levels of these clonally variant genes, resulting in phenotypic diversity. Changes in environmental conditions can select for preexisting parasites with transcriptional profiles that are favorable under those conditions (Rovira-Graells et al. 2012). Instead of responding to environmental changes by transcriptional reprogramming, *P. falciparum* thus seems to mainly use an epigenetic approach to generate phenotypic heterogeneity to increase its chances of survival.

In addition to the epigenetic regulation of clonally variant gene families, other features of chromatin structure also contribute to gene regulation. While these are perhaps not epigenetic mechanisms in the strictest definition in that they can transmit different gene expression profiles to genetically identical progeny, they do contribute to gene regulation of both clonally variant genes and other parasite genes and will therefore be discussed as well.

11.3 Epigenetic Regulation of Virulence Genes

11.3.1 P. falciparum *Virulence Genes*

The most extensively studied clonally variant gene family in *P. falciparum* is the family of *var* genes, which encode variants of Erythrocyte Membrane Protein 1 (PfEMP1). These proteins are expressed on the surface of infected red blood cells and mediate adherence of the cell to the microvasculature, preventing clearance by the spleen and contributing to severe disease by obstructing blood flow in major organs, such as the brain (Miller et al. 1994). Each *P. falciparum* parasite carries approximately 60 different *var* genes, only one of which is expressed at any time (Scherf et al. 2008). Switching *var* gene expression allows the parasite to escape from host immune responses in a process called antigenic variation. All chromosomes contain clusters of *var* genes in the subtelomeric regions, and several chromosomes harbor clusters of *var* genes internally. The subtelomeric regions contain several other variant surface antigen gene families, including *rifin*, *stevor*, *pfmc-2tm*, and *phist*, which are not expressed in the same strictly clonal fashion as the *var* genes are, but that are largely regulated by the same mechanisms.

11.3.2 Histone Posttranslational Modifications

Among the several layers of epigenetic control that are involved in silencing of *var* genes, histone posttranslational modifications (PTMs) are extremely important (Fig. 11.2). Silent *var* genes are localized in one or multiple clusters of H3K9me3/PfHP1-marked heterochromatin at the nuclear periphery (Freitas-Junior et al. 2000, 2005; Crowley et al. 2011; Chookajorn et al. 2007; Lopez-Rubio et al. 2009; Salcedo-Amaya et al. 2009; Perez-Toledo et al. 2009; Flueck et al. 2009; Ay et al. 2014). The repressive environment

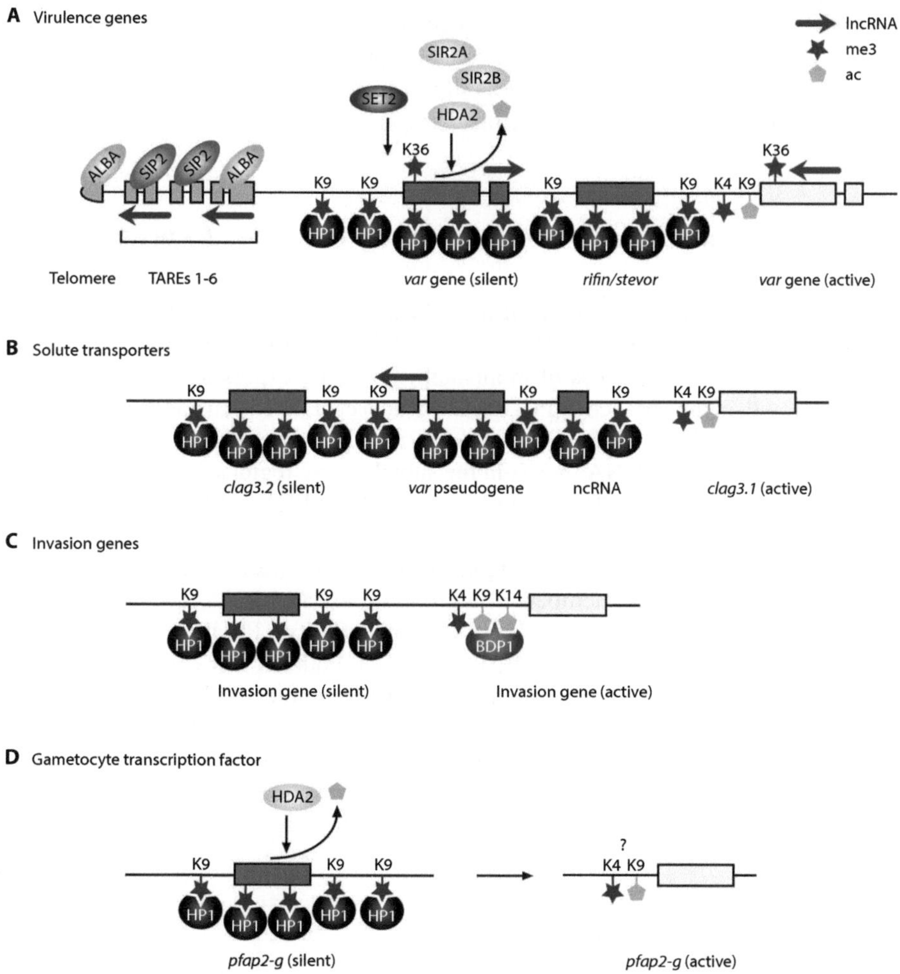

Fig. 11.2 *P. falciparum* genes regulated through epigenetic mechanisms. (**a**) Virulence gene families, including *var*, *rifin*, and *stevor* that are predominantly located in subtelomeric regions, although the same mechanisms also regulate internally located *var* genes. (**b**) The genes *clag3.1* and *clag3.2* that encode transporters of ion, nutrients, and other solutes. (**c**) Invasion genes. (**d**) The transcription factor that is thought to drive gametocyte differentiation

is maintained by multiple histone deacetylases (HDAC), including the NAD^+-dependent class III HDAC silent information regulator 2 proteins PfSIR2A and PfSIR2B, as well as the class II HDAC PfHDA2 (Duraisingh et al. 2005; Tonkin et al. 2009; Lopez-Rubio et al. 2009; Coleman et al. 2014). Removal of these proteins from the parasite results in loss of monoallelic *var* gene expression. *Var* gene silencing, and to a lesser extent *rifin* and *stevor* gene expression, is also influenced by H3K36me3. *P. falciparum* encodes histone lysine methyltransferase (HKMT) PfSET2 (also called PfSETvs for variant silencing) that specifically marks the virulence genes and does not act in other regions of the genome (Jiang et al. 2013; Ukaegbu et al. 2014). While both the active and silent *var* genes carry H3K36me3, overexpression of a dominant negative mutant of PfSET2 resulted in a higher *var* gene switching rate (Ukaegbu et al. 2014), while mutually exclusive *var* gene expression is completely disrupted in the absence of PfSET2 (Jiang et al. 2013). These results suggest that H3K36me3 is involved in both maintenance of monoallelic expression and epigenetic memory.

11.3.3 Proteins Involved in Regulating var *Gene Expression*

Many different proteins are involved in maintaining the repressive heterochromatin that harbors the silent *var* genes (Fig. 11.2). PfHP1 is an essential structural component, demonstrated by the loss of monoallelic *var* gene expression and an arrest in parasite growth in the absence of PfHP1 (Brancucci et al. 2014). In addition, one of the members of the ApiAP2 TF family, *P. falciparum* SPE2-interacting protein (PfSIP2), binds to arrays of SPE2 DNA motifs in the subtelomeric regions upstream of the *var* genes and colocalizes with PfHP1 (Flueck et al. 2010). SPE2 motifs are also present in the telomere-associated repeat elements (TARE), suggesting that PfSIP2 is also involved in the maintenance and organization of chromosome ends. The TARE regions are also bound by members of a group of archaeal DNA/RNA binding proteins, termed Alba for "Acetylation lowers binding affinity." PfAlba1, 2, and 4 bind to TARE6 (Chene et al. 2012), while PfAlba3 binds to both telomeric and subtelomeric regions and is additionally found in *var* gene promoters (Goyal et al. 2012). Deacetylation of lysine residues in the N-terminus of PfAlba3 by PfSIR2A is thought to increase its affinity for DNA binding (Goyal et al. 2012). In addition to the proteins mentioned above, many other factors are likely to be involved in the recruitment and organization of subtelomeric heterochromatin and are the subject of active research.

11.3.4 var *Gene Switching*

The single active *var* gene is physically separated from these repressive heterochromatin regions and is enriched for H3K4me3 and H3K9ac in its 5′ flanking region (Duraisingh et al. 2005; Lopez-Rubio et al. 2007, 2009) (Fig. 11.2). This single active *var* gene is only actively transcribed during the ring stage. In trophozoite and schizont stages, H3K4 methylation in the active *var* gene is controlled by the HKMT PfSET10, allowing re-activation of the same gene in the next replication cycle (Volz et al. 2012). In addition, the histone acetyltransferase (HAT) is recruited to the active *var* gene and acetylates H4 at multiple residues (Miao et al. 2010). During *var* gene switching, the activated *var* gene is repositioned from the heterochromatin compartment to a transcriptionally permissive environment in an actin-dependent mechanism that requires the binding of an actin protein complex to an 18-bp nuclear protein-binding element in the *var* gene intron (Zhang et al. 2011). In the absence of immune pressure, approximately 2% of parasites will spontaneously switch *var* gene activation with every replication cycle (Roberts et al. 1992). The switch rate is thought to be inherently low to ensure that an infected host will not develop immunity to a wide variety of surface antigens, which would reduce the chances of parasite survival and the ability to reinfect the same host. It has been shown that different *var* genes have different intrinsic switch rates (Horrocks et al. 2004) and that *var* genes that are located on internal chromosomal loci are more likely to be activated than subtelomeric *var* genes (Frank et al. 2007; Noble et al. 2013). In addition, *var* gene expression is more likely to switch towards highly divergent *var* genes than to genes that are relatively conserved among parasite strains (Noble et al. 2013). This switching hierarchy may have an evolutionary advantage by selecting for surface antigens that are more likely to evade existing host immune responses. However, the molecular basis for *var* gene switching and its inherent patterns are still largely unknown.

11.3.5 Long Noncoding RNAs

Another distinct mechanism controlling *var* gene expression is the transcription of two types of long noncoding RNAs (lncRNAs) from the *var* genes (Epp et al. 2009; Amit-Avraham et al. 2015) (Fig. 11.2). Both lncRNAs are transcribed from a bidirectional promoter in the *var* intron, are capped, are not polyadenylated, and are incorporated into chromatin. During the late trophozoite and schizont stage, most or all *var* genes transcribe a transcript in the sense direction that spans the entire second exon. Transcription of an antisense transcript that extends into the first exon is restricted to the active *var* genes during the ring stage. Transcription of the sense lncRNA by RNA polymerase II may play a role in depositing H3K36me3 at *var* gene loci by PfSET2 (Ukaegbu et al. 2014). This lncRNA may thus be required to mark *var* genes as members of a clonally variant gene family to recruit other epigenetic factors necessary for the maintenance of monoallelic expression. In contrast, the antisense transcript seems to be involved in regulating the active *var* gene, since expression of the antisense transcript can trigger the activation of the corresponding silent gene and downregulation of the antisense transcript can deactivate the corresponding active *var* gene (Amit-Avraham et al. 2015). Histone PTMs, proteins that organize the perinuclear heterochromatin compartments, and lncRNAs together thus create a complex regulatory network that maintains epigenetic memory of *var* gene expression and silencing and allows *var* gene switching.

11.4 Epigenetic Regulation of Other Clonally Variant Genes

11.4.1 Solute Transporters

To be able to transport ions, nutrients, and other solutes across the erythrocyte membrane, *P. falciparum* expresses a plasmodial surface anion channel (PSAC) on the surface of the infected red blood cell (Nguitragool et al. 2011; Pillai et al. 2012). These PSACs are encoded by two highly homologous genes, *clag3.1* and *clag3.2*, which are located approximately 10 kb from each other in a subtelomeric region of the parasite genome. The *clag* genes are expressed at the schizont stage, but the proteins are not expressed at the erythrocyte surface until the ring stage of the next cycle (Nguitragool et al. 2011). Until their essential function in solute transport was discovered, these genes were thought to play a role in cytoadherence and invasion of host cells, and it has not been completely ruled out that they may have a dual biological function.

Together with the *var* genes, these are the only gene families in *P. falciparum* that are expressed in a strict mutually exclusive fashion (Cortes et al. 2007). The silenced gene is marked by H3K9me3, while the active gene is enriched in H3K9ac and H3K4me3, and has an open nucleosome organization (Crowley et al. 2011; Comeaux et al. 2011) (Fig. 11.2). These epigenetics marks are heritable, and the activation or repression status of each gene is consequently transmitted to the next parasite generation (Cortes et al. 2007; Comeaux et al. 2011). In addition, a lncRNA of over 1 kb in length is transcribed in the schizont stage from a *var* pseudogene located on the opposite strand of the region between the two *clag3* genes (Otto et al. 2010; Siegel et al. 2014). Expression of this lncRNA correlates with expression of *clag3.1*, indicating that multiple layers of transcriptional regulation are involved in control of the *clag* genes, similar to the *var* genes.

Although mutually exclusive expression of these genes is highly favored, both genes can be silenced or activated simultaneously under conditions of extreme selective pressure. Several water-soluble antimalarial drugs are dependent upon PSACs to cross the erythrocyte membrane. *In vitro* cultures of *P. falciparum* in the presence of such drugs were shown to decrease the expression of both *clag3* genes, resulting in decreased intracellular drug concentrations and a drug-resistant phenotype, in combination with a reduced growth rate (Sharma et al. 2013; Mira-Martinez et al. 2013). In addition, in vitro selective pressure can also result in activation of both genes (Rovira-Graells et al. 2015). While these experiments were performed in vitro under somewhat artificial conditions, these results point to a potential new mechanism of drug resistance and are therefore important to further explore.

It is not well understood why the parasite encodes two different solute transporters and why these are expressed in a mutually exclusive way. The transporters show functional differences (Mira-Martinez et al. 2013), suggesting that certain growth conditions may favor one transporter over the other. In addition, it has been postulated that mutually exclusive expression of these surface proteins may enable the parasite to escape from immune responses directed against the transporter that is being expressed (Cortes 2008). However, these proteins are not thought to be a major target for antibody responses, and it is thus unclear whether this is a plausible explanation.

11.4.2 Invasion Genes

Invasion of a new red blood cell is mediated by the binding of parasite ligands to specific receptors on the erythrocyte surface (Cowman and Crabb 2006). Several multigene families are involved in the invasion process, including *eba*, *rhoph1/clag*, *acbp*, and *pfRH*. Since many members of these families are not essential for parasite growth, the function of individual proteins is not entirely clear, and there seems to be considerable redundancy in host–ligand interactions among these proteins. These genes can be either active or silent, giving rise to differential expression profiles in related parasite lines, which are thought to be partially regulated through epigenetic mechanisms (Cortes et al. 2007).

For two closely related parasite strains with different capacities for erythrocyte invasion, it was shown that activation of *pfRH4* conferred the ability to invade red blood cells that do not express sialic acid as a receptor (Jiang et al. 2010). Activation of *pfRH4* was associated with increased acetylation levels of histones H3 and H4, as well as higher levels of H3K4me3, while the 5′ untranslated region of the silent genes was enriched in H3K9me3 (Fig. 11.2). In addition, the active *pfRH4* gene showed lower occupancy of the +1 nucleosome at the time of gene transcription. Recently, a bromodomain protein, PfBDP1, was shown to bind to acetylated H3 (H3K9ac and H3K14ac) and to positively regulate transcription of invasion genes, among others (Josling et al. 2015). The epigenetic mechanisms involved in expression of invasion genes are thus comparable to the regulatory network used to control expression of the *var* and *clag* gene families, as described above. However, little is known about how individual variant invasion genes are regulated and which environmental clues may influence the expression of different subsets of invasion genes.

11.5 Epigenetic Regulation of Gametocytogenesis

During the IDC, most parasites will continue to replicate asexually, but with every replication cycle, a small percentage of parasites will commit to gametocyte differentiation. This commitment is thought to be made during the schizont stage, followed by egress and reinvasion of daughter parasites that will all continue on the path towards gametocyte formation. It is not very well understood what triggers this differentiation process. The activation of the gametocyte differentiation pathway seems to be mostly stochastic, but can also be prompted by certain environmental conditions, such as the presence of antimalarial drugs or nutrient starvation.

One of the master regulators of gametocyte differentiation is an AP2 transcription factor located on chromosome 12 in *P. falciparum* (*pfap2-g*) (Kafsack et al. 2014; Sinha et al. 2014). In asexual parasites, the locus encoding this TF is epigenetically silenced by H3K9me3, bound by PfHP1, and localized at the nuclear periphery (Lopez-Rubio et al. 2009; Flueck et al. 2009), similar to the *var* gene family (Fig. 11.2). It is unknown what triggers activation of the *pfap2-g* locus in vivo, but in vitro experiments have shown that downregulation of HDAC PfHDA2 results in activation of *pfap2-g* and an induction of gametocyte formation (Coleman et al. 2014). This HDAC is present in perinuclear foci and presumably functions by removing acetyl groups on histones, which makes the chromatin accessible to H3K9me3 deposition and PfHP1 binding, resulting in heterochromatin formation. PfHDA2 also acts as a silencer of *var* genes, and a "leaky" repression of *var* genes and the *pfap2-g* locus may result in both low-frequency *var* gene switching and lowrate gametocyte conversion (Coleman et al. 2014). In a similar fashion, depletion of PfHP1 from the parasite leads to activation of PfAP2-G and an increased rate of gametocyte differentiation, although this process is most likely to be downstream from PfHDA2 regulation (Brancucci et al. 2014).

Upon activation, PfAP2-G promotes the expression of early gametocyte genes that drive sexual differentiation and generates a positive feedback loop that ensures high transcriptional activity of the *pfap2-g* locus (Kafsack et al. 2014). In a rodent malaria parasite, *P. berghei*, a second AP2 TF (PfAP2-G2) assists in gametocyte formation by repressing approximately one-third of the genes that are expressed during the IDC (Yuda et al. 2015). This TF acts at an early time point during differentiation, before male and female gametocyte characteristics are noticeable. A homolog of this gene is present in *P. falciparum*, but its role in gametocyte differentiation awaits confirmation. Together, these two TFs are likely to be responsible for transcriptionally reprogramming the parasite gene expression network, leading to differentiation of the parasite into the stage that is transmissible to mosquitoes. Other genes have been implicated in gametocytogenesis, including PfGDV1 and early gametocyte-specific genes (Ikadai et al. 2013; Eksi et al. 2012; Silvestrini et al. 2010), but it remains to be determined where these genes reside within the transcriptional network that regulates the induction and progression of gametocyte differentiation.

11.6 Local Chromatin Structure

Several features of the local chromatin structure in *P. falciparum* are distinct from how chromatin is organized in other eukaryotes. Heterochromatin in *P. falciparum* is essentially limited to regions of the genome that harbor the virulence gene families, the *pfag2-g* locus, and several other genes that are repressed during the IDC, as described in more detail in the preceding sections. These are the only regions in the genome that are enriched for the repressive marks H3K9me3, H3K36me3, and H4K20me3 (Lopez-Rubio et al. 2009; Jiang et al. 2013). In contrast, the majority of the genome is present in a euchromatin compartment with an abundance of histone acetylation (Lopez-Rubio et al. 2009; Miao et al. 2006), as is also observed for other unicellular organisms (Garcia et al. 2007). Intergenic regions are marked by activating histone modifications H3K9ac and H3K4me3 (Bartfai et al. 2010; Salcedo-Amaya et al. 2009; Trelle et al. 2009), which are typically only found in active promoters in mammalian genomes (Barski et al. 2007; Bernstein et al. 2005; Kim et al. 2005; Nishida et al. 2006; Wang et al. 2008) and thus have a much broader distribution in *P falciparum*. In addition, H3K9ac is enriched in the promoter and 5′ coding regions of highly transcribed genes in *P. falciparum* (Cui et al. 2007b; Bartfai et al. 2010). Thus, the majority of the *P. falciparum* genome seems to be permanently in a transcriptionally permissive state, while only a subset of genes is strictly repressed.

A second interesting feature of local chromatin structure in *P. falciparum* is the localization of histone variants. Histone variant H2A.Z is highly enriched in +1 and −1 nucleosomes on either site of the promoter region in other eukaryotes (Guillemette et al. 2005; Raisner et al. 2005; Tolstorukov et al. 2009), where they promote assembly of transcription machinery. Instead, the entire intergenic regions in *P. falciparum* are lined by nucleosomes that contain both H2A.Z and H2B.Z (Bartfai et al. 2010; Hoeijmakers et al. 2013; Petter et al. 2013). It is thought that these histone variants have evolved a specialized function to facilitate nucleosome deposition in the highly AT-rich intergenic regions of *P. falciparum*. Of all eukaryotic organisms that have been sequenced to date, *P. falciparum* has the highest AT content, with an average of 80.7% AT in the entire genome, but reaching 90–95% AT in intergenic regions (Gardner et al. 2002). AT-rich DNA is thought to be more rigid and therefore less prone to wrap around the histone core and form a stable nucleosome (Segal and Widom 2009; Tillo and Hughes 2009), necessitating a specialized histone variant to enable nucleosome positioning in the intergenic regions of *P. falciparum*.

Finally, the nucleosome landscape around *P. falciparum* genes is also divergent from that in other eukaryotes. Eukaryotes generally have strongly positioned −1 and +1 nucleosomes bordering the promoter, followed by an array of nucleosomes with increasing variation in their positioning (called "fuzziness") until the 3′ end of the gene. Such nucleosomal arrays have not been observed in *P. falciparum* (Bunnik et al. 2014; Kensche et al. 2015; Ponts et al. 2010), which could be due to the high AT content of its genome that hampers the positioning of nucleosomes at fixed intervals. However, *Tetrahymena thermophila* and *Dictyostelium discoideum*, both with an average AT content of 78%, do show an array of nucleosomes downstream of the transcription start site (TSS) (Beh et al. 2015; Chang et al. 2012), suggesting that a high AT content alone may not fully explain this phenomenon in *P. falciparum*. It is possible that chromatin remodelers function differently in establishing nucleosome positioning in *P. falciparum* or that the signal from nucleosome arrays with different intervals in various subsets of genes is averaged out in metagenomic analyses.

The *P. falciparum* genome shows characteristic nucleosome features around the TSS, the start and end of the coding region, the transcription termination site, and at splice donor and acceptor sites (Bunnik et al. 2014; Kensche et al. 2015; Ponts et al. 2010). As in other eukaryotic organisms, the TSS is preceded by a nucleosome-depleted region that is assumed to allow binding of transcription factors and recruitment of the transcription machinery. In accordance, genes with higher mRNA transcription levels show a more open nucleosome organization in their TSSs (Bunnik et al. 2014; Kensche et al. 2015). It remains to be determined whether this lower nucleosome occupancy in the TSS drives transcriptional activity, or whether it is a consequence of the presence of RNA polymerase II and other transcription-associated proteins.

Two aspects of the *P. falciparum* nucleosome landscape are still the subject of debate. First, the nucleosome level was reported to be lower in intergenic regions than in coding regions (Bunnik et al. 2014; Ponts et al. 2010, 2011 Westenberger et al. 2009). Others dispute this finding and attribute it to experimental and sequencing biases introduced by the high AT content in these regions (Kensche et al. 2015). However, a lower density of nucleosomes is also observed in intergenic regions of other eukaryotes (Pokholok et al. 2005; Valouev et al. 2008; Chang et al. 2012; Wu et al. 2014), including the AT-rich ciliate *Tetrahymena thermophile* (Beh et al. 2015). In addition, it has been shown that certain DNA elements, in particular poly(dA:dT) tracts and AT repeats that are both highly abundant in the intergenic regions of the *P. falciparum* genome, strongly disfavor nucleosome binding (Anderson and Widom 2001; Field et al. 2008; Peckham et al. 2007; Raveh-Sadka et al. 2012). Recent accurate in vitro and in vivo nucleosome positioning study in AT- and GC-rich eukaryotic genomes confirms that AT-rich regions are unfavorable to nucleosome formation and validates that the variation in GC content may guide the establishment of nucleosome patterns within coding regions of eukaryotic genomes (Beh et al. 2015). Nucleosome sequence preferences may act in concert with epigenetic marks to establish the distinctive nucleosome pattern observed in genes and contribute to gene regulation (Beh et al. 2015). While *P. falciparum* seems to have evolved the specialized histone variants H2A.Z and H2B.Z for nucleosome binding to highly AT-rich DNA sequences, these nucleosomes are likely to be less tightly bound to the rigid AT-rich intergenic regions, resulting in a more open nucleosome conformation.

A second controversial finding is that nucleosome levels change as the parasite progresses through the IDC. Various experimental methodologies, including western blotting (Le Roch et al. 2004), mass spectrometry (Oehring et al. 2012; Bunnik et al. 2014), MNase-Seq and its complementary approach FAIRE-Seq (Ponts et al. 2010), as well as ChIP-Seq (Bunnik et al. 2014), showed that histone levels are lower during the transcriptionally most active trophozoite stage. This has led to a model in which nucleosome eviction drives the efficient transcription of 70–80% of the *P. falciparum* genome at the trophozoite stage, followed by repackaging of the genome at the schizont stage in preparation for egress and reinvasion. This model provides an explanation for the limited need for specific TFs, since the majority of the genome is transcribed simultaneously, and places more importance on post-transcriptional mechanisms of gene regulation (Balu et al. 2011; Bunnik et al. 2013; Eshar et al. 2015; Vembar et al. 2015).

11.7 Nuclear Architecture

Chromosomes are long-chain molecules that due to their size are unable to freely mingle in the nucleus. They are therefore arranged in a nonrandom fashion into distinct chromosome territories, with little intermingling (Rosa and Everaers 2008). The conformation of chromosomes and their organization within the nucleus play important roles in regulating gene expression [reviewed in (Dekker et al. 2013; Gibcus and Dekker 2013)]. The nucleus is divided into compartments that separate the chromatin into different transcriptional environments. In addition, looping of chromosomes can bring together distal loci, such as enhancer and promoter elements, and give rise to interchromosomal interactions that influence gene expression.

In *P. falciparum*, this compartmentalization of the nucleus is best exemplified by the localization of clonally variant gene families, such as subtelomeric and internally located *var* genes, into one or multiple heterochromatin clusters at the nuclear periphery (Freitas-Junior et al. 2000; Lopez-Rubio et al. 2009; Ay et al. 2014), while the center of the nucleus contains a large euchromatin compartment. These observations were initially made using fluorescent in situ hybridization experiments and were recently confirmed using an advanced next-generation sequencing application, called Hi-C (Ay et al. 2014; Lemieux et al. 2013). The Hi-C technology generates a genome-wide map of millions of intra- and interchromosomal interactions, comparable to a FISH experiment with probes that highlight all or most genomic loci. Hi-C experiments have been performed on the single cell level in a mammalian genome (Nagano et al. 2013), but thus far only population level interaction maps are available for *P. falciparum*. Nevertheless, these results have greatly increased our understanding of nuclear organization in the parasite.

P. falciparum chromosomes are positioned in a folded conformation, in which the chromosomes are anchored at the centromere and the two arms of the chromosome are folded back over each other in a parallel arrangement (Ay et al. 2014). The clustering of subtelomeric and internal *var* genes adds structural complexity to this organization, by introducing loops in chromosomes with internal *var* genes that are not observed in chromosomes that do not harbor internally located *var* genes (Ay et al. 2014). As a result, the *P. falciparum* genome has a more complex arrangement than similarly sized organisms, such as budding and fission yeast (Ay et al. 2014; Duan et al. 2010; Tanizawa et al. 2010). Other hallmarks of nuclear organization include the colocalization of centromeres at the nuclear periphery (Ay et al. 2014; Hoeijmakers et al. 2012), colocalization of active rDNA loci in the ring stage (Ay et al. 2014; Lemieux et al. 2013), and a strict organization of chromosome territories in the ring and schizont stages (Ay et al. 2014).

The organization of the nucleus and its content undergoes several distinct changes during the IDC that are most likely necessary to accommodate the high level of transcriptional activity observed during this stage of the parasite life cycle (Fig. 11.3). First, the size of the nucleus expands dramatically, from an average volume of approximately 20 μm^3 in the ring stage to a maximum volume

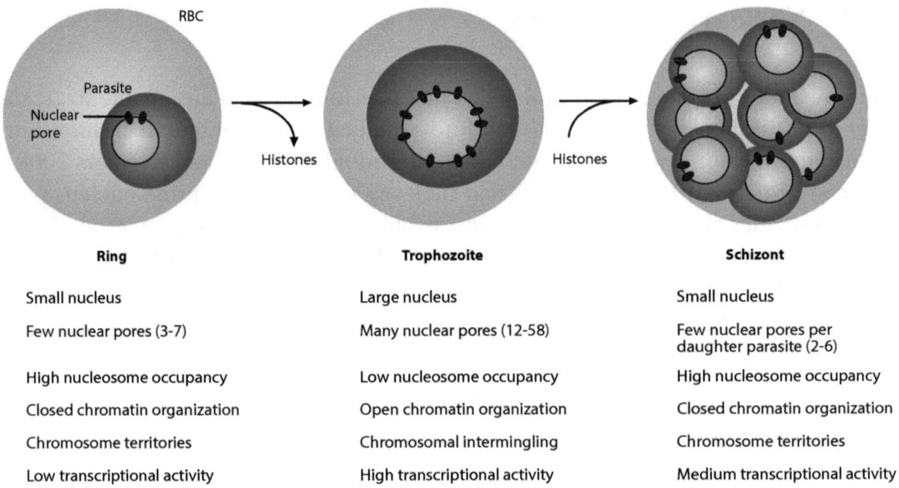

Fig. 11.3 Dynamic remodeling of the parasite nucleus during the IDC. Schematic overview of the changes that occur during the transitions from ring to trophozoite to schizont and that are likely to be associated with differences in transcriptional activity in these three main IDC stages

of 50 μm^3 in the trophozoite stage (Bannister et al. 2005; Weiner et al. 2011). In parallel with the increase in nuclear size, the number of nuclear pores increases from 3–7 pores in rings to 12–58 pores in trophozoites, which are uniformly distributed in the nuclear membrane (Weiner et al. 2011). The nuclear pores are located adjacent to areas of euchromatin, indicating that the increased number of nuclear pores facilitates export of messenger RNA into the cytosol. These changes revert when the parasite undergoes nuclear division and the nuclear content and its pores are distributed over all daughter parasites. During the transcriptionally more active trophozoite stage, chromosome territories are partially lost (Ay et al. 2014), suggesting an association between chromosome compaction and transcriptional activity. In addition, genes that are located more closely together have more similar expression profiles, and colocalization of genes that are most active during other stages of the parasite life cycle is also observed (Ay et al. 2014). Overall, these observations point towards an important role of dynamic nuclear rearrangements in control of gene expression.

11.8 Epigenetic Mechanisms as Drug Targets

Given the importance of epigenetics mechanisms for regulating gene expression in *P. falciparum*, the molecular components involved in these processes are currently being explored as potential antimalarial drug targets. Interfering with histone acetylation and methylation levels has been shown to often result in decreased parasite survival. For example, six out of the ten HKMTs encoded by *P. falciparum* are essential for parasite development during the IDC. Thus far, two related small molecules with strong HKMT inhibitory activity in *P. falciparum* and *P. vivax* have been identified (Malmquist et al. 2012, 2015). Likewise, inhibition of PfGNC5, a HAT with a preference for acetylation of H3K9 and H3K14, by curcumin results in hypoacetylation of these residues and induces parasite death (Cui et al. 2007a). Inhibition of HDACs also deregulates parasite development by inducing hyperacetylation, resulting in large changes in gene expression (Chaal et al. 2010). Several compounds with selective HDAC inhibitory activity against multiple life cycle stages of *P. falciparum* have been discovered (Hansen et al. 2014; Marfurt et al. 2011), giving promise for the development of this class of drugs as antimalarial therapy. However, the catalytic sites of histone-modifying enzymes tend to be well conserved across eukaryotic species. The potential toxicity of inhibitors of histone-modifying enzymes in mammalian cells will thus need to be carefully characterized. In addition, improved compounds displaying increased antiplasmodial activity and enhanced pharmacokinetic properties will need to be developed before they can be considered as novel therapeutic agents.

The integrity of the repressive heterochromatin compartment that harbors the *var* genes and other clonally variant expressed genes is also vital for parasite survival. As described earlier, disruption of PfHP1 or PfHDA1 interferes with the maintenance of this compartment, resulting in an arrest in parasite development (Brancucci et al. 2014; Coleman et al. 2014). In addition, it is likely that many other, as yet uncharacterized, proteins are involved in regulating the structure and dynamic remodeling of the nucleus and its contents. Some of these proteins may exhibit parasite-specific features that will make them attractive drug targets. An example of a chromatin remodeling protein with a domain architecture that is unique to apicomplexan parasites is the *P. falciparum* homolog of the Snf2/Swi2 ATPase ISWI, which contains additional five PHD domains around its core helicase domain in comparison to the mammalian version (Aravind et al. 2003; Templeton et al. 2004). Further characterization of such parasite-specific proteins could be highly informative for drug development. In addition, it will be very interesting to map the pathways involved in biogenesis of the nuclear membrane and the nuclear pore complex. Considering the large changes in nuclear size and number of nuclear pores, disruption of these processes may have a dramatic effect on parasite survival.

Finally, little is known about the role of epigenetic regulation in other life cycle stages. Studying these stages is complicated due to the lack of in vitro culture systems, but has received much attention, in particular with an eye to discovering novel therapies. In particular, much effort is being devoted to understanding the establishment and reactivation of hypnozoites, the dormant forms of *P. vivax* and *P. ovale*, which can survive for weeks to years in hepatocytes and can cause malaria relapses. Only one antimalarial drug can activate hypnozoites, but administration of this compound can be fatal in persons with glucose-6-phosphate dehydrogenase deficiency. A recent study described an increased rate of hypnozoite activation upon treatment with a HKMT inhibitor that reduces H3K9me3 levels (Dembele et al. 2014), indicating that epigenetic repression and heterochromatin formation play important roles in maintaining hypnozoites. These results suggest that inhibitors of histone-modifying enzymes may also be a promising class of therapeutics for other stages of the malaria parasite life cycle.

11.9 Conclusions

The development of the human malaria parasite *P. falciparum* during the IDC is highly dependent on epigenetic mechanisms of gene regulation. In particular, the clonally variant expression of virulence genes and other parasite-specific genes and gene families are tightly controlled at the chromatin level by repressive and activating histone PTMs, heterochromatin formation, nuclear localization, and long noncoding RNAs. In addition, drastic changes in both local chromatin structure and global genome organization are observed during the IDC. Disruption of proteins involved in the maintenance of the heterochromatin compartment has been shown to efficiently arrest parasite development. Epigenetic mechanisms are also likely to play important roles during other stages of the parasite life cycle. Inhibitors of the molecular components involved in epigenetic regulatory mechanisms thus are a potentially interesting novel class of antimalarial compounds with cross-stage activity that should be further explored.

References

Amit-Avraham I, Pozner G, Eshar S et al (2015) Antisense long noncoding RNAs regulate var gene activation in the malaria parasite Plasmodium falciparum. Proc Natl Acad Sci U S A 112 (9):E982–E991. doi:10.1073/pnas.1420855112

Anderson JD, Widom J (2001) Poly(dA-dT) promoter elements increase the equilibrium accessibility of nucleosomal DNA target sites. Mol Cell Biol 21(11):3830–3839. doi:10.1128/MCB.21.11.3830-3839.2001

Aravind L, Iyer LM, Wellems TE et al (2003) Plasmodium biology: genomic gleanings. Cell 115 (7):771–785

Ay F, Bunnik EM, Varoquaux N et al (2014) Three-dimensional modeling of the P. falciparum genome during the erythrocytic cycle reveals a strong connection between genome architecture and gene expression. Genome Res. doi:10.1101/gr.169417.113

Balaji S, Babu MM, Iyer LM et al (2005) Discovery of the principal specific transcription factors of Apicomplexa and their implication for the evolution of the AP2-integrase DNA binding domains. Nucleic Acids Res 33(13):3994–4006. doi:10.1093/nar/gki709

Balu B, Maher SP, Pance A et al (2011) CCR4-associated factor 1 coordinates the expression of Plasmodium falciparum egress and invasion proteins. Eukaryot Cell 10(9):1257–1263. doi:10.1128/EC.05099-11

Bannister LH, Margos G, Hopkins JM (2005) Making a home for *Plasmodium* post-genomics: ultrastructural organization of the blood stages. In: Sherman IW (ed) Molecular approaches to Malaria. ASM Press, Washington, DC, pp 24–49

Barski A, Cuddapah S, Cui K et al (2007) High-resolution profiling of histone methylations in the human genome. Cell 129(4):823–837. doi:10.1016/j.cell.2007.05.009

Bartfai R, Hoeijmakers WA, Salcedo-Amaya AM et al (2010) H2A.Z demarcates intergenic regions of the plasmodium falciparum epigenome that are dynamically marked by H3K9ac and H3K4me3. PLoS Path 6(12):e1001223. doi:10.1371/journal.ppat.1001223

Beh LY, Muller MM, Muir TW et al (2015) DNA-guided establishment of nucleosome patterns within coding regions of a eukaryotic genome. Genome Res 25(11):1727–1738. doi:10.1101/gr.188516.114

Bernstein BE, Kamal M, Lindblad-Toh K et al (2005) Genomic maps and comparative analysis of histone modifications in human and mouse. Cell 120(2):169–181. doi:10.1016/j.cell.2005.01.001

Brancucci NM, Bertschi NL, Zhu L et al (2014) Heterochromatin protein 1 secures survival and transmission of malaria parasites. Cell Host Microbe 16(2):165–176. doi:10.1016/j.chom.2014.07.004

Bunnik EM, Chung DW, Hamilton M et al (2013) Polysome profiling reveals translational control of gene expression in the human malaria parasite Plasmodium falciparum. Genome Biol 14 (11):R128. doi:10.1186/gb-2013-14-11-r128

Bunnik EM, Polishko A, Prudhomme J et al (2014) DNA-encoded nucleosome occupancy is associated with transcription levels in the human malaria parasite Plasmodium falciparum. BMC Genomics 15:347. doi:10.1186/1471-2164-15-347

Chaal BK, Gupta AP, Wastuwidyaningtyas BD et al (2010) Histone deacetylases play a major role in the transcriptional regulation of the Plasmodium falciparum life cycle. PLoS Pathog 6(1): e1000737. doi:10.1371/journal.ppat.1000737

Chang GS, Noegel AA, Mavrich TN et al (2012) Unusual combinatorial involvement of poly-A/T tracts in organizing genes and chromatin in Dictyostelium. Genome Res 22(6):1098–1106. doi:10.1101/gr.131649.111

Chene A, Vembar SS, Riviere L et al (2012) PfAlbas constitute a new eukaryotic DNA/RNA-binding protein family in malaria parasites. Nucleic Acids Res 40(7):3066–3077. doi:10.1093/nar/gkr1215

Chookajorn T, Dzikowski R, Frank M et al (2007) Epigenetic memory at malaria virulence genes. Proc Natl Acad Sci U S A 104(3):899–902. doi:10.1073/pnas.0609084103

Coleman BI, Skillman KM, Jiang RH et al (2014) A Plasmodium falciparum histone deacetylase regulates antigenic variation and gametocyte conversion. Cell Host Microbe 16(2):177–186. doi:10.1016/j.chom.2014.06.014

Comeaux CA, Coleman BI, Bei AK et al (2011) Functional analysis of epigenetic regulation of tandem RhopH1/clag genes reveals a role in Plasmodium falciparum growth. Mol Microbiol 80 (2):378–390. doi:10.1111/j.1365-2958.2011.07572.x

Cortes A (2008) Switching Plasmodium falciparum genes on and off for erythrocyte invasion. Trends Parasitol 24(11):517–524. doi:10.1016/j.pt.2008.08.005

Cortes A, Carret C, Kaneko O et al (2007) Epigenetic silencing of Plasmodium falciparum genes linked to erythrocyte invasion. PLoS Pathog 3(8):e107. doi:10.1371/journal.ppat.0030107

Cowman AF, Crabb BS (2006) Invasion of red blood cells by malaria parasites. Cell 124 (4):755–766. doi:10.1016/j.cell.2006.02.006

Crowley VM, Rovira-Graells N, Ribas de Pouplana L et al (2011) Heterochromatin formation in bistable chromatin domains controls the epigenetic repression of clonally variant Plasmodium falciparum genes linked to erythrocyte invasion. Mol Microbiol 80(2):391–406. doi:10.1111/j.1365-2958.2011.07574.x

Cui L, Miao J, Cui L (2007a) Cytotoxic effect of curcumin on malaria parasite Plasmodium falciparum: inhibition of histone acetylation and generation of reactive oxygen species. Antimicrob Agents Chemother 51(2):488–494. doi:10.1128/AAC.01238-06

Cui L, Miao J, Furuya T et al (2007b) PfGCN5-mediated histone H3 acetylation plays a key role in gene expression in Plasmodium falciparum. Eukaryot Cell 6(7):1219–1227. doi:10.1128/EC.00062-07

Dekker J, Marti-Renom MA, Mirny LA (2013) Exploring the three-dimensional organization of genomes: interpreting chromatin interaction data. Nat Rev Genet 14(6):390–403. doi:10.1038/nrg3454

Dembele L, Franetich JF, Lorthiois A et al (2014) Persistence and activation of malaria hypnozoites in long-term primary hepatocyte cultures. Nat Med 20(3):307–312. doi:10.1038/nm.3461

Duan Z, Andronescu M, Schutz K et al (2010) A three-dimensional model of the yeast genome. Nature 465(7296):363–367. doi:10.1038/nature08973

Duraisingh MT, Voss TS, Marty AJ et al (2005) Heterochromatin silencing and locus repositioning linked to regulation of virulence genes in Plasmodium falciparum. Cell 121 (1):13–24. doi:10.1016/j.cell.2005.01.036

Eksi S, Morahan BJ, Haile Y et al (2012) Plasmodium falciparum gametocyte development 1 (Pfgdv1) and gametocytogenesis early gene identification and commitment to sexual development. PLoS Pathog 8(10):e1002964. doi:10.1371/journal.ppat.1002964

Epp C, Li F, Howitt CA et al (2009) Chromatin associated sense and antisense noncoding RNAs are transcribed from the var gene family of virulence genes of the malaria parasite Plasmodium falciparum. RNA 15(1):116–127. doi:10.1261/rna.1080109

Eshar S, Altenhofen L, Rabner A et al (2015) PfSR1 controls alternative splicing and steady-state RNA levels in Plasmodium falciparum through preferential recognition of specific RNA motifs. Mol Microbiol 96(6):1283–1297. doi:10.1111/mmi.13007

Field Y, Kaplan N, Fondufe-Mittendorf Y et al (2008) Distinct modes of regulation by chromatin encoded through nucleosome positioning signals. PLoS Comp Biol 4(11):e1000216. doi:10.1371/journal.pcbi.1000216

Flueck C, Bartfai R, Niederwieser I et al (2010) A major role for the Plasmodium falciparum ApiAP2 protein PfSIP2 in chromosome end biology. PLoS Path 6(2):e1000784. doi:10.1371/journal.ppat.1000784

Flueck C, Bartfai R, Volz J et al (2009) Plasmodium falciparum heterochromatin protein 1 marks genomic loci linked to phenotypic variation of exported virulence factors. PLoS Pathog 5(9): e1000569. doi:10.1371/journal.ppat.1000569

Frank M, Dzikowski R, Amulic B et al (2007) Variable switching rates of malaria virulence genes are associated with chromosomal position. Mol Microbiol 64(6):1486–1498. doi:10.1111/j.1365-2958.2007.05736.x

Freitas-Junior LH, Bottius E, Pirrit LA et al (2000) Frequent ectopic recombination of virulence factor genes in telomeric chromosome clusters of P. falciparum. Nature 407(6807):1018–1022. doi:10.1038/35039531

Freitas-Junior LH, Hernandez-Rivas R, Ralph SA et al (2005) Telomeric heterochromatin propagation and histone acetylation control mutually exclusive expression of antigenic variation genes in malaria parasites. Cell 121(1):25–36. doi:10.1016/j.cell.2005.01.037

Garcia BA, Hake SB, Diaz RL et al (2007) Organismal differences in post-translational modifications in histones H3 and H4. J Biol Chem 282(10):7641–7655. doi:10.1074/jbc.M607900200

Gardner MJ, Hall N, Fung E et al (2002) Genome sequence of the human malaria parasite Plasmodium falciparum. Nature 419(6906):498–511. doi:10.1038/nature01097

Gibcus JH, Dekker J (2013) The hierarchy of the 3D genome. Mol Cell 49(5):773–782. doi:10.1016/j.molcel.2013.02.011

Goyal M, Alam A, Iqbal MS et al (2012) Identification and molecular characterization of an Alba-family protein from human malaria parasite Plasmodium falciparum. Nucleic Acids Res 40 (3):1174–1190. doi:10.1093/nar/gkr821

Guillemette B, Bataille AR, Gevry N et al (2005) Variant histone H2A.Z is globally localized to the promoters of inactive yeast genes and regulates nucleosome positioning. PLoS Biol 3(12): e384. doi:10.1371/journal.pbio.0030384

Hansen FK, Sumanadasa SD, Stenzel K et al (2014) Discovery of HDAC inhibitors with potent activity against multiple malaria parasite life cycle stages. Eur J Med Chem 82:204–213. doi:10.1016/j.ejmech.2014.05.050

Hoeijmakers WA, Flueck C, Francoijs KJ et al (2012) Plasmodium falciparum centromeres display a unique epigenetic makeup and cluster prior to and during schizogony. Cell Microbiol 14(9):1391–1401. doi:10.1111/j.1462-5822.2012.01803.x

Hoeijmakers WA, Salcedo-Amaya AM, Smits AH et al (2013) H2A.Z/H2B.Z double-variant nucleosomes inhabit the AT-rich promoter regions of the Plasmodium falciparum genome. Mol Microbiol 87(5):1061–1073. doi:10.1111/mmi.12151

Horrocks P, Pinches R, Christodoulou Z et al (2004) Variable var transition rates underlie antigenic variation in malaria. Proc Natl Acad Sci U S A 101(30):11129–11134. doi:10.1073/pnas.0402347101

Ikadai H, Shaw Saliba K, Kanzok SM et al (2013) Transposon mutagenesis identifies genes essential for Plasmodium falciparum gametocytogenesis. Proc Natl Acad Sci U S A 110(18): E1676–E1684. doi:10.1073/pnas.1217712110

Iwanaga S, Kaneko I, Kato T et al (2012) Identification of an AP2-family protein that is critical for malaria liver stage development. PloS One 7(11):e47557. doi:10.1371/journal.pone.0047557

Jiang L, Lopez-Barragan MJ, Jiang H et al (2010) Epigenetic control of the variable expression of a Plasmodium falciparum receptor protein for erythrocyte invasion. Proc Natl Acad Sci U S A 107(5):2224–2229. doi:10.1073/pnas.0913396107

Jiang L, Mu J, Zhang Q et al (2013) PfSETvs methylation of histone H3K36 represses virulence genes in Plasmodium falciparum. Nature 499(7457):223–227. doi:10.1038/nature12361

Josling GA, Petter M, Oehring SC et al (2015) A Plasmodium falciparum bromodomain protein regulates invasion gene expression. Cell Host Microbe 17(6):741–751. doi:10.1016/j.chom.2015.05.009

Kafsack BF, Rovira-Graells N, Clark TG et al (2014) A transcriptional switch underlies commitment to sexual development in malaria parasites. Nature 507(7491):248–252. doi:10.1038/nature12920

Kensche PR, Hoeijmakers WA, Toenhake CG et al (2015) The nucleosome landscape of Plasmodium falciparum reveals chromatin architecture and dynamics of regulatory sequences. Nucleic Acids Res. doi:10.1093/nar/gkv1214

Kim TH, Barrera LO, Zheng M et al (2005) A high-resolution map of active promoters in the human genome. Nature 436(7052):876–880. doi:10.1038/nature03877

Le Roch KG, Johnson JR, Florens L et al (2004) Global analysis of transcript and protein levels across the Plasmodium falciparum life cycle. Genome Res 14(11):2308–2318. doi:10.1101/gr.2523904

Lemieux JE, Kyes SA, Otto TD et al (2013) Genome-wide profiling of chromosome interactions in Plasmodium falciparum characterizes nuclear architecture and reconfigurations associated with antigenic variation. Mol Microbiol. doi:10.1111/mmi.12381

Lopez-Rubio JJ, Gontijo AM, Nunes MC et al (2007) 5′ flanking region of var genes nucleate histone modification patterns linked to phenotypic inheritance of virulence traits in malaria parasites. Mol Microbiol 66(6):1296–1305. doi:10.1111/j.1365-2958.2007.06009.x

Lopez-Rubio JJ, Mancio-Silva L, Scherf A (2009) Genome-wide analysis of heterochromatin associates clonally variant gene regulation with perinuclear repressive centers in malaria parasites. Cell Host Microbe 5(2):179–190. doi:10.1016/j.chom.2008.12.012

Malmquist NA, Moss TA, Mecheri S et al (2012) Small-molecule histone methyltransferase inhibitors display rapid antimalarial activity against all blood stage forms in Plasmodium falciparum. Proc Natl Acad Sci U S A 109(41):16708–16713. doi:10.1073/pnas.1205414109

Malmquist NA, Sundriyal S, Caron J et al (2015) Histone methyltransferase inhibitors are orally bioavailable, fast-acting molecules with activity against different species causing malaria in humans. Antimicrob Agents Chemother 59(2):950–959. doi:10.1128/AAC.04419-14

Marfurt J, Chalfein F, Prayoga P et al (2011) Ex vivo activity of histone deacetylase inhibitors against multidrug-resistant clinical isolates of Plasmodium falciparum and P. vivax. Antimicrob Agents Chemother 55(3):961–966. doi:10.1128/AAC.01220-10

Miao J, Fan Q, Cui L et al (2006) The malaria parasite Plasmodium falciparum histones: organization, expression, and acetylation. Gene 369:53–65. doi:10.1016/j.gene.2005.10.022

Miao J, Fan Q, Cui L et al (2010) The MYST family histone acetyltransferase regulates gene expression and cell cycle in malaria parasite Plasmodium falciparum. Mol Microbiol 78(4):883–902. doi:10.1111/j.1365-2958.2010.07371.x

Miller LH, Good MF, Milon G (1994) Malaria pathogenesis. Science 264(5167):1878–1883
Mira-Martinez S, Rovira-Graells N, Crowley VM et al (2013) Epigenetic switches in clag3 genes mediate blasticidin S resistance in malaria parasites. Cell Microbiol 15(11):1913–1923. doi:10.1111/cmi.12162
Nagano T, Lubling Y, Stevens TJ et al (2013) Single-cell Hi-C reveals cell-to-cell variability in chromosome structure. Nature 502(7469):59–64. doi:10.1038/nature12593
Nguitragool W, Bokhari AA, Pillai AD et al (2011) Malaria parasite clag3 genes determine channel-mediated nutrient uptake by infected red blood cells. Cell 145(5):665–677. doi:10.1016/j.cell.2011.05.002
Nishida H, Suzuki T, Kondo S et al (2006) Histone H3 acetylated at lysine 9 in promoter is associated with low nucleosome density in the vicinity of transcription start site in human cell. Chromosome Res 14(2):203–211. doi:10.1007/s10577-006-1036-7
Noble R, Christodoulou Z, Kyes S et al (2013) The antigenic switching network of Plasmodium falciparum and its implications for the immuno-epidemiology of malaria. eLife 2:e01074. doi:10.7554/eLife.01074
Oehring SC, Woodcroft BJ, Moes S et al (2012) Organellar proteomics reveals hundreds of novel nuclear proteins in the malaria parasite Plasmodium falciparum. Genome Biol 13(11):R108. doi:10.1186/gb-2012-13-11-r108
Otto TD, Wilinski D, Assefa S et al (2010) New insights into the blood-stage transcriptome of Plasmodium falciparum using RNA-Seq. Mol Microbiol 76(1):12–24. doi:10.1111/j.1365-2958.2009.07026.x
Peckham HE, Thurman RE, Fu Y et al (2007) Nucleosome positioning signals in genomic DNA. Genome Res 17(8):1170–1177. doi:10.1101/gr.6101007
Perez-Toledo K, Rojas-Meza AP, Mancio-Silva L et al (2009) Plasmodium falciparum heterochromatin protein 1 binds to tri-methylated histone 3 lysine 9 and is linked to mutually exclusive expression of var genes. Nucleic Acids Res 37(8):2596–2606. doi:10.1093/nar/gkp115
Petter M, Selvarajah SA, Lee CC et al (2013) H2A.Z and H2B.Z double-variant nucleosomes define intergenic regions and dynamically occupy var gene promoters in the malaria parasite Plasmodium falciparum. Mol Microbiol 87(6):1167–1182. doi:10.1111/mmi.12154
Pillai AD, Nguitragool W, Lyko B et al (2012) Solute restriction reveals an essential role for clag3-associated channels in malaria parasite nutrient acquisition. Mol Pharmacol 82(6):1104–1114. doi:10.1124/mol.112.081224
Pokholok DK, Harbison CT, Levine S et al (2005) Genome-wide map of nucleosome acetylation and methylation in yeast. Cell 122(4):517–527. doi:10.1016/j.cell.2005.06.026
Ponts N, Harris EY, Lonardi S et al (2011) Nucleosome occupancy at transcription start sites in the human malaria parasite: a hard-wired evolution of virulence? Infect Genet Evol 11(4):716–724. doi:10.1016/j.meegid.2010.08.002
Ponts N, Harris EY, Prudhomme J et al (2010) Nucleosome landscape and control of transcription in the human malaria parasite. Genome Res 20(2):228–238. doi:10.1101/gr.101063.109
Raisner RM, Hartley PD, Meneghini MD et al (2005) Histone variant H2A.Z marks the 5′ ends of both active and inactive genes in euchromatin. Cell 123(2):233–248. doi:10.1016/j.cell.2005.10.002
Raveh-Sadka T, Levo M, Shabi U et al (2012) Manipulating nucleosome disfavoring sequences allows fine-tune regulation of gene expression in yeast. Nat Genet 44(7):743–750. doi:10.1038/ng.2305
Roberts DJ, Craig AG, Berendt AR et al (1992) Rapid switching to multiple antigenic and adhesive phenotypes in malaria. Nature 357(6380):689–692. doi:10.1038/357689a0
Rosa A, Everaers R (2008) Structure and dynamics of interphase chromosomes. PLoS Comput Biol 4(8):e1000153. doi:10.1371/journal.pcbi.1000153
Rovira-Graells N, Crowley VM, Bancells C et al (2015) Deciphering the principles that govern mutually exclusive expression of Plasmodium falciparum clag3 genes. Nucleic Acids Res 43(17):8243–8257. doi:10.1093/nar/gkv730

Rovira-Graells N, Gupta AP, Planet E et al (2012) Transcriptional variation in the malaria parasite Plasmodium falciparum. Genome Res 22(5):925–938. doi:10.1101/gr.129692.111

Salcedo-Amaya AM, van Driel MA, Alako BT et al (2009) Dynamic histone H3 epigenome marking during the intraerythrocytic cycle of Plasmodium falciparum. Proc Natl Acad Sci U S A 106(24):9655–9660. doi:10.1073/pnas.0902515106

Scherf A, Lopez-Rubio JJ, Riviere L (2008) Antigenic variation in Plasmodium falciparum. Annu Rev Microbiol 62:445–470. doi:10.1146/annurev.micro.61.080706.093134

Segal E, Widom J (2009) Poly(dA:dT) tracts: major determinants of nucleosome organization. Curr Opin Struct Biol 19(1):65–71. doi:10.1016/j.sbi.2009.01.004

Sharma P, Wollenberg K, Sellers M et al (2013) An epigenetic antimalarial resistance mechanism involving parasite genes linked to nutrient uptake. J Biol Chem 288(27):19429–19440. doi:10.1074/jbc.M113.468371

Siegel TN, Hon CC, Zhang Q et al (2014) Strand-specific RNA-Seq reveals widespread and developmentally regulated transcription of natural antisense transcripts in Plasmodium falciparum. BMC Genomics 15:150. doi:10.1186/1471-2164-15-150

Silvestrini F, Lasonder E, Olivieri A et al (2010) Protein export marks the early phase of gametocytogenesis of the human malaria parasite Plasmodium falciparum. Mol Cell Proteomics 9(7):1437–1448. doi:10.1074/mcp.M900479-MCP200

Sinha A, Hughes KR, Modrzynska KK et al (2014) A cascade of DNA-binding proteins for sexual commitment and development in Plasmodium. Nature 507(7491):253–257. doi:10.1038/nature12970

Tanizawa H, Iwasaki O, Tanaka A et al (2010) Mapping of long-range associations throughout the fission yeast genome reveals global genome organization linked to transcriptional regulation. Nucleic Acids Res 38(22):8164–8177. doi:10.1093/nar/gkq955

Templeton TJ, Iyer LM, Anantharaman V et al (2004) Comparative analysis of apicomplexa and genomic diversity in eukaryotes. Genome Res 14(9):1686–1695. doi:10.1101/gr.2615304

Tillo D, Hughes TR (2009) G+C content dominates intrinsic nucleosome occupancy. BMC Bioinformatics 10:442. doi:10.1186/1471-2105-10-442

Tolstorukov MY, Kharchenko PV, Goldman JA et al (2009) Comparative analysis of H2A.Z nucleosome organization in the human and yeast genomes. Genome Res 19(6):967–977. doi:10.1101/gr.084830.108

Tonkin CJ, Carret CK, Duraisingh MT et al (2009) Sir2 paralogues cooperate to regulate virulence genes and antigenic variation in Plasmodium falciparum. PLoS Biol 7(4):e84. doi:10.1371/journal.pbio.1000084

Trelle MB, Salcedo-Amaya AM, Cohen AM et al (2009) Global histone analysis by mass spectrometry reveals a high content of acetylated lysine residues in the malaria parasite Plasmodium falciparum. J Proteome Res 8(7):3439–3450. doi:10.1021/pr9000898

Ukaegbu UE, Kishore SP, Kwiatkowski DL et al (2014) Recruitment of PfSET2 by RNA polymerase II to variant antigen encoding loci contributes to antigenic variation in P. falciparum. PLoS Pathog 10(1):e1003854. doi:10.1371/journal.ppat.1003854

Valouev A, Ichikawa J, Tonthat T et al (2008) A high-resolution, nucleosome position map of C. elegans reveals a lack of universal sequence-dictated positioning. Genome Res 18 (7):1051–1063. doi:10.1101/gr.076463.108

Vembar SS, Macpherson CR, Sismeiro O et al (2015) The PfAlba1 RNA-binding protein is an important regulator of translational timing in Plasmodium falciparum blood stages. Genome Biol 16:212. doi:10.1186/s13059-015-0771-5

Volz JC, Bartfai R, Petter M et al (2012) PfSET10, a Plasmodium falciparum methyltransferase, maintains the active var gene in a poised state during parasite division. Cell Host Microbe 11 (1):7–18. doi:10.1016/j.chom.2011.11.011

Wang Z, Zang C, Rosenfeld JA et al (2008) Combinatorial patterns of histone acetylations and methylations in the human genome. Nat Genet 40(7):897–903. doi:10.1038/ng.154

Weiner A, Dahan-Pasternak N, Shimoni E et al (2011) 3D nuclear architecture reveals coupled cell cycle dynamics of chromatin and nuclear pores in the malaria parasite Plasmodium falciparum. Cell Microbiol 13(7):967–977. doi:10.1111/j.1462-5822.2011.01592.x

Westenberger SJ, Cui L, Dharia N et al (2009) Genome-wide nucleosome mapping of Plasmodium falciparum reveals histone-rich coding and histone-poor intergenic regions and chromatin remodeling of core and subtelomeric genes. BMC Genomics 10:610. doi:10.1186/1471-2164-10-610

WHO (2015) The World Malaria Report. http://www.who.int/malaria/publications/world-malaria-report-2015/report/en/

Wu Y, Zhang W, Jiang J (2014) Genome-wide nucleosome positioning is orchestrated by genomic regions associated with DNase I hypersensitivity in rice. PLoS Genet 10(5):e1004378. doi:10.1371/journal.pgen.1004378

Yuda M, Iwanaga S, Kaneko I et al (2015) Global transcriptional repression: an initial and essential step for Plasmodium sexual development. Proc Natl Acad Sci U S A 112 (41):12824–12829. doi:10.1073/pnas.1504389112

Yuda M, Iwanaga S, Shigenobu S et al (2009) Identification of a transcription factor in the mosquito-invasive stage of malaria parasites. Mol Microbiol 71(6):1402–1414. doi:10.1111/j.1365-2958.2009.06609.x

Yuda M, Iwanaga S, Shigenobu S et al (2010) Transcription factor AP2-Sp and its target genes in malarial sporozoites. Mol Microbiol 75(4):854–863. doi:10.1111/j.1365-2958.2009.07005.x

Zhang Q, Huang Y, Zhang Y et al (2011) A critical role of perinuclear filamentous actin in spatial repositioning and mutually exclusive expression of virulence genes in malaria parasites. Cell Host Microbe 10(5):451–463. doi:10.1016/j.chom.2011.09.013

Glossary of Terms for Epigenetics of Infectious Diseases

Adenine methylation A type of base modification found in bacterial DNA, involving the addition of a methyl group to the N^6 position of adenosine moieties embedded in specific DNA motifs. DNA adenine methylation is performed by DNA methyltransferases belonging to restriction-modification systems and by "orphan" DNA methylases like Dam in Gamma-proteobacteria and CcrM in Alpha-proteobacteria.

Acetylation The introduction, via an enzymatic reaction, of an acetyl group to an organic compound, for instance to *histones* or other proteins.

Adenoviruses (human) This group of DNA viruses with genome sizes of between 30 and 35 kb pairs is one of the major causative agents for upper respiratory infections, particularly in childhood. In the 1960s to 1990s and beyond, adenoviruses have served a major role as tools in the study of the molecular biology of human (mammalian) cells and in tumor biology. Adenoviruses can transform cells in culture to tumor-like cells. Some of the adenoviruses (adenovirus type 12 and others) can directly induce tumors in newborn hamsters. More recently, adenoviruses have been used as vectors in gene technology. Attempts to apply such manipulated adenovirus genomes as vectors in human gene therapeutic regimes have led to fatal accidents. In experimental research, however, adenoviruses continue to be very useful gene transfer agents.

Agouti gene The agouti gene (A) controls fur color through the deposition of a yellow pigment in developing hairs. Several variants of the gene exist, and for one of these (Agouti Variable Yellow, A^{vy}), the expression levels can be heritably modified by *DNA methylation*.

Alleles Different variants or copies of a gene. For most genes on the chromosomes, there are two copies: one copy inherited from the mother and the other from the father. The DNA sequence of each of these copies may be different because of genetic polymorphisms.

Arginine vasopressin Human peptide hormone with antidiuretic activity.

Assisted reproduction technologies (ART) The combination of approaches that are being applied in the fertility clinic, including *IVF* and *ICSI*.

W. Doerfler, J. Casadesús (eds.), *Epigenetics of Infectious Diseases*, Epigenetics and Human Health, DOI 10.1007/978-3-319-55021-3

Astrocyte Glial cell with star shape found in the brain and the spinal cord.

5-Azacytidine A cytidine analogue in which the five carbon of the cytosine ring has been replaced with nitrogen. It integrates in DNA and RNA, while the analogue 5-aza-2-deoxycitidine is specific for DNA. Both drugs are potent inhibitors of mammalian *DNA methyltransferases*.

Bacteremia Presence of bacteria in the bloodstream.

Biofilm Microbial community made of cells stuck to each other, often adhered to a surface.

Bisphenol A Chemical used in the production of polycarbonate plastics and epoxy resins.

Bivalent chromatin A chromatin region that is modified by a combination of histone modifications such that it represses gene transcription but at the same time retains the potential of acquiring gene expression.

Bisulfite genomic sequencing A procedure in which bisulfite is used to deaminate cytosine to uracil in genomic DNA. Conditions are chosen so that 5-methylcytosine is not changed. PCR amplification and subsequent DNA sequencing reveals the exact position of cytosines which are methylated in genomic DNA.

Bromodomain Protein motif found in a variety of nuclear proteins including transcription factors and HATs involved in transcriptional activation. Bromodomains bind to histone tails carrying acetylated lysine residues.

Brno nomenclature Regulation of the nomenclature of specific histone modifications formulated at the Brno meeting of the NoE in 2004. Rules are: <Histone><amino-acid position><modification type><type of modification>. Example: H3K4me3 = trimethylated lysine-4 on histone H3.

Brucellosis Infection caused by bacteria belonging to the genus *Brucella*, which colonizes the reproductive tract of animals causing infertility and abortion.

CGG repeat expansions CGG repeats have been found at several locations in the human genome. The naturally occurring CGG repeat in the first untranslated exon of the FMR1 (*fragile X* mental retardation 1) gene has been studied in detail, because its expansion due to an instability of unknown origin is causally related to the *Fragile X Syndrome*, the most frequent cause of mental retardation in human males. With a few exceptions, this expansion is linked to *methylation* of the FMR1 promoter region which causes the shutdown of the gene and the Fragile X Syndrome.

Cholangiocyte Epithelial cell of the bile duct.

Cell fate The programmed path of differentiation of a cell. Although all cells have the same DNA, their cell fate can be different. For instance, some cells develop into brain, whereas others are the precursors of blood. Cell fate is determined in part by the organization of *chromatin*—DNA and the histone proteins—in the nucleus.

Cellular memory (epigenetic) Specific active and repressive organizations of chromatin can be maintained from one cell to its daughter cells. This is called *epigenetic inheritance* and ensures that specific states of gene expression are inherited over many cell generations.

ChIP See chromatin *immunoprecipitation.*

ChIP on chip After chromatin immunoprecipitation, DNA is purified from the immunoprecipitated chromatin fraction and used to hybridize arrays of short DNA fragments representing specific regions of the genome.

ChIP Seq Sequencing of the totality of DNA fragments obtained by ChIP to determine their position on the genome. Sequencing is usually preceded by PCR amplification of ChIP-derived DNA to increase its amount.

Chromatid In each somatic cell generation, the genomic DNA is replicated in order to make two copies of each individual chromosome. During M phase of the cell cycle, these copies—called chromatids—are microscopically visible one next to the other, before they get distributed to the daughter cells.

Chromatin The nucleoprotein complex constituting the chromosomes in eukaryotic cells. The structural organization of chromatin is complex and involves different levels of compaction. The lowest level of compaction is represented by an extended array of nucleosomes.

Chromatin remodeling Locally, the organization and compaction of chromatin can be altered by different enzymatic machineries. This is called chromatin remodeling. Chromatin remodeling protein complexes move *nucleosomes* along the DNA and require ATP for their action.

Chromodomain (chromatin organization modifier domain) Protein–protein interaction motif first identified in *Drosophila melanogaster HP1* and *Polycomb group proteins*. Also found in other nuclear proteins involved in transcriptional silencing and heterochromatin formation. Chromodomains consist of approximately 50 amino acids and bind to histone tails that are methylated at certain lysine residues.

Chromosomal domain In higher eukaryotes, it is often observed that in a specific cell type, chromatin is organized (e.g., by DNA or *histone methylation*) the same way across hundreds to thousands of kilobases of DNA. These "chromosomal domains" can comprise multiple genes that are similarly expressed. Some chromosomal domains are controlled by *genomic imprinting*.

Chromatin immunoprecipitation (ChIP) Incubation of chromatin fragments comprising one to several nucleosomes, with an antiserum directed against particular proteins bound to DNA (histones, posttranslational modified histones, transcription factors, etc.). After ChIP, the genomic DNA is purified from the chromatin fragments brought down by the antiserum and analyzed.

Cohesin Protein complex that mediates cohesion between sister chromatids during chromosome segregation in eukaryotic cells.

CpG dinucleotide A cytosine followed by a guanine in the sequence of bases of DNA. *Cytosine methylation* in mammals occurs predominantly at CpG dinucleotides.

CpG island A small stretch of DNA, of several hundred up to several kilobases in size, that is particularly rich in *CpG dinucleotides* and is also relatively enriched in cytosines and guanines. CpG islands occur often, but not exclusively, in promoter sequences which control the expression of genes. About one-third of all promoters in the human genome lack such islands.

Crown gall disease Tumor-like disease of plants caused by the bacterium *Agrobacterium tumefaciens*.

Curcumin Polyphenol present in turmeric, with anti-inflammatory and antioxidant properties.

Cyclin Member of a family of proteins involved in control of the eukaryotic cell cycle.

Cytosine methylation In mammals, DNA methylation occurs predominantly at cytosines that are part of *CpG dinucleotides*. As a consequence of the palindromic nature of the CpG sequence, methylation is symmetrical, i.e., affects both strands of DNA at a methylated target site. Methylated *CpG* sequences present in promoters, are frequently associated with transcriptional repression. Cytosine methylation was the first epigenetic signal whose function was recognized.

Deacetylation The removal of acetyl groups from proteins. Deacetylation of histones is often associated with gene repression and is mediated by histone deacetylases (HDACs).

Dendritic cell Antigen-presenting cell of the human immune system.

DNA demethylation Removal of methyl groups from DNA. This can occur "actively," i.e., by an enzymatically mediated process, or "passively," when methylation is not maintained after DNA replication. Active and passive methylation are known to occur in higher eukaryotes while prokaryotic DNA demethylation is only passive.

"de novo" DNA methylation Addition of methyl groups to a stretch of DNA which is not yet methylated (acquisition of "new" DNA methylation). Generation of novel patterns of DNA methylation, frequently found in foreign DNA integrated into an established (mammalian) genome.

Disomy The occurrence in the cell of two copies of a chromosome, or part of a chromosome, that are identical and of the same parental origin (uniparental disomy).

DNA methylation A biochemical modification of DNA resulting from the addition of a methyl group to cytosine bases, and in prokaryotes also to adenine bases. In mammals, methylation is frequently found in cytosines that are in *CpG dinucleotides*. Methyl groups can be removed from DNA by DNA demethylation.

DNA methyltransferase Enzyme which adds a methyl group ($-CH_3$) onto the DNA, either de novo or maintaining the existing patterns of DNA methylation after DNA replication.

Dosage compensation The X chromosome is present in two copies in one sex and in one copy in the other. Dosage compensation ensures that in spite of the copy number difference, X-linked genes are expressed at the same level in males and females. In mammals, dosage compensation occurs by inactivation of one of the X chromosomes in females. This inactivation does, however, not involve all segments of the "inactivated" X chromosome. The traditional term of dosage compensation does not take into account the actual complexity of *X chromosome inactivation*.

E-cadherin Glycoprotein involved in calcium-dependent cell adhesion. Loss of cadherin is associated with multiple types of cancer.

Embryonic stem (ES) cells Cultured cells obtained from the inner cell mass of the blastocyst, and for human ES cells possibly also from the epiblast. These cells are totipotent; they can be differentiated into every somatic cell lineage. ES-like cells can be obtained by dedifferentiation in vitro of somatic cells (see *iPS cells*). The iPS cells, however, differ significantly from the naturally occurring embryonic stem cells. There has been much publicity about iPS cells.

Endocrine disruptor A chemical component that can have an antagonistic effect on the action of a hormone (such as an estrogen) with which it shares structural properties. Some pesticides act as endocrine disruptors and have been found to have adverse effects on animal development and (for some of them) to induce altered *DNA methylation* at specific loci. A well-characterized endocrine disruptor is *Bisphenol A*, a chemical used for the production of certain plastics.

Enhancer A small, specialized sequence of DNA which, when recognized by specific regulatory proteins, can enhance the activity of the promoter of a gene(s) located in close or even quite distant (several 1000 nucleotides) vicinity. The mechanism of action of an enhancer is not really understood.

Epi-alleles Copies of a DNA sequence or a gene differing in their epigenetic and/or expression states without the occurrence of a genetic mutation.

Epigenesis The development of an organism from fertilization through a sequence of steps leading to a gradual increase in complexity through differentiation of cells and formation of organs. This term was originally introduced in the 1940s by Conrad Hal Waddington (1905–1975) in Edinburgh. At the time, Waddington had a quite different perception of this complex term.

Epigenetics The study of changes in gene function that can be transmitted through cell division (in eukaryotes, mitosis or meiosis) and arise without an apparent change in the genomic DNA sequence. Epigenetic mechanisms are involved in the formation and maintenance of cell lineages during development, and, in mammals, in *X-inactivation* and *genomic imprinting*, and are frequently perturbed in diseases. Epigenetics, an overly popularized term, is just a different way of looking at genetics.

Epigenetic code Patterns of DNA methylation and histone modifications can modify the way genes on the chromosomes are expressed. This has led to the idea that combinations of epigenetic modifications can constitute a code on top of the genetic code which modulates gene expression. We prefer to exert caution about introducing the term "code" here.

Epigenetic inheritance Somatic inheritance, or inheritance through the germ line, of epigenetic information (changes that affect gene function, without the occurrence of an alteration in the DNA sequence). At present, it has not been proven beyond doubt whether, and if by what mechanisms, epigenetic modifications are truly inheritable through the germ line.

Epigenetic marks Regional modifications of DNA and chromatin proteins, including *DNA methylation* and histone methylation, which can be maintained from one cell generation to the next and which may affect the way genes are expressed.

Epigenetic reprogramming Resetting of *epigenetic marks* on the genome so that these become like those of another cell type, or of another developmental stage. Epigenetic reprogramming occurs for instance in *primordial germ cells*, to bring them back in a "ground state." Epigenetic reprogramming and dedifferentiation also occur after *somatic cell nuclear transfer*.

Epigenome The epigenome is the overall epigenetic state of a particular cell. In the developing embryo, each cell type has a different epigenome. Epigenome maps represent the presence of DNA methylation, histone modification, and other chromatin modifications along the chromosomes.

Epigenotype The totality of epigenetic marks that are found along the DNA sequence of the genome in a particular cell lineage or at a particular developmental stage.

Epimutation A change in the normal epigenetic marking of a gene or a regulatory DNA sequence (e.g., a change in DNA methylation) which affects gene expression.

Epstein–Achong–Barr Virus (EBV also termed Human Herpes-Virus 4, HHV 4) is an enveloped virion of the Herpesviridae group and carries a double-stranded linear DNA genome of about 172 kbp. The virion DNA is not methylated. The virus was isolated by Epstein et al. in 1964 from the B-lymphocytes of an African Burkitt lymphoma patient. [The original description of EBV was by MA Epstein, BG Achong, and YM Barr. Virus particles in cultured lymphoblasts from Burkitt lymphoma. The Lancet 283:702–703, 1964. For unknown reasons, the documented contribution by BG Achong has not survived in the conventional designation of EBV.] Most people are innocuously infected with EBV in early childhood. In later stages of life, the first encounter with EBV frequently causes infectious mononucleosis. The virus persists throughout life. EBV infections have been implicated to play a role in Morbus Hodgkin, Burkitt lymphoma, and other lymphomas and tumors in humans. EBV can replicate in epithelial cells. Infection of B lymphocytes, however, leads to a latent viral life cycle with specifically limited gene expression profiles. In latency, the EBV genome persists in an episomal form as double-stranded circular DNA which can become methylated in specific patterns that regulate viral gene expression. Several of the viral proteins—EBNA (EBV nuclear antigens) and LMPs (latent membrane proteins)—are among the regulators of limited viral gene expression in latency.

Euchromatin A type of chromatin which is lightly stained when observed through the microscope at interphase. Euchromatic *chromosomal domains* are loosely compacted and relatively rich in genes, usually actively transcribed. The opposite type of chromatin organization is *heterochromatin*.

Exosome Vesicle present in all mammalian fluids, containing proteins and RNA from a eukaryotic cell.

Foreign DNA DNA molecule which is introduced into a species different from the species of origin. Foreign DNA is often brought into mammalian cells by viral infections (human immunodeficiency virus = HIV, adenovirus, human papillomavirus, and others). Foreign DNA is also introduced into cells artificially by a number of techniques in order to manipulate cells for experimental or commercial purposes. In gene therapeutic regimens, foreign DNA is thought to alter the cell's functional repertoire in the hope to influence pathological processes in the cell, e.g., in a cancer cell or in genetic disease. The unsolved problem in such procedures, however, remains that the introduction of foreign DNA per se elicits a number of defensive mechanisms in the cell which may do more harm to the cell or the organisms subject to the regimen than the intended therapeutic effort can have a chance of achieving. These unintended problems arise from the only partly understood epigenetic consequences in the wake of foreign DNA introductions.

Fragile X syndrome The fragile X syndrome is caused by a fragile chromosomal site at Xq27.3 and results in intellectual disability, attention deficit/hyperactivity disorder, and macroorchidism after puberty, plus facial and skeletal dysmorphisms. At the molecular level, the *expansion of a CGG repeat* located in the 5′-untranslated region of the first exon of the *FMR1* (fragile X mental retardation) gene and the hypermethylation of its promoter region inactivate the *FMR1* gene early in human development. Inactivation or very rare mutations of the *FMR1* gene determines the lack of its gene product during development and leads to the Fragile X Syndrome.

Genomic imprinting An epigenetic phenomenon which affects a small subset of genes in the genome and results in mono-allelic gene expression in a parent-of-origin-dependent way (for a given pair of alleles, either the maternally or paternally derived copy is active). The extent of genomic imprinting across the entire mammalian genome is only incompletely understood. We do not know whether at certain times, many different parts of the genome might be intermittently imprinted.

Germ line-specific stem cells Cells derived from undifferentiated germ cells that can be maintained through many cell divisions without alterations in their characteristics.

Gram-negative Type of bacterium that is not stained by crystal violet due to the thinness of the peptidoglycan layer in the cell wall.

Gram-positive Type of bacterium that is stained by crystal violet due to the thickness of the peptidoglycan layer in the cell wall.

Heterochromatin A type of chromatin which is darkly stained when observed under the microscope at interphase. Heterochromatic chromosomal domains, found in all cell types, are highly compacted, are rich in repeat sequences, and show little or no gene expression. Extended regions of heterochromatin are found close to centromeres and at telomeres. A historic term that fails to describe the subtleties of functional chromosomal organization.

Histone acetylation Posttranslational modification of the ε-amino group of lysine residues in histones catalyzed by a family of enzymes called *histone acetyltransferases (HATs)*. Acetylation contributes to the formation of decondensed, transcriptionally permissive chromatin structures and facilitates interaction with proteins containing *bromodomains*.

Histone acetyltransferase (HAT) An enzyme that acetylates (specific) lysine amino acids on histone proteins.

Histone code Hypothesis sustaining that distinct chromatin states of condensation and function are marked by specific histone modifications or specific combinatorial codes (see also epigenetic code). A popular but incompletely supported idea.

Histone deacetylase (HDAC) Member of a group of enzymes that remove acetyl groups from histone proteins. Deacetylation increases the positive charge of histones and enhances their attraction to the negatively charged phosphate groups in DNA. Deacetylation of histones often leads to gene inactivation.

Histone demethylase (HDM) Member of a group of proteins that catalyze enzymatic removal of methyl groups from either lysine or arginine residues of histones. Prominent examples are LSD1 and Jumonji proteins.

Histone methylation Posttranslational methylation of amino acid residues in histones, catalyzed by *histone methyltransferases (HMTs)*. Histone methylation is found at arginine as mono- or dimethylation and in lysine as mono-, di-, or trimethylation. Modifications are described depending on the position and type of methylation (mono-, di-, trimethylation) according to the *Brno nomenclature*. Different types of methylation can be found in either open transcriptionally active or silent (repressive) chromatin (*histone code*). Methylated lysine residues are recognized by proteins containing *chromodomains*.

Histone methyltransferase (HMT) Enzyme that catalyzes the transfer of methyl groups from S-adenosyl-methionine (SAM) to lysine or arginine residues in histones.

Homeodomain protein Family of transcription factors involved in anatomical development in fungi, plants, and animals.

Human papillomaviruses (HPV) Are a group of more than 100 types of non-enveloped icosahedral viruses containing double-stranded circular DNA of about 8 kbp with highly conserved genome organization. HPVs preferentially infect squamous, multilayered, and differentiating epithelial cells but are able to replicate only in cells of the basal epithelial cell layers. Infection is frequently asymptomatic. The products of the early viral genes E6 and E7 have been shown to transform cells in culture. There is a sizable literature that relates HPV infections to different human tumor diseases, depending on the HPV type. The possible role of HPV16 and HPV18 persistence in human cervix carcinoma has been most intensely studied. The viral genomes are integrated into the cellular genomes in these tumor cells. The methylation profile of HPV DNA changes during the viral life cycle. HPV DNA integrated in cervix carcinoma cells is hypermethylated as expected for integrated (foreign) viral DNA. The question of to what extent these integrates of HPV DNA lead to genome-wide epigenetic alterations in the infected cells has not been investigated. Only a small percentage of women infected with HPV types 16 or 18 actually develop tumors with a latency period of decades. There must be many hitherto unknown factors to explain the cause of cervix carcinoma or other presumably HPV-related tumors in humans.

Integration Insertion by covalent linkage of a foreign DNA molecule into the established genome of a cell.

Intracytoplasmic sperm injection (ICSI) Capillary-mediated injection of a single sperm into the cytoplasm of an oocyte, followed by activation to promote directed fertilization.

Imprinted genes Genes showing a parent-of-origin-specific gene expression pattern controlled by epigenetic marks that originate from the germ line. See *genomic imprinting*.

Imprinting See *genomic imprinting*.

Imprinted X-inactivation Preferential inactivation of the paternal X-chromosome in rodents (presumably also in humans) during early embryogenesis and in the placenta of mammals.

Imprinting control region (ICR) Region that shows germ line-derived, parent-of-origin-dependent epigenetic marking which controls the imprinted expression of neighboring imprinted genes. Its mechanism of action is not known.

Inner cell mass (ICM) Cells of the inner part of the blastocyst. Inner mass cells are the source for ES cells.

Induced pluripotent stem cells (iPS) Cells derived from differentiated somatic cells by in vitro reprogramming. Reprogramming is triggered by activation of pluripotency factor genes and cultivation in ES cell medium. iPS cells are able to generate all cell types of an embryo. Since the "triggering" involves the insertion of (foreign) genes into the genome, the patterns of methylation and transcription of the cells might be fundamentally altered. Hence, the iPS cells are in no way identical to naturally occurring pluripotent stem cells.

In vitro fertilization (IVF) Fertilization of a surgically retrieved oocyte in the laboratory, followed by a short period of in vitro cultivation before the embryo is transferred back into the uterus to allow development to term.

Isoschizomers Restriction enzymes that recognize the same target sequence in DNA. These enzymes often respond differently to methylation of bases within their target sequence, which makes them useful tools in DNA methylation analysis. Thus, *Msp*I cuts both CCGG and C5mCGG, whereas *Hpa*II cuts only the unmethylated sequence.

Kaposi's sarcoma-associated Herpesviruses (KSHV/HHV-8) Viruses belonging to the ubiquitous Gamma herpesviruses which infect a broad range of animal species and follow a biphasic life cycle. The viral genome is a linear double-stranded DNA molecule of about 165 kbp and carries a considerable number of genes with homology to cellular genes. Infection alternates between latent and lytic cell interactions. Persistent infections can be of long duration with the viral genome in an episomal state and a strictly controlled program of viral gene activities. The latent state of the viral genome is characterized by specific profiles of DNA methylation and histone modifications in the cell. There is evidence that the episomal viral population can be heterogeneous with respect to genome regulation. By various stimuli, persistently infected cells can be induced to the lytic viral life cycle. Gamma herpesviruses are lymphotropic and can lead to tumor diseases, particularly in immune-compromised patients like in HIV-infected individuals. HHV8 was originally isolated in 1994 from an AIDS-associated Kaposi sarcoma. HHV8 was also found in primary effusion lymphomas and in multicentric Castleman disease, a complex syndrome linked to the frequent development of lymphomas.

Lamin Family of architectural proteins lining the inside of the nuclear membrane in the eukaryotic nucleus.

Lipopolysaccharide Large molecule composed of a lipid and a polysaccharide, found in the outer membrane of Gram-negative bacteria.

Locus control region (LCR) Region marked by insulator functions and DNase-hypersensitive sites. LCRs contain binding sites for insulator proteins and enhancer binding proteins. LCRs control the domain specific, developmentally regulated expression of genes by long-range interactions with gene promoters.

Lymphosarcoma Malignant tumor of lymphatic tissue, especially common in cats and dogs.

MAP kinase (mitogen-activated protein kinase) Family of eukaryotic kinases involved in the control of cellular responses to mitogens, inflammatory cytokines, and multiple types of stress.

Maternal effects Long-term effects on the development of the embryo triggered by factors in the cytoplasm of the oocyte.

M cell Epithelial cell of the intestinal lymphoid tissue, specialized in the transport of antigens, soluble macromolecules, small particles, and microorganisms.

Meningoencephalitis Inflammation of the brain and the meninges, caused by viral and protozoan infections.

Methyl-binding domain (MBD) Protein domain in Methyl-CpG-binding proteins (MBPs) responsible for recognizing and binding to methylated cytosine residues in DNA. Proteins containing MBDs form a specific family of proteins with various molecular functions.

Methyl-CpG-binding proteins (MBPs) Proteins containing domains (such as MBD) binding to 5-methyl-cytosine in the context of CpG dinucleotides. MBPs mostly act as mediators for molecular functions such as transcriptional control or DNA repair.

Microsporidia Phylum of eukaryotic, unicellular parasites whose spore forms a polar tube to infiltrate host cells in both vertebrates and invertebrates.

N-acetylglucosamine Monosaccharide present in biopolymers such as chitin and bacterial and fungal cell walls as well as in the surface proteins of enveloped viruses, like the influenza virus.

Neuroepithelioma Tumor derived from neuroepithelium in the retina or in a peripheral nerve.

Neutrophil Mammalian white blood cell belonging to the polymorphonuclear family and involved in innate immunity.

Nitrogen fixation Conversion of atmospheric nitrogen into ammonia performed by Cyanobacteria and by soil bacteria such as *Rhizobium* and *Azotobacter*.

Noncoding RNA (ncRNA) Any RNA transcript that does not code for a protein. ncRNA generation frequently involves RNA processing.

Non-Mendelian inheritance Inheritance of genetic traits that do not follow Mendelian rules and/or cannot be explained in simple mathematically modeled traits.

Nuclear periphery Region around the nuclear membrane characterized by contacts of the chromosomes with the nuclear lamina.

Nuclear (chromosomal) territory Cell type-specific areas within the nucleus occupied by specific chromosomes during interphase (G1).

Nucleolus Compartment(s) within the nucleus formed by rDNA repeat domains. Nucleoli are marked by specific heterochromatic structures and active gene expression.

Nucleosome Fundamental organizational unit of chromatin consisting of 147 base pairs of DNA wound around a histone octamer.

O-antigen Glycan polymer present in the lipopolysaccharide of Gram-negative bacteria, which constitutes a target recognized by host antibodies and serves as a receptor for certain bacteriophages.

Ookinete Motile zygote of the malaria parasite and other sporozoans, able to form an oocyst in the stomach of *Anopheles* females.

Parasitophorous vacuole Compartment formed by certain parasites (e.g., *Plasmodium* and *Toxoplasma*) in the cytoplasm of host cells, which protects the parasite from host defense mechanisms.

Phagolysosome Cytoplasmic body formed upon fusion of a phagosome and a lysosome.

Phase variation Reversible switching of gene expression at high frequency, generating ON and OFF subpopulations.

Pili and fimbriae Bacterial appendages that enable bacterial cells to adhere to specific surfaces, including bacterial cells and eukaryotic tissues.

Pleomorphic Term used to describe cell types showing variability in their shape and their size.

Pluripotency Capacity of stem cells to form all cell types of an embryo including germ cells.

Polycomb group proteins Epigenetic regulatory proteins forming multiprotein complexes (PRCs = polycomb repressive complexes). Polycomb group proteins possess enzymatic properties to control the maintenance of a suppressed state of developmentally regulated genes, mainly through histone methylation and ubiquitination. Family of proteins that are part of complexes that remodel chromatin, usually causing gene silencing.

Position effect variegation (PEV) Cell/tissue-specific variability of gene expression controlled by the temporal inheritance of certain epigenetic states. PEV is a consequence of variable expression patterns across the respective gene. A classical example of PEV is found in certain mutations leading to variegated eye pigmentation in *Drosophila* eyes.

Primordial germ cell During early embryogenesis, mammalian cells are set aside which migrate through the hindgut of the developing mammalian embryo into the "Gonadenanlagen" to form founder cells of the latter germ line.

Protamines Small, arginine-rich proteins that replace histones late in the haploid phase of spermatogenesis. Protamines are thought to be essential for sperm head condensation and DNA stabilization. After fertilization, protamines are removed from paternal chromosomes in the mammalian zygote.

Pyelonephritis Inflammation of the upper urinary tract, usually caused by bacterial infection.

Restriction-modification Pair of enzymes made of a DNA endonuclease and a DNA methyltransferase involved in bacterial defense against foreign DNA invasion.

RNA interference (RNAi) Posttranscriptional regulatory effects on mRNAs (control of translation and/or stability) triggered by processed double-stranded and single-stranded small RNA (si-, mi-, piRNAs) molecules. Effects are propagated by enzymatic complexes such as RISC containing the small RNAs bound by Argonaute proteins.

SAHA Suberoylanilide hydroxamic acid, an inhibitor of certain histone deactylases, leading to enhanced levels of histone acetylation. See also *TSA*.

S-adenosyl homocysteine (SAH) Hydrolyzed product formed after the methylation reaction catalyzed by DNA and *histone methyltransferases* using SAM as methyl group donor. SAH is a competitive inhibitor of SAM for most methyltransferases.

S-adenosyl methionine (SAM) A cofactor for all DNA (DNMTs) and histone methyltransferases (HMTs) providing the methyl group added to either DNA or histones (arginine or lysine).

SET domain A domain found in virtually all lysine-specific *histone methyltransferases (HMTs)*. A protein–protein interaction domain required for HMT activity and modulation of chromatin structure, frequently associated with cysteine-rich Pre-SET and Post-SET domains.

Silencer DNA element to which proteins bind inhibiting transcription of a nearby promoter. Silencer elements are recognized and bound by silencer proteins.

siRNAs Small interfering RNAs in the size range of 21–24 nucleotides, derived from double-stranded long RNAs cleaved by Dicer. siRNAs are incorporated into the RISC complex to be targeted to complementary RNAs to promote cleavage of these mRNAs.

Somatic cell nuclear transfer (SCNT) Transfer of the nucleus of a somatic cell into an enucleated oocyte using a glass capillary to form an SCNT- zygote. After activation of the zygote, the genome of the nucleus derived from the somatic cells becomes reprogrammed to start development.

Spermatogenesis The process by which spermatogonia develop into mature sper matozoa. Spermatozoa (sperm) are the mature male gametes. Thus, spermatogenesis is the male version of gametogenesis.

Spermiogenesis The final stage of spermatogenesis which involves the maturation of spermatids into mature, motile spermatozoa (sperm). During this stage, cells no longer divide and undergo a major morphological transformation. In addition, at most of the genome, histone proteins are replaced by the more basic *protamines*.

Sporozoite Motile stage of the malaria parasite and other sporozoans, which is often the infective agent introduced into a host.

Stem cell Noncommitted cell which has the capacity to self-renew and divide many times giving rise to daughter cells that maintain the stem cell function. Stem cells have the property to differentiate into specialized cells.

Toll receptor Protein located in the membrane of macrophages and dendritic cells, involved in activation of the innate immune system upon detection of microbial antigens.

Totipotency Capacity of stem cells to produce all cell types required to form a mammalian embryo, i.e., embryonic and extraembryonic cells *(see Pluripotency)*. Totipotent cells are formed during the first cleavages of the embryo.

TSA Trichostatin-A, an inhibitor of certain types of histone deacetylases.

Trithorax group proteins Proteins containing a Trithorax-like bromodomain. They are usually involved in recognizing histone modifications marking transcriptionally active regions and contribute to the maintenance of activity.

Trophoblast Cells of the blastoderm which form the placental tissues in mammals.

Typhoid fever Infection caused by *Salmonella* Typhi and *Salmonella* Paratyphi, acquired by ingestion of contaminated food or water.

Urothelium Epithelium made of 3–5 cell layers and a glycoprotein surface, found in the urinary tract.

X chromosome inactivation Epigenetically controlled form of *dosage compensation* in female mammals resulting in transcriptional silencing of genes on surplus X-chromosome. X-chromosome inactivation is triggered by the noncoding RNA Xist and manifested by various epigenetic modifications including histone methylation, histone deacetylation, and DNA methylation. This inactivation, however, does not affect all regions of the "inactivated" X chromosome. The traditional term of dosage compensation does not take into account the actual complexity of "X chromosome inactivation."

MIX
Papier aus verantwortungsvollen Quellen
Paper from responsible sources
FSC® C105338

If you have any concerns about our products,
you can contact us on
ProductSafety@springernature.com

In case Publisher is established outside the EU,
the EU authorized representative is:
Springer Nature Customer Service Center GmbH
Europaplatz 3, 69115 Heidelberg, Germany

Printed by Libri Plureos GmbH
in Hamburg, Germany